PRODUCTION AND NEUTRALIZATION OF NEGATIVE IONS AND BEAMS

Eighth International Symposium

PRODUCTION AND APPLICATION OF LIGHT NEGATIVE IONS

Seventh European Workshop

A Joint Meeting

PRODUCTION AND NEUTRALIZATION OF NEGATIVE IONS AND BEAMS

Eighth International Symposium

PRODUCTION AND APPLICATION OF LIGHT NEGATIVE IONS

Seventh European Workshop

A Joint Meeting
Villagium of Giens, France   September 1997

EDITOR
Claude Jacquot
Centre d'Etudes de Cadarache, France

American Institute of Physics

AIP CONFERENCE PROCEEDINGS 439

Woodbury, New York

Editor:

Claude Jacquot
Department of the Research on Controlled Thermonuclear Fusion
Centre d'Etudes de Cadarache
F-13108 Saint Paul lez Durance
FRANCE

Email: jacquot@drfc.cad.cea.fr

Authorization to photocopy items for internal or personal use, beyond the free copying permitted under the 1978 U.S. Copyright Law (see statement below), is granted by the American Institute of Physics for users registered with the Copyright Clearance Center (CCC) Transactional Reporting Service, provided that the base fee of $15.00 per copy is paid directly to CCC, 222 Rosewood Drive, Danvers, MA 01923. For those organizations that have been granted a photocopy license by CCC, a separate system of payment has been arranged. The fee code for users of the Transactional Reporting Service is: 1-56396-737-5/ 98 /$15.00.

© 1998 American Institute of Physics

Individual readers of this volume and nonprofit libraries, acting for them, are permitted to make fair use of the material in it, such as copying an article for use in teaching or research. Permission is granted to quote from this volume in scientific work with the customary acknowledgment of the source. To reprint a figure, table, or other excerpt requires the consent of one of the original authors and notification to AIP. Republication or systematic or multiple reproduction of any material in this volume is permitted only under license from AIP. Address inquiries to Office of Rights and Permissions, 500 Sunnyside Boulevard, Woodbury, NY 11797-2999; phone: 516-576-2268; fax: 516-576-2499; e-mail: rights@aip.org.

L.C. Catalog Card No. 98-72695
ISBN 1-56396-737-5
ISSN 0094-243X
DOE CONF- 9709134

Printed in the United States of America

CONTENTS

Preface ix
Welcome Address xii
Committees xv

FUNDAMENTAL PROCESSES

Effect of Cesium Seeding on Hydrogen Negative Ion Volume Production 3
    M. Bacal, F. El Balghiti-Sube, L. I. Elizarov, and A. J. Tontegode
Creation of Highly Excited H_2 Molecules in a Hydrogen Flux
Penetrative through Cesium-Hydrogen Discharge 12
    M. Bacal, F. G. Baksht, and V. G. Ivanov
Negative Ion Yields in Hydrogen Scattering from Graphite Surfaces 37
    M. A. Gleeson, W. R. Koppers, K. Tsumori, and A. W. Kleyn
First Results from a Double Vlasov Model for Negative Ion Extraction
from Volume Sources—The Possibility of an Enhanced Transverse
Space Charge Limit 41
    J. H. Whealton, D. K. Olsen, and R. J. Raridon
Modeling Negative Ion Production in Volume Sources 54
    O. Fukumasa and K. Yoshino
Modeling of Negative Ion Transport in a Plasma Source 62
    D. Riz and J. Paméla

DIAGNOSTICS

Measurement of the Temporal Behaviour of H^- Density in a Pulsed
Hydrogen Multipole using the Photodetachment Technique 77
    T. Mosbach, H. M. Katsch, and H. F. Döbele
Measurement of Negative Ion Beam Emittance 81
    C. Michaut, J. Bucalossi, and D. Riz

SOURCES/H^-, D^-

Development of Large Scale Negative Ion Source for LHD-NBI 93
    K. Tsumori, T. Takanashi, S. Asano, M. Osakabe, Y. Takeiri, Y. Oka,
    E. Asano, T. Kawamoto, R. Akiyama, T. Okuyama, Y. Suzuki, and O. Kaneko
Long Pulse Operation of the Kamaboko Negative Ion Source
on the MANTIS Test Bed 105
    R. Trainham, C. Jacquot, D. Riz, K. Miyamoto, Y. Fujiwara, and Y. Okumura
Status and Objectives of the DENISE Ion Source Project at DCU 113
    P. McNeely, D. Boilson, N. Curran, M. B. Hopkins, and D. Vender
A Large-Area RF Source for Negative Hydrogen Ions 119
    P. Frank, J. H. Feist, W. Kraus, E. Speth, B. Heinemann, F. Probst,
    R. Trainham, and C. Jacquot

Efficient Production of H⁻ in High Density Helicon Plasmas 123
    D. Hayashi, K. Sasaki, K. Kadota, Y. Oka, K. Tsumori, and O. Kaneko

Development of a High Current H⁻ Source for ESS 133
    A. Maaser, P. Beller, H. Klein, K. Volk, and M. Weber

Studying of Negative Ions Source Based on Reflective Discharge in Regimes with Cesium Added and Without Cesium 139
    I. A. Soloshenko, A. I. Shchedrin, and A. V. Ryabtsev

Negative Ion Beams from Compact Surface Plasma Sources and their Merits for New Applications 158
    S. K. Guharay

Production and Control of ECR Plasmas for Negative Ion Sources 165
    O. Fukumasa and N. Tsuda

The DESY RF Driven H⁻ Volume Source 172
    J. Peters

NEGATIVE IONS ACCELERATION

Multi-Stage, Multi-Aperture Electrostatic Accelerator for H⁻ Beams 179
    K. Watanabe, Y. Fujiwara, M. Hanada, T. Inoue, K. Miyamoto, N. Miyamoto, Y. Ohara, and Y. Okumura

Results of the 1 MV SINGAP Experiment 187
    C. Desgranges, J. Bucalossi, M. Fumelli, R. S. Hemsworth, P. Massmann, J. Paméla, and A. Simonin

Beam Transmission in the ITER Neutral Beam Injectors 199
    A. Krylov, E. Di Pietro, E. Dlugach, T. Inoue, M. Hanada, R. Hemsworth, K. Miyamoto, and A. Panasenkov

Radiation Induced Conductivity and Voltage Holding Characteristics of Insulation Gas for the ITER-NBI 205
    Y. Fujiwara, M. Hanada, T. Inoue, K. Miyamoto, N. Miyamoto, Y. Ohara, Y. Okumura, and K. Watanabe

Surface Charging of Insulated Materials by Negative Ion Beam Bombardment 217
    J. Ishikawa, H. Tsuji, and Y. Gotoh

Secondary Electron Emission from Insulators 221
    H. J. Hopman and J. Verhoeven

APPLICATIONS AND SYSTEMS

Update of Neutral Beams for ITER 227
    R. S. Hemsworth, E. Di Pietro, T. Inoue, A. Krylov, Y. Okumura, M. Tanii, and M. Watson

Negative Ion Beam Source Development in Europe 237
    R. S. Hemsworth, J. Bucalossi, C. Desgranges, P. Frank, M. Fumelli, B. Heinemann, E. R. Hodgson, C. Jacquot, W. Krause, P. Massmann, C. Michaut, W. Ott, H.-P. Penningsfeld, D. Riz, A. Simonin, E. Speth, A. Stäbler, R. Trainham, and O. Volmer

**Separation of Beam and Electrons in the Spallation Neutron Source
H⁻ Ion Source** 244
        J. H. Whealton, R. J. Raridon, and K. N. Leung
Historical Perspective of the H⁻ Ion Source Symposia 254
        C. W. Schmidt

Summary and Closing Remarks 259
List of Participants 277
Author Index 281

Preface

The joint meeting of the Eighth International Symposium on the Production and Neutralization of Negative Ions and Beams and the Seventh European Workshop on the Production and Application of Light Negative Ions was held at the Villagium of Giens, France, from September 14 to 19, 1997. Future joint meetings are planned to alternate between sites in Europe and BNL every three years based on a decision made by the Conference participants during the meeting. The next joint meeting will be organized by BNL in 2000.

The BNL Symposia series has been held at Brookhaven every three years since 1977. Similarly, the European workshop series, whose inception was at Ecole Polytechnique, France, in 1984, has been held every two or three years at different locations in western Europe. Activities of world-wide research, as well as progress in negative ion sources, negative ion beams, and neutral beams, was well reflected in presentations at these meetings and their published proceedings.

The world-wide activity level in negative ion research has dramatically reduced. Considering this reduction in large scale negative development programs, it seemed prudent to hold a combined BNL-European meeting and to draw a larger portion of the contributions from small programs in which a variety of negative ion sources are being developed in industrial accelerators and basic research applications. Impressive progress reported from the Japanese fusion program, together with the interest in neutral beam heating (ITER and European fusion machines), have resuscitated hope for large-scale fusion programs.

We would like to thank the head of the Departement de Recherches sur la Fusion Controlee, Jérôme Pamela, for supporting this Symposium. Thanks to the Commissariat à l'Energie Atomique for ensuring success of this meeting. Members of the International Program Committee have provided valuable advice. We would like to thank the Local Organizing Committee, the Centre d'Etudes de Cadarache Staff Services Division, and the Villagium of Giens for handling a lot of the details involved in organizing the meeting. Special thanks to Michel Chatelier, Deputy Director of the DRFC for the welcome address, and Peter Massmann for his concluding remarks. Veronique Poli and Anne Bertin Maghit's large effort in putting this meeting together have guaranteed its success.

Finally, we thank the participants for their invaluable contribution which made it a very stimulating meeting with the spirit of Brookhaven . We look forward with optimism to the next meeting at the beginning of the next century.

Claude Jacquot
DRFC/STID, CEA Cadarache
EURATOM-Association
13108 Saint Paul lez Durance
FRANCE

Welcome Address

M. Chatelier
Deputy Head of Fusion Department, Cadarache

Ladies and gentlemen, good morning:

It is my pleasure to welcome you here in the Villagium de Giens, on behalf of our Fusion Department at Cadarache. Unfortunately, Jérôme Pamela the Head of our Department, who, as you know, belongs or belonged to your negative ion community cannot attend the meeting since he is participating in another meeting with the European Consultative Committee of the Fusion Programme in Brussels. I want to welcome particularly those of you coming from Japan, USA, Russia and Ukraine who made a long trip to participate.

Before giving the floor to the organisers of the meeting, I would like to take this opportunity to convey some thoughts to you about the present situation of magnetic fusion. Magnetic fusion is now, and for some time, crossing a difficult period as far as strategic options have to be determined for its near future (10-20 years) and far future (50 years or more).

There is absolutely no doubt in my view that magnetic fusion is, and will remain, definitely one of the most important options (not to say the most important) for energy supply in the far future. There are many reasons for that, among which:

- first, the availability of fusion fuel over many centuries (deuterium and lithium, considering the sea reservoir),

- second, the good prospects in terms of safety : no thermal excursion of the reactor, low residual heat to remove in case of an accident, no population evacuation in case of an accident and low radiotoxicity waste with human time scale decay.

- third, the fact that it does not contribute to CO_2 production (as other nuclear plants).

These characters of magnetic fusion have been established in the past and revisited recently by different expert groups which reached still positive conclusions on these aspects.

The main argument, counterbalancing these assets of fusion energy are:

- first, the difficulty of realizing a fusion reactor (both scientifically and technically) and

- second, the direct consequence of this difficulty, i.e. that the conception and construction of a commercial fusion reactor will require a long time (>50 years from now) and substantial amounts of money.

- third, the cost of electricity from a fusion plant which is not expected to be very much competitive today with the present other forms of production available (e.g. fossile)

Therefore, fusion is considered neither a guaranteed solution nor an urgent task by those in charge of providing funds to energy research. Renewable energies seem to have a better visibility (solar cells, wind mills, etc), due to the short time scale over which some pay-off is expected and a better public acceptance. However, it is also recognized that these forms of energy production will not provide in the future with more than some tens of percent of the overall energy consumption (say 20-30) which is far insufficient.

Now, in this context, what are the kinds of arguments that can be developed for the promotion of fusion energy? In my view, there are two ways of communicating some enthusiasm to the public or to deciders about magnetic fusion:

- first, our human society has to develop not only a model but also a strategy regarding the evolution of worldwide demography and energy consumption over the next century and beyond. Sharing energy resources fairly around the world is probably the starting point of healthy economy and therefore of long-lasting peace. Then, resource availibility might be the prime argument instead of a cheap kilowatt at some time. (It should be mentioned here that the CO_2 produced by fossil fuels, at least, seems to be fairly distributed around the earth in spite of an uneven production). Socioeconomic studies, by expert groups, are presently underway to forecast the evolution of the world in front of this question (USA, EU). We may hope that these studies will somewhat clarify the question.

- second, the production of solar-like conditions on the earth with the promise of inexhaustible energy production, namely fusion energy, remains the kind of venture that should fascinate the public, although it is true that some suspicion exists today about the ability of science to bring happiness in general. It is therefore important that we can reach a high standard of communication to the public associating reliable technical information together with good pedagogy.

I am personally confident that, on the long term, fusion energy, and particularly magnetic fusion energy, will remain a fundamental option whatever the turbulences (not only plasma turbulences) that we will have to face during the forthcoming years.

Finally, coming back to the subject of this meeting, I would like to point out that additional power for plasma heating and current drive remains a hot issue, since both physics scenarios contemplated for next step tokamaks include extremely high power levels. Indeed, the expected threshold of the H mode, the relevant confinement mode to reach ignition in ITER, lays in the range of a 100MW. Steady-state advanced tokamak scenarios based on the control of the current density profile are also extremely demanding (2-4MW/m^3), owing to the generally modest efficiency of the non inductive current generation methods.

I do not know what the eventual choice for current drive and heating will be in the next step machines (waves or/and beams). Several options have probably to remain open for some time since none of them can pretend gathering all virtues alone. Negative ion beams are certainly one of them. This method is very actively studied by your community and I should certainly mention the results obtained recently on JT60U which certainly open the way to the use of this technology on other fusion devices (in this respect stellarators are also interested, not only tokamaks).

Many specific questions still require progress for negative ion based NBI to be applied reliably on next step devices, namely MV power supplies and accelerators, large insulators, source optimization, long pulse operation with cesium, etc. I understand that some of these aspects will be tackled during the meeting. Another kind of question is linked to the implementation of neutral beam injectors on a future reactor: penetration of tritium into the beam lines, magnetic shielding, life time of the components, remote maintainability, etc. These aspects should be cautiously kept in mind since they could contribute heavily in the eventual choice for one or the other method.

Finally I would like to congratulate you on merging both the European Workshop and the International Symposium into a single effective meeting, although, in my view, its title should be revisited for the sake of brevity.

I wish you a successful and fruitful meeting as well as an enjoyable visit to Tore Supra and the fusion installations of Cadarache next Wednesday.

Thank you for your attention.

LOCAL ORGANIZING COMMITTEE

C. Jacquot, Chairman
V. Poli
A. Bertin Maghit

INTERNATIONAL PROGRAMME COMMITTEE

C. Jacquot	DRFC, Centre de Cadarache, Saint Paul lez Durance, France
A. Hershcovitch	Brookhaven National Laboratory, Upton, New York, USA
M. Bacal	Ecole Polytechnique, Palaiseau, France
Y. I. Belchenko	Siberian branch of the Academy of Science, Novosibirsk, Russia
R. S. Hemsworth	DRFC, Centre de Cadarache, Saint Paul lez Durance, France
M. Hopkins	Physics Dept, Dublin City University, Glasnevin Dublin, Ireland
N. Ishikawa	Department of Electronics, Kyoto University, Kyoto, Japan
O. W. Kleyn	FOM Institute for Atomic and Molecular Physics, Amsterdam, The Netherlands
V. Kuligyn	Kurchatov Institute of Atomic Energy, Moscow, Russia
T. Kuroda	National Institute for Fusion Science, Nagoya, Japan
U. Okumura	Japan Atomic Energy Research Institute, Nakamachi, Japan
J. Sherman	Los Alamos National Laboratory, Los Alamos, New Mexico, USA

FUNDAMENTAL PROCESSES

Effect of Cesium Seeding on Hydrogen Negative Ion Volume Production

M. Bacal, F. El Balghiti-Sube, L.I. Elizarov* and A.J. Tontegode**

Laboratoire de Physique des Milieux Ionisés, Laboratoire du C.N.R.S., Ecole Polytechnique 91128 Palaiseau, France

Abstract. The effect of cesium vapor partial pressure on the plasma parameters has been studied in the dc hybrid negative ion source "CAMEMBERT III". The cesium vapor pressure was varied up to 10^{-5} Torr and was determined by a surface ionization gauge in the absence of the discharge. The negative ion relative density measured by laser photodetachment in the center of the plasma extraction region increases by a factor of four when the plasma is seeded with cesium. However the plasma density and the electron temperature (determined using a cylindrical electrostatic probe) are reduced by the cesium seeding. As a result, the negative ion density goes up by a factor of two at the lowest hydrogen pressure studied. The velocity of the directed negative ion flow to the plasma electrode, determined from two-laser beam photodetachment experiments, appears to be affected by the cesium seeding. The variation of the extracted negative ion and electron currents versus the plasma electrode bias will also be reported for pure hydrogen and cesium seeded plasmas. The cesium seeding leads to a dramatic reduction of the electron component, which is consistent with the reduced electron density and temperature. The negative ion current is enhanced and a goes through a maximum at plasma electrode bias lower than 1 V.

These observations lead to the conclusion that the enhancement of pure volume production occurs in this type of plasma. Possible mechanisms for this type of volume process will be discussed.

INTRODUCTION

In hydrogen volume negative ion sources the introduction of cesium considerably enhances the extracted negative ion current and reduces the extracted electron current. This effect was observed in 1988[1] but its nature remained oscure because the change of the plasma properties due to cesium seeding was not studied. A considerable difficulty is due to the fact that the cesium partial pressure present in the plasma is not known. If we refer to the sources constructed in view of neutral beam injection into ITER or LHD, a new uncertainty appears due to the fact that these sources are pulsed, and are studied with pulses of different length. If the discharge pulse is long enough, the cesium deposited on surfaces can be evaporated due to the discharge and is condensed again after its end. Cesium evaporation in the plasma volume can lead to the enhancement of volume production of H⁻ ions. Until now the existence of such effects was not demonstrated, but various scenarios were proposed.[2]

We have initiated the study of the effects of cesium seeding in 1995 in the hydrogen and deuterium plasmas produced in the hybrid multicusp source[3] CAMEMBERT III. We found[4] that the negative ion density at the center of the plasma extraction region is enhanced due to cesium seeding, while the electron density goes down. The purpose of this work is to correlate the observations of the negative ion and electron densities with the density of neutral cesium. We observed the enhancement of volume production of H⁻ ions and the decrease of the electron density and the electron temperature in the plasma extraction region but we do not exclude that the change of the extracted currents is related to the effect of cesium ions on the sheath present in front of the extraction opening.

EXPERIMENTAL SETUP AND TECHNIQUE

The work with cesium seeding to H_2 and D_2 plasma was effected in the large hybrid multicusp H⁻ ion source Camembert III, developped at Ecole Polytechnique[3]. In this source the magnetic filter is represented by the multicusp magnetic field. It separates the driver region, located on the source border, and the extraction region, located in the center of the source. The cesium metal was introduced into the source cesium oven in a sealed glass container, which was broken when the oven was heated to a temperature of about 200°C. After this initial introduction of cesium vapor, the cesium reservoir was heated again, to obtain higher cesium pressure. We had no trouble with the electrostatic probes or with operating the filaments and the discharge. We determined the cesium partial pressure with a surface ionization gauge, which has been installed on the source axis opposite to the extraction opening. The maximum cesium pressure studied was 1.1×10^{-5} Torr.

The plasma characteristics were measured both in the center of Camembert III, i.e. in the center of its extraction region, and near the extraction opening in the plasma electrode. The negative ion relative density, n_-/n_e, was measured by photodetachment[5,6]. The Nd-YAG laser beam diameter was 6 mm; the probe, coaxial with the laser beam, was biased positive at 50 V. In all the experiments the discharge voltage was 50 V. The electron density and temperature were measured using the same cylindrical electrostatic probe which was used for the measurement by photodetachment of n_-/n_e. The principal results for hydrogen will be reported.

We have observed the spectral lines of the neutral cesium. Their intensity is going down when the hydrogen pressure goes up. The intensity of hydrogen lines goes up with the hydrogen pressure.

EFFECT OF CESIUM SEEDING ON THE PLASMA PARAMETERS

Study of the center of the extraction region.

The relative negative ion density goes up when the cesium additive is seeded into the hydrogen plasma. The variation of n^-/n_e with hydrogen pressure is illustrated on Fig. 1 for pure hydrogen and for two cesium partial pressures, for a constant value of the discharge current (I_d). Note the large enhancement of n^-/n_e, particularly at low pressure (a factor of four in H_2 at 1 mTorr). At 1 mTorr n^-/n_e attains a maximum value (0.3) at the intermediate (5.5×10^{-5} Torr) cesium partial pressure and goes down when the cesium pressure is further enhanced.

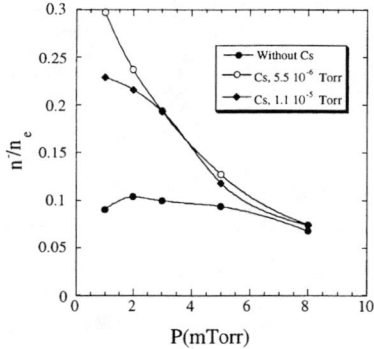

FIGURE 1. Dependence of the ratio n^-/n_e on gas pressure in hydrogen plasma for pure hydrogen and two cesium partial pressures. The discharge current is 50 A.

Figure 2 presents the variation of the electron density, n_e, the electron temperature, T_e, and plasma potential, V_p, versus the hydrogen pressure, for the same partial pressures of cesium as in Figure 1. The cesium seeding first leads to a decrease of the electron density (as reported in Ref. 4), but the further increase of the cesium partial pressure leads to a small increase of n_e. A similar effect is observed on T_e. The plasma potential V_p exhibits first a large reduction, followed by a considerable recovery when the cesium partial pressure is enhanced. If both the probe and the inner wall surfaces are equally covered by cesium, the measured change of plasma potential corresponds to an actual change of this potential. Note that an increase of the plasma potential due to cesium seeding was observed in our earlier work[4].

Figure 3 shows that the increase of the cesium partial pressure leads to a monotonous enhancement of the negative ion density for hydrogen pressures below 4 mTorr. There

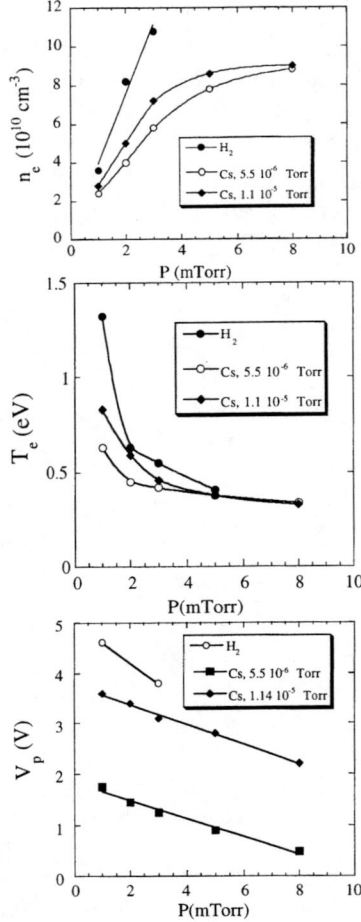

FIGURE 2. The variation with the hydrogen pressure of the electron density and temperature for two partial pressures of cesium. For comparison the values measured earlier without cesium are also shown.

may be a maximum of n^- at the lowest cesium partial pressure, when the hydrogen pressure is very low.

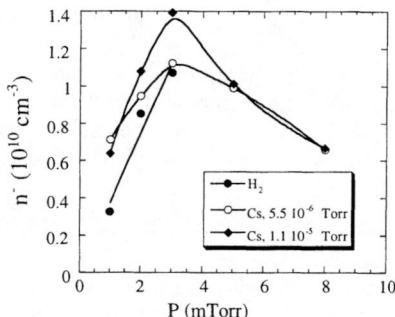

FIGURE 3. The variation of the negative ion density with the hydrogen pressure for two partial pressures of cesium. For comparison the values measured earlier without cesium are also shown.

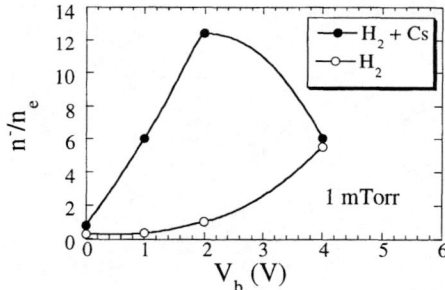

FIGURE 4. Dependence of the ratio n^-/n_e on the plasma electrode bias V_b in hydrogen plasma with cesium seeding. Partial pressure of cesium : 5.5×10^{-6} Torr.

Study of the plasma near the extraction opening

In the plasma region near the extraction opening, a weak magnetic field essentially parallel to the plasma electrode is produced due to a leak from the magnets located in the extractor in order to separate the electrons from the negative ions in the extracted beam. This field profoundly modifies the plasma properties.[7]

Figure 4 presents the variation of n^-/n_e versus the plasma electrode bias V_b near the extraction opening, for a partial pressure of cesium of 5.5×10^{-6} Torr. At the lowest hydrogen pressure, the observed enhancement is dramatic, a maximum being identified for $V_b = 2$ V. For the higher hydrogen pressure of 3 mTorr (not shown), the enhancement of the maximum value of n^-/n_e is modest, because its value without cesium seeding is very high (approximately 10). One can note a shift of the maximum value towards lower V_b.

Experiments effected using the two-laser-beam technique in this region[8] have shown that the cesium seeding leads to the enhancement of the velocity of the directed negative ion flow to the positively biased plasma electrode. This is leads to an increase of the extracted negative ion current.

Study of the extracted currents

Figure 5 shows the variation with V_b of the extracted electron and negative ion currents for the same discharge conditions and for the hydrogen pressure of 3 mTorr. The maximum partial pressure of cesium was 1.1×10^{-5} Torr. For comparison results without cesium are also shown. Hydrogen pressure was 3 mTorr. A partial pressure of cesium 1.1×10^{-5} Torr was studied here. The cesium seeding leads to a modification of the shapt of the dependence of the negative current on the plasma electrode bias: while in pure hydrogen this dependence is monotonous in the studied range, a pronounced maximum appears in the seeded plasma. An enhancement of the negative ion current by a factor 2.5 is observed due to cesium seeding, a maximum being observed for $V_b = 0.75$ V (Fig. 5b).

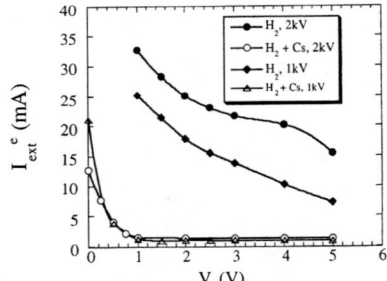

FIGURE 5a. Variation of the extracted electron current with the plasma electrode bias at two values of extaction voltage. The partial pressure of cesium was 1.1×10^{-5} Torr. For comparison results without cesium are also shown. Hydrogen pressure was 3 mTorr.

The reduction of the electron extracted current is larger than the increase of the extracted negative ion current: a factor of 30 at the optimum V_b.

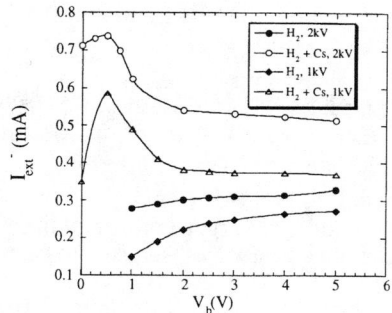

FIGURE 5b. Variation of the extracted negative ions current with the plasma electrode bias at two values of extaction voltage. For comparison results without cesium are also shown.

DISCUSSION

Several causes can be proposed for explaining the increased negative ion production in cesium seeded hydrogen plasma:

1. <u>Electron cooling.</u> The introduction of Cesium into the plasma should reduce the electron temperature, due to excitation and ionization collisions. The cross sections of excitation and ionization of Cesium by electron impact are large. There are no comparable inelastic cross sections in H_2 below 10 eV. This should: (a) increase the dissociative attachment rates; (b) reduce the H^- loss by electron collisional detachment.

2. <u>Gettering.</u> The atomic hydrogen density is reduced by its absorption by Cesium (gettering). The atomic Hydrogen destroys the H^- ions by associative detachment:
$$H^- + H^0 \rightarrow H_2(v) + e$$
The continuous absorption of atomic hydrogen by cesium can substantially increase the H^- ion density.

3. <u>Formation of CsH.</u> It is known[8,9] that CsH may be formed in hydrogen discharges containing Cesium. Calculations of the ground state potential energy curves for CsH and CsH^- have been effected by Stevens, Karo and Hiskes[10]. More work is however required for the theoretical evaluation of the possibility of dissociative attachment to this molecule. To our knowledge, this cross section has not been measured yet.

Ziesel and Herzenberg[11] have attempted to measure the cross section for dissociative attachment in LiH, because it was thought that this cross section was large compared to the corresponding cross section of vibrationally non-excited ground state. However the vapor of LiH which had to be produced by heating the corresponding powder was dissociated in the process of heating; the LiH fraction (10^{-4})) remaining in the molecular beam was then too low for H^-/LiH detection.

4. <u>Enhanced density of high Rydberg state molecules</u>. Studies effected in ArF laser-produced plasma indicate the possibility of H^- formation from the dissociative electron attachment to High Rydberg state molecules (HR)[12]:

$$H_2 (HR) + e \rightarrow H^- + H (n=2)$$

Such molecules could be produced in larger number in the presence of cesium, because of the interaction of metastable hydrogen molecules with cesium atoms in excited states.

The causes of the reduction of the plasma density, following Cesium seeding are as follows:
1. The increase of n^- at constant total density $n^+ = n_e + n^-$ leads to lower n_e.
2. Due to the increase in n^-/n_e (up to 0.4 at 1 mTorr) the positive ion loss due to mutual neutralization with negative ions goes up, leading to reduced n^+ and also n_e.

Earlier, one thought that the efficient ionization of cesium in the discharge would lead to a much higher plasma density for a given discharge current. This was in contradiction with the lower electron current extracted from cesium seeded plasma.

In conclusion we found an enhancement of the negative ion density and a reduction of the electron density and electron temperature due to cesium seeding. We do not exclude, however, that the change of the extracted currents is related to the effect of cesium ions on the sheath present in front of the extraction opening.

ACKNOWLEDGEMENTS

This work was supported by Direction des Recherches, Etudes et Techniques (France) and by INTAS Contract N° 94-0316. Enlightening discussions with Dr. J.P. Ziesel are gratefully acknowledged.

REFERENCES

* Permanent address: RRC Kurchatov Institute, Moscow (Russia)
**Permanent address: A.F. Ioffe Physical-Technical Institute, St. Petersburg, Russia.

1. Walther, S.R., Leung, K.N. and Kunkel, W.B., *J. Appl. Phys.*, **64**, 3424 (1988)
2. Peterson, J.R., *AIP Conf. Proc. N° 158,* AIP, New York, 1987, p. 113.
3. Courteille, C., Bruneteau, A.M. and Bacal, M., *Rev. Sci. Instr.*, **66**, 2533-2540 (1995)
4. Bacal, M., Michaut, C., Elizarov, L.I. and F. El Balghiti, *Rev. Sci. Instrum.*, **67**, 1138-1143 (1996)
5. Bacal, M., *Plasma Sources Sci.&Techn.*, **2**, 190-197 (1993)
6. El Balghiti-Sube, F., Baksht, F.G. and Bacal, M., *Rev. Sci. Instrum.*, **67**, 2221-2227 (1996)
7. Bacal, M., Bruneteau, J. and Devynck, P., *Rev. Sci. Instrum.*, **59**, 2152-2157 (1986)
8. Ivanov, A.A., Sionov, A.B., El Balghiti-Sube, F. and Bacal, M., *Phys. Rev. E*, **55**, 956 (1997)
8. Tam, A., Moe, G. and Happer, W., Phys. Rev. Lett., **35**, 1630 (1975)
9. Picqué, J.L., Vergès J. and Vetter, R., J. Physique-LETTRES, **41**, L-305 (1980)
10. Walther J. Stevens, Arnold M. Karo and John R. Hiskes, J. Chem. Phys., 74, 3989 (1981)
11. Ziesel, J.P. and Herzenberg, A., Private communication
12. L.A. Pinnaduwage, L.A. and Christophorou, L.G., *J. Appl. Phys.*, **76**, 46 (1994)

Creation of Highly Excited H_2 Molecules in a Hydrogen Flux Penetrative through Cesium-Hydrogen Discharge

M. Bacal[1], F.G. Baksht[2], V.G. Ivanov[2]

[1] *Laboratoire de Physique des Millieux Ionisés, Ecole Polytechnique, 91128, Palaiseau, France*

[2] *A.F. Ioffe Physical-Technical Institute, 26 Polytechnicheskaya, St. Petersburg, 194021, Russia.*

Abstract. There is a considerable interest in the creation of vibrationally excited H_2 molecules because of their application to H^- volume- plasma sources, plasma chemistry etc. . It was shown previously that a low-voltage discharge in Cs-H_2 mixture might be used as an effective source of vibrationally excited H_2 molecules. In the present communication, it is shown that vibrational distribution function (VDF) of H_2 molecules may be very significantly improved if vibrational excitation occurs in a moving flux of H_2 molecules penetrative through the discharge.

The following theoretical model is considered. Hydrogen flux flows in a flat channel and passes through two sections one after another. The first section (I) is a discharge one, where low-voltage Cs-H_2 discharge occurs and vibrationally excited H_2 molecules are created. In the next section (II), the excited hydrogen flows in a plane channel, which walls are unheated. Because of a large value of the mean vibrational energy of H_2 molecule, very significant intensification of H_2 high vibrational level population takes place in the section II. This intensification is of the order of 10^3 - 10^4 for several vibrational levels. The increase of vibrational level population occurs due to the diffusion of vibrational quanta, by v-v exchange, up to high vibrational levels. VDF, obtained at the exit of a section II, is significantly improved for the purposes of H^- generation by means of dissociative attachment.

1. Interest in vibrationally excited H_2 molecules is stimulated by numerouos practical applications of excited hydropgen molecules: e. g. hydrogen ion sources, plasmachemistry etc. Vibrationally excited H_2 molecules can be obtained in various kinds of the discharge, which takes place in pure hydrogen plasma or in hydrogen plasma with some easily ionized additives, in particular with cesium additives. Depending on concrete applications, various methods of H_2 vibrational excitation are used. Investigation of the methods of creation of vibrationally excited H_2

molecules is very essential in conformity with H^- volume-plasma sources, where H^- generation is due to the dissociative attachment (DA) of thermal electrons to vibrationally excited H_2 molecules in principal electronic $X^1\Sigma_g^+$ state [1]. In existent volume sources, H_2 vibrational excitation often occurs by caskade way - due to the radiative deexcitation of singlet excited electronic H_2 states $B^1\Sigma_u^+$, $C^1\Pi_u$ etc. [2]. In H^- sources, which use such two-step scheme of generation of vibrationally excited $H_2(X^1\Sigma_g^+)$, the processes of H_2 vibrational excitation and DA are usually separated in the volume (e.g. see [3-5]). The separation is achieved by using the special discharge chamber, where excited H_2 molecules are created (tandem source), or by means of utilizing of so called hybrid source [3,4], where fast cathode electrons, which excite H_2 molecules, are confined near the walls of the camera because of the electric drift in crossed $\vec{E}$, $\vec{H}$ fields. In hybrid sources, H^- ions are created mainly in the central part of the discharge volume, where fast or heated thermal electrons, which destroy H^- ions, are absent and electron temperature T_e is close to 1 eV, this T_e value being optimum for DA [6]. Among a number of the investigations of tandem and hybrid sources, we point to the work [4], where the optimum discharge current and electron temperature for H^- generation were calculated and comparison between calculated and experimental data was fulfilled.

The volume separation between the processes of H_2 vibrational excitation and H^- generation due to the DA may be also desirable because of possible application of low-voltage (LV) cesium-hydrogen discharge to the creation of vibrationally excited H_2 molecules [7-9]. Utilizing of this kind of the discharge for H_2 vibrational excitation is the most perspective in a comparatively dense plasma. In such a plasma, a good vibrational H_2 distribution function (VDF) can be achieved due to e-v and v-v exchanges in a principal $X^1\Sigma_g^+$ electronic state. But for some purposes, e.g. for H^- sources for thermonuclear application, it is desirable to have a considerably low hydrogen pressure in the volume, from which H^- ions are extracted. For this reason, it is also desirable to separate the processes of a generation of vibrationally excited H_2 molecules (the first chamber with a considerably high pressure) and the processes of H^- creation and subsequent extraction (the second chamber with a low hydrogen pressure). The required pressure difference between two chambers can be obtained because of hydrogen outflow with a sound velocity V_s from the first chamber into the second one.

In present communication, it will be shown that the VDF of H_2 molecules, which flow out of the first chamber, may be significantly improved if the outflow will be organized by the proper way.

2. The model, which is considered in the present investigation, is illustrated by Fig. 1a. Molecular hydrogen flows in a channel between two parallel flat walls and passes through two sections I and II one after the other, the lengths of the sections being h_1 and h_2 respectively. In section I, LV cesium-hydrogen discharge exists between two flat electrodes. Here the initial vibrational excitation of molecular hydrogen occurs. After it, the hydrogen flux penetrates into the section II, which is isolated from section I. At the exit of the channel of section II ($x = h_2$) molecular hydrogen flows out with the sound velocity V_s. The temperature of the walls of section II is close to the room temperature. In this section, an intensification of the vibrational excitation of H_2 molecules occurs due to the non-resonance v-v exchange of the molecules in a cold gas.

The distribution of the molecular hydrogen pressure $p(x)$ and concentration $N_{H_2}(x)$ along the length of the channel of section II can be defined approximately in analogy with viscous flow of a gas in a round channel [11]. If the channel is long enough ($h_2/L \gg 1$), the following expression for a mean drift velocity of the gas flow can be obtained by means of the averaging of the gas dynamic velocity [12] over the cross-section of a flat channel:

$$V = -\frac{L^2}{12\eta}\frac{dp}{dx}, \quad (1)$$

where η is a viscosity of a molecular hydrogen. Taking into account $V = V_s$ at the exit of a channel, we obtain the molecular hydrogen pressure distribution along the channel:

$$p(x) = \left[p_0^2 - (p_0^2 - p_s^2)x/h_2\right]^{1/2} \quad (2)$$

Here $p_0 = p(0)$ is the hydrogen pressure at the inlet of the channel, p_s is a pressure at the exit of the channel. The length h_2 of the channel and its width L satisfy the relation:

$$h_2/L = (R_s/24)(c_p/c_v)^{-1}[(p_0/p_s)^2 - 1], \quad (3)$$

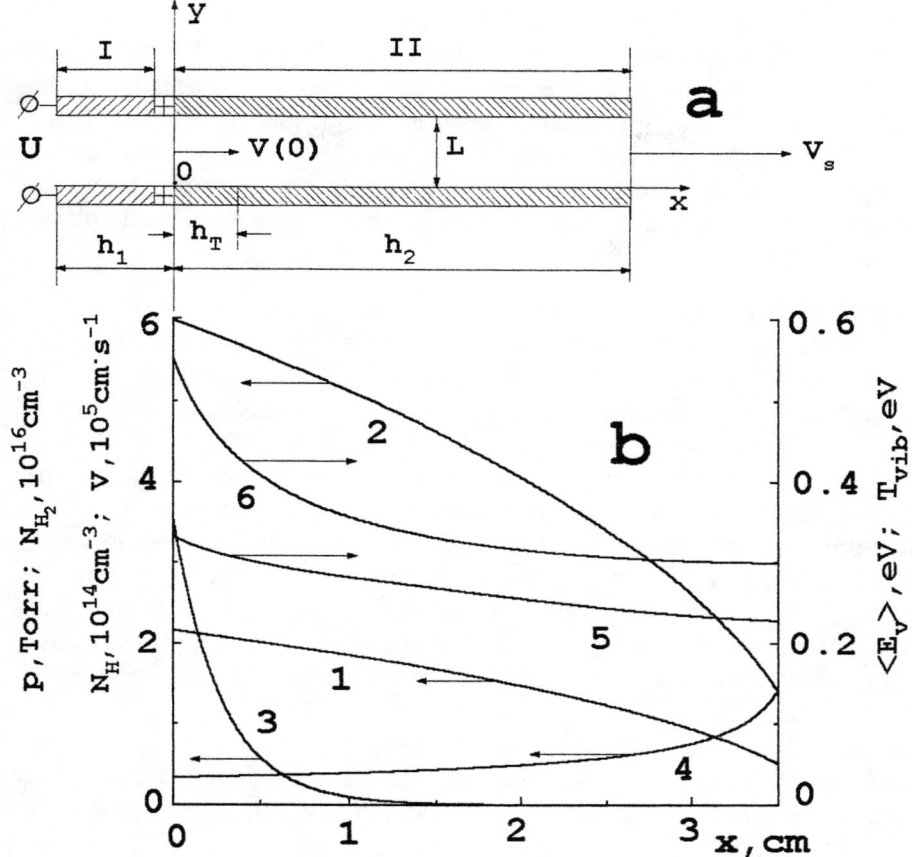

Fig.1a. Scheme of the hydrogen flow in a channel.
I - discharge zone. II - flow of the vibrationally excited gas in a channel.

Fig.1b. Distribution of the main parameters of the hydrogen flux along the channel: 1 - p, 2 - N_{H_2}, 3 - N_H, 4 - V, 5 - $\langle E_v \rangle$, 6 - T_{vib}.
Discharge parameters: L = 0.3 cm, h_1 > 1 cm, $N_{H_2}^{(0)}$ = 3·10^{16} cm^{-3}, $N_H^{(0)}$ = 1.62·10^{14} cm^{-3}, $N_{Cs}^{(0)}$ = 7·10^{13} cm^{-3}, $N_{H^-}^{(0)}$ = 5.1·10^{12} cm^{-3}, n_e = 2.07·10^{13} cm^{-3}, T_e = 0.88eV, $\langle E_v \rangle$ = 0.366 eV, T_{vib} = 0.59 eV.
Gas temperature in the discharge - T_0 = 0.08 eV.
Gas temperature in the channel - T_0 = 0.04 eV.

where $R_s = \rho_s L V_s/\eta$ is Reynolds number, which is calculated by means of channel transverse dimension L and gas density $\rho_s = M_{H2}\, p_s/kT$ at the exit of the channel ($x = h_2$), T being a gas temperature in the channel. Plasma parameters in the initial LV discharge (section I) and gas parameters in the channel (section II) are considered to be homogeneous over the cross-section of the channel. As an a example, in Fig. 1b the distribution of the main parameters of the gas flow along the channel of section II are shown: here hydrogen pressure p, concentration N_{H_2}, mean velocity V, hydrogen atom concentration N_H (see Appendix), and mean vibrational energy $<E_v>$ of H_2 molecules are depicted together with the vibrational temperature $T_{vib} = (E_1/k)/\ln(N_0/N_1)$. Here and later E_v and N_v are the energies and concentrations of H_2 molecules at the vibrational level numbered v.

3. Let as consider the main peculiarities of the calculations of the plasma composition and H_2 VDF in the initial LV discharge. The calculation of the plasma parameters of LV $Cs-H_2$ discharge was performed by the method, which was elaborated in [13] for homogeneous gas discharge gap. A comparatively dense discharge plasma is considered. Therefore it is supposed that the cathode beam relaxation takes place due to the pair electron-electron collisions and the beam energy is spent for the heating of thermal plasma electrons [14].

The calculations were performed in two stages [9]. At the first stage, the system of the equations, which describe the electron-vibration kinetics in the discharge, was solved, an electron temperature T_e being considered as a given value at this stage. All the plasma parameters, including VDF of H_2 molecules and ion fluxes j_i from plasma to the electrodes, were determined by the solution of these equations.

At the second stage, the system of the boundary conditions, which describes the particle and electron energy balance on plasma boundaries, was solwed. The cathode φ_1 and anode φ_2 potential barriers and electron fluxes, j_{e1} и j_{e2}, from plasma to the electrodes were determined by the solution of this system of the equations. As a result, the discharge voltage, $U = \varphi_1 - \varphi_2$, and the discharge current, $j = j_{e2} - j_i$, which correspond to the given electron temperature T_e, were determined etc.

The characteristic peculiarity of the discharge considered here is a comparatively small value of an emission current j_s ($j_s < 10$ A/cm^2). This j_s value is significantly less than in analogous modes of the LV discharge in a dense collisional plasma, which were considered previously (e.g. see [13]). This peculiarity simplifies significantly a practical realization of this kind of the discharge. This feature of the discharge is illustrated by Fig. 2, where the

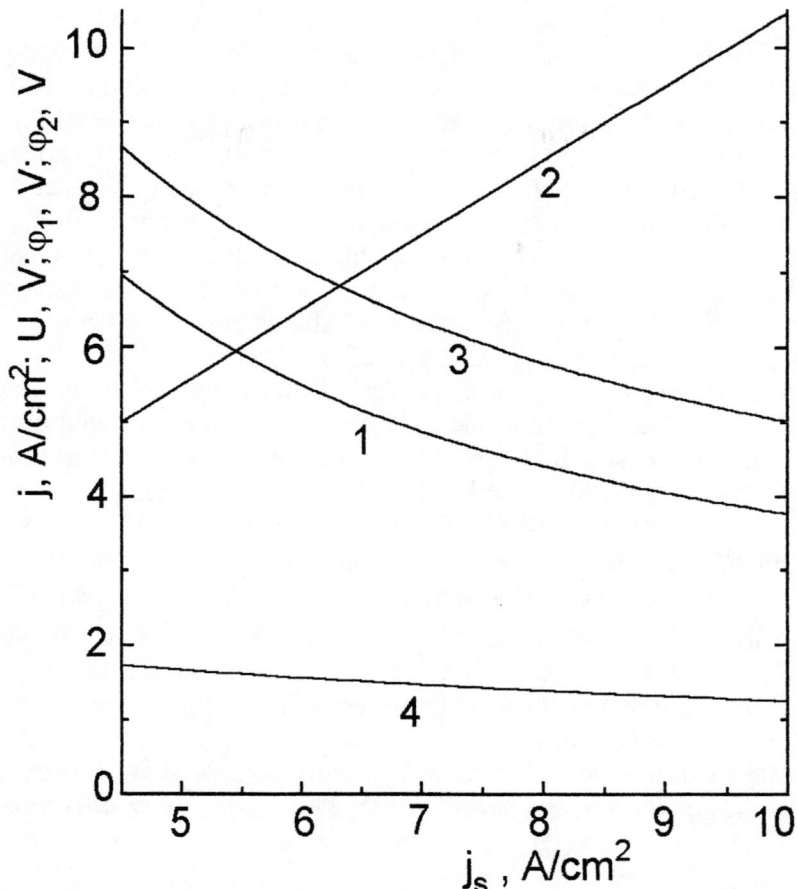

Fig.2. The dependence of the voltage, discharge current and near-electrode potential barriers in Langmuir sheaths upon cathode emission current:

1 - U, 2 - j, 3 - φ_1, 4 - φ_2.

Discharge parameters: $L = 0.3$ cm, $h_1 > 1$ cm, $N_{H_2}^{(0)} = 3 \cdot 10^{16}$ cm^{-3}, $N_H^{(0)} = 1.8 \cdot 10^{14}$ cm^{-3}, $N_{Cs}^{(0)} = 10^{14}$ cm^{-3}, $N_{H^-}^{(0)} = 3.8 \cdot 10^{12}$ cm^{-3}, $n_e = 3.26 \cdot 10^{13}$ cm^{-3}, $T_e = 0.65$ eV, $\langle E_v \rangle = 0.335$ eV, $T_{vib} = 0.55$ eV.

dependences of φ_1, φ_2, U and j upon j_s are shown. One can see that typical LV mode, where $\varphi_1 < E_d/q$, is realized at the comparatively moderate current densities: 5 A/cm^2 < j_s < 10 A/cm^2, which can be obtained without difficulties in practice. Here $E_d \cong 8.8$ eV is a threshold of the direct H$_2$ dissociation by electron impact. Of course, in this current range, a vibrational H$_2$ excitation of high v levels is less than it occurs in corresponding discharge modes, which were considered previously (e.g. see [7-8], [13]). But now, the needed level of high vibrational level population will be achieved by means of the significant additional vibrational excitation in a proper organized hydrogen flow in a cold channel of the section II.

Because of a considerably small value of the gas flow velocity V_0 in a discharge zone (section I), the calculation of the discharge plasma properties was performed by the method, which was elaborated for a plasma at rest. But two circumstances, which are specific for a discharge in a moving plasma, have to be analized specially.

Firstly, it is a comparatively high value of a time of the vibrational relaxation of H$_2$ molecules. It results in a comparatively slow creation of H$_2$ VDF $f_v^{(0)}$ in the discharge. Therefore the establishment of a vibrational molecular distribution was considered at the plasma parameters, which are typical for the considered mode. The method of calculations was described in [15]. In addition to [15] in present investigation the time dependences of hydrogen atom concentration N_H and negative hydrogen ion concentration N_{H^-} were taken into account, because of a considerably slow relaxation of these concentrations in the discharge. In Fig. 3 the results of calculations are shown. Here the values $f_v(t)/ f_v(\infty)$ are depicted, were $f_v(t)$ is an instant value of VDF during the process of vibrational relaxation and, $f_v(\infty) = f_v^{(0)}$, is a final value of VDF, which is established at the end of the relaxation process. The calculations are performed for two values of translational gas temperature T_0. One can see that the process of vibrational relaxation depends significantly on T_0 (compare Fig. 3a and 3b). At a comparatively high value of T_0, the maximums of $f_v(t)/ f_v(\infty)$ appear at the time dpendences of VDF, i.e. the vibrational relaxation of the concentrations $N_v(t)$ of vibrationally excited H$_2$ levels is non-monotonous in this case. This is the result of v-t exchange, which is more essential at high T_0 value and which diminishes the populations N_v of high v levels at the end of vibrational relaxation. Just for this reason, the final value $N_v^{(0)} \equiv N_v(\infty)$ of high vibrational level population and the final value $N_H^{(0)}$ of the hydrogen atom concentration, which are established in the discharge at high translational temperature T_0 ($T_0 = 0.08$ eV), are remarkably less than at low T_0 ($T_0 = 0.026$ eV). The time of the whole vibrational relaxation and the

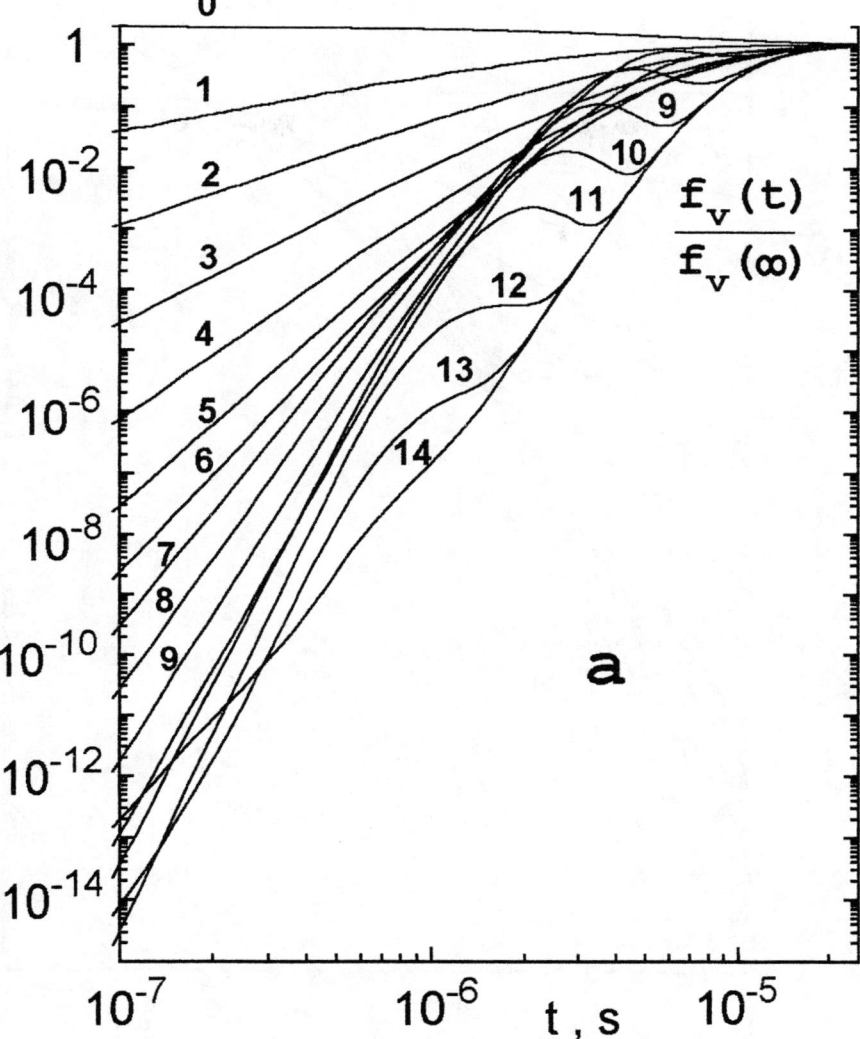

Fig.3a. Vibrational relaxation of H_2 molecules in a low-voltage discharge plasma. The numbers of vibrational levels are shown near the curves.

$N_{H_2}^{(0)} = 3 \cdot 10^{16}$ cm^{-3}, $N_{Cs}^{(0)} = 10^{14}$ cm^{-3}, $T_e = 0.8$ eV.
$n_e = 5.8 \cdot 10^{13}$ cm^{-3}, $N_H^{(0)} = 7.8 \cdot 10^{14}$ cm^{-3},
$N_{H^-}^{(0)} = 6.6 \cdot 10^{12}$ cm^{-3}, $T_0 = 0.026$ eV.

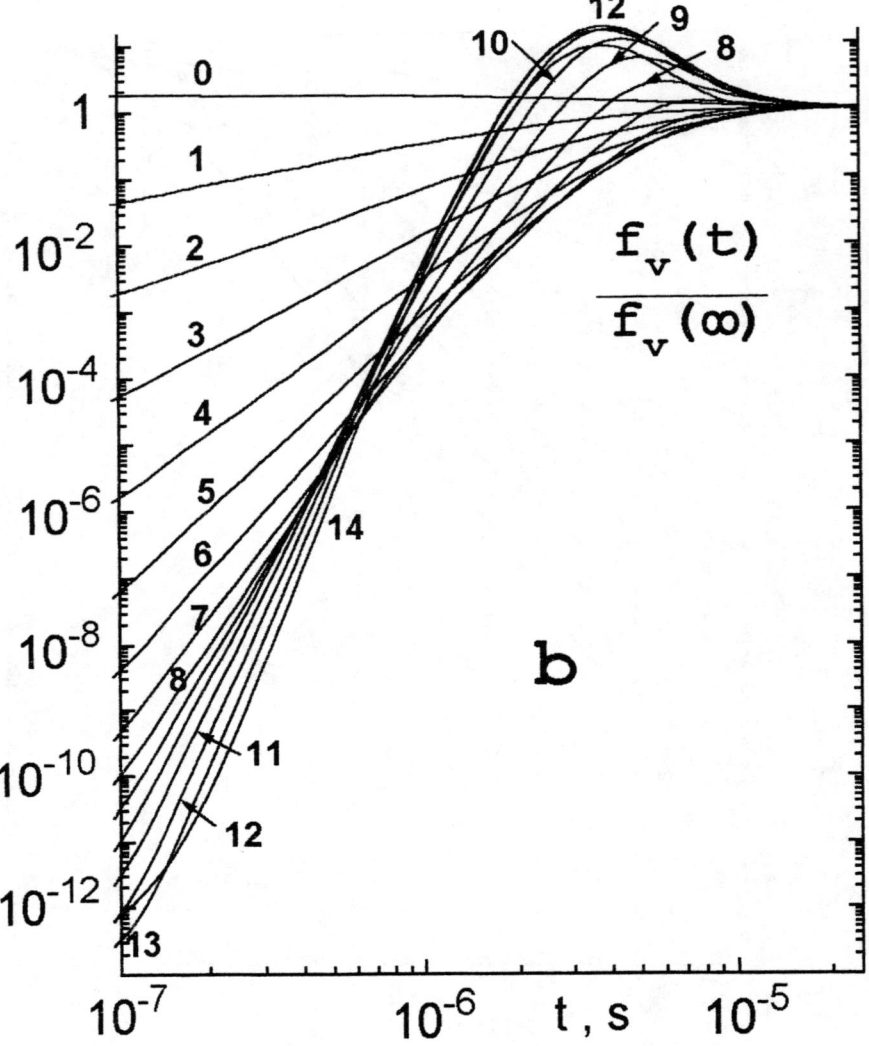

Fig.3b. Vibrational relaxation of H_2 molecules in a low-voltage discharge plasma. The numbers of vibrational levels are shown near the curves.

$N_{H_2}^{(0)} = 3 \cdot 10^{16}$ cm^{-3}, $N_{Cs}^{(0)} = 10^{14}$ cm^{-3}, $T_e = 0.8$ eV.
$n_e = 5.4 \cdot 10^{13}$ cm^{-3}, $N_H^{(0)} = 3.3 \cdot 10^{14}$ cm^{-3},
$N_{H^-}^{(0)} = 7 \cdot 10^{12}$ cm^{-3}, $T_0 = 0.08$ eV.

corresponding length of vibrational relaxation are equal to $\tau_v^{(0)} \cong 15$ mcs, and $l_v^{(0)} = V_0 \tau_v^{(0)} \cong 1$ cm for both two cases considered before.

Secondly, a specific peculiarity of the discharge in a moving plasma is a considerably slow relaxation of a translational gas temperature in the discharge. This peculiarity is essential because, as it was mentioned above, the translational temperature T_0 influences significantly upon the process of vibrational relaxation and upon the VDF, which is created in the discharge. In the conditions, which are considered here, the gas heating in the discharge gap is caused mainly by the gas contact with a hot emitter, but not by the collisions between the electrons and the heavy particles or by v-t or non-resonance v-v exchange. Because of the gas contact with a hot emitter, an expanding layer takes place near the emitter surface. Estimating the length h_T of the emitter, at which the width δ_T of the thermal boundary layer is equal to the interelectrode distance L, one can obtain $h_T = \theta \cdot \text{Pr} \cdot R_0 L$ [16], where $\text{Pr} = \eta_0 c_p / \kappa_0$ is Prandtl number; $R_0 = \rho_0 L V_0 / \eta_0$ is Reynolds number, which is calculated according to the hydrogen flux parameters in a discharge zone (section I); κ_0, η_0 and ρ_0 are the heat conductivity, viscosity and density of the molecular hydrogen in a discharge gap; c_p is a hydrogen specific heat at a constant pressure; $\theta \sim 0.1$ is a numerical factor. Using this estimation, one can obtain that the temperature T_0, which is typical for stationary discharge, has to be established in a discharge zone at the distance $h_T \cong 0.2$ cm, along the hydrogen flux. This distance is less than $l_v^{(0)}$. The other typical times and corresponding lengths (e.g. length of electron maxwellization or cesium atom ionization [17] etc.) are shorter than the times and lengths considered above. So, in the case considered here it is sufficient to use the length h_1 of the discharge zone, which is greater than 1 cm, for plasma to relax to the conditions, which are characteristic of the stationary LV cesium-hydrogen discharge. The condition $h_1 > 1$ cm is supposed to be fulfilled in subsequent calculations.

The definition of gas temperature T in the channel, i.e. in section II, is essential for further consideration. Because the probabilities of excitation of high vibrationally excited levels due to the non-resonance v-v exchange increase significantly (in comparison with corresponding deexcitation probabilities) at lower translational temperature T, it is reasonable to use a sufficiently low value of gas temperature in order to receive high concentration of vibrationally excited H_2 molecules in the channel. In further calculations, we shall suppose that the walls of the channel will be at room temperature, but the gas temperature T in the channel will be slightly greater than the room temperature. Approximately, it takes into account gas heating in the channel because of v-t and non-resonance v-v exchange. In the majority of calculations, we shall suppose that $T \cong (0.03 - 0.04)$ eV.

Because the length h_T of gas temperature relaxation is small in comparison with the length h_2 of the channel (see Fig. 1a), we shall neglect the variation of the gas pressure along the length $\sim h_T$. At the same time we shall introduce a jump of gas temperature $\Delta T = T_0 - T$ at the boundary (x = 0) between the gas discharge gap (I) and the channel (II). Because pressure and neutral flux are continuous at this boundary, the jumps of the concentrations N_{H_2} and N_H and of the flux velocity V must take place at the boundary. The H_2 VDF f_v and the degree of H_2 dissociation are continuous at the boundary. The values of N_{H_2}, N_H and V in the discharge zone, i.e. before the jump, are denoted as $N_{H_2}^{(0)}$, $N_H^{(0)}$ and V_0. The corresponding values behind the jump are denoted as $N_{H_2}(0)$, $N_H(0)$ and $V(0)$.

4. The VDF of H_2 molecules in the channel of section II was obtained by means of the solution of the following system of the equations:

$$\frac{d}{dx}(N_v V) = I_v^{(vv)}\{N_v\} + I_{vM}^{(vt)}\{N_v\} + I_{vA}^{(vt)}\{N_v\} + I_v^{(w)}\{N_v\}, \quad (4)$$

$$(v = 0,1,2,...,14)$$

The collision terms on the right side of the equation (4) correspond to v-v exchange, v-t exchanges with hydrogen molecules and hydrogen atoms and H_2 vibrational relaxation on the walls of the channel. The collision terms, which describe v-v and v-t exchanges, are formulated in [8,13].

Now we shall consider the collision term $I_v^{(w)}\{N_v\}$, which describes the interaction between H_2 molecules and the walls of the channel. As opposed to the discharge section I, in the channel II, where e-v exchange does not occur, the concrete form of $I_v^{(w)}$ terms is essential because it influences significantly on the VDF $f_v(x)$, which is created in the channel. It is convenient to describe the losses of vibrationally excited H_2 molecules on the walls by introducing of corresponding effective times of life τ_v, so $I_v^{(w)}\{N_v\} = -N_v/\tau_v$. For the gas, which is at rest in a gap with parallel plane surfaces, τ_v can be written in the analogy of [18]:

$$\tau_v = \Lambda^2 / D_{sd} + (L/\bar{v}) \cdot (2-\gamma_v)/\gamma_v, \quad (5)$$

where $\Lambda = L/\pi$ is an effective diffusion length for a plane geometry; $D_{sd} = (3\pi/16\sqrt{2})\bar{v}\,l_{H_2}$ is a self-diffusion coefficient of H_2 molecules [19]; $\bar{v} = \sqrt{8kT/\pi \cdot M_{H_2}}$; $l_{H_2} = 1/N_{H_2}\bar{\sigma}^{(1)}$ is a free path of H_2 molecule, γ_v is a probability of the wall loss of H_2 molecule, which is vibrationally excited at v level. The first term on the right side of (5) is significantly greater than the second one, if $\gamma_v >> (3\pi^2/8\sqrt{2})(l_{H_2}/\Lambda)$. In this case, $\tau_v \cong \tau_d \equiv \Lambda^2/D_{sd}$, where τ_d is the time of the diffusion escape of H_2 molecule, which does not depend on γ_v. In the opposite limiting case, where $\gamma_v << (3\pi^2/8\sqrt{2})(l_{H_2}/\Lambda)$, i.e. γ_v is a very small value (note that $l_{H_2}/\Lambda << 1$), $\tau_v \cong 2L/(\gamma_v \cdot \bar{v})$. In this case τ_v is significantly greater than τ_d. In Appendix, it is shown that, under certain conditions, the expression of the type of $I_v^{(w)}\{N_v\} = -N_v/\tau_v$ describes correctly the escape of particles (molecules, atoms etc.) from the gas flux to the walls of a channel, if τ_v is expressed by (5). Because the second term in (5) is rather evident in the limit of small γ_v, the consideration is carried out in Appendix conformably to calculation of the losses of particles in the opposite limit case, where $\gamma_v \sim 1$ and $\tau = \tau_d$. The concrete calculations are performed in Appendix conformably to computation of the diffusional losses of atomic hydrogen. The diffusion coefficient D_{12} of atomic hydrogen in the molecular one is adopted from [20].

5. Let us consider the calculation of γ_v probabilities. The calculations, which were performed in present communication by means of the method considered above, show strong dependence of the VDF $f_v(h_2)$ of H_2 molecules, excited at high levels, upon the value of γ_v. Therefore it is necessary to consider a concrete mechanism of the interaction between vibrationally excited H_2 molecules and the walls in the theoretical consideration of the flow of vibrationally excited H_2 molecules in the channel.

Naturally, it is reasonable to have a wall material with a large value of potential barrier for surface adsorption of molecular hydrogen. Further, the copper will be considered as such a material because the values of the refered barrier of (0.8-1.5) eV were revealed for Cu (e.g. see [21-24]). The most complete, accessible for us data about the probabilities of the destruction of vibrationally excited H_2 molecules on the copper surface are contained in [23]. It was shown in [23] that surface desactivation of the vibrationally excited H_2 molecules was caused by the tunneling of H_2 molecules through the surface potential barrier, if the kinetic energy of H_2 molecule is comparatively small ($E_{kin} \leq 0.1$ eV). The tunneling probability $w_v(E_{kin})$ was determined in [23] as the function of a vibrational quantum number v for $v \leq 10$ and as the function of kinetic energy of molecule for kinetic energies $E_{kin} \geq 0.1$ eV. In present communication, it was supposed that $\gamma_v = w_v(E_{kin})$. The data for $v = 11-14$ were obtained by extrapolation of the data obtained for $v \leq 10$. Because the extrapolation of the data obtained for $E_{kin} \geq 0.1$ eV is difficult, in the present communication, the data obtained for $E_{kin} = 0.1$ eV were used for $E_{kin} < 0.1$ eV. According to the last assumption, a noticeable increase of the wall losses γ_v of vibrationally excited H_2 molecules takes place in a number of cases. It was supposed also that the interaction between H_2 molecules and the channel walls did not change the whole number of H_2 molecules, i.e. the diminution of the number of vibrationally excited molecules because of their interaction with the walls was compensated by the desorption of H_2 molecules in the principal state (v=0). In [23] two sets of the data were obtained. These sets of the data correspond to two directions (parallel or perpendicular) of the molecular axis with regard to the surface, when molecule is disposed at the top of the potential barrier for the surface adsorption. The calculations, which were performed in the present communication, show that the results depend weakly on this factor. In further calculations, the data, which correspond to the perpendicular direction of the molecular axis with respect to the surface, were used.

6. Now we shall consider the results of the calculations. In the calculations, the H_2 VDF was determined by means of the equations (4). Simultaneously, distribution of hydrogen atoms along the channel was calculated by means of the solution of the equation

$$\frac{d}{dx}(N_H V) = - N_H/\tau_d^{(H)} , \qquad (6)$$

which is a consequence of the relation (A.8) (see Appendix). The initial conditions for the equations (4) and (6) were formulated at $x = 0$, i.e. at the boundary between the discharge section (I) and the channel (II). The known H_2 VDF $f_v^{(0)}$, which was created in the discharge, was used in the initial conditions as well as the values of H_2 and H concentrations:

$N_{H_2}(0) = N_{H_2}^{(0)} T_0/T$, $N_H(0) = N_H^{(0)} T_0/T$, which are created just after the temperature jump at the inlet of the channel. The results of the calculations are shown in Fig. 1b and Fig. 4-6.

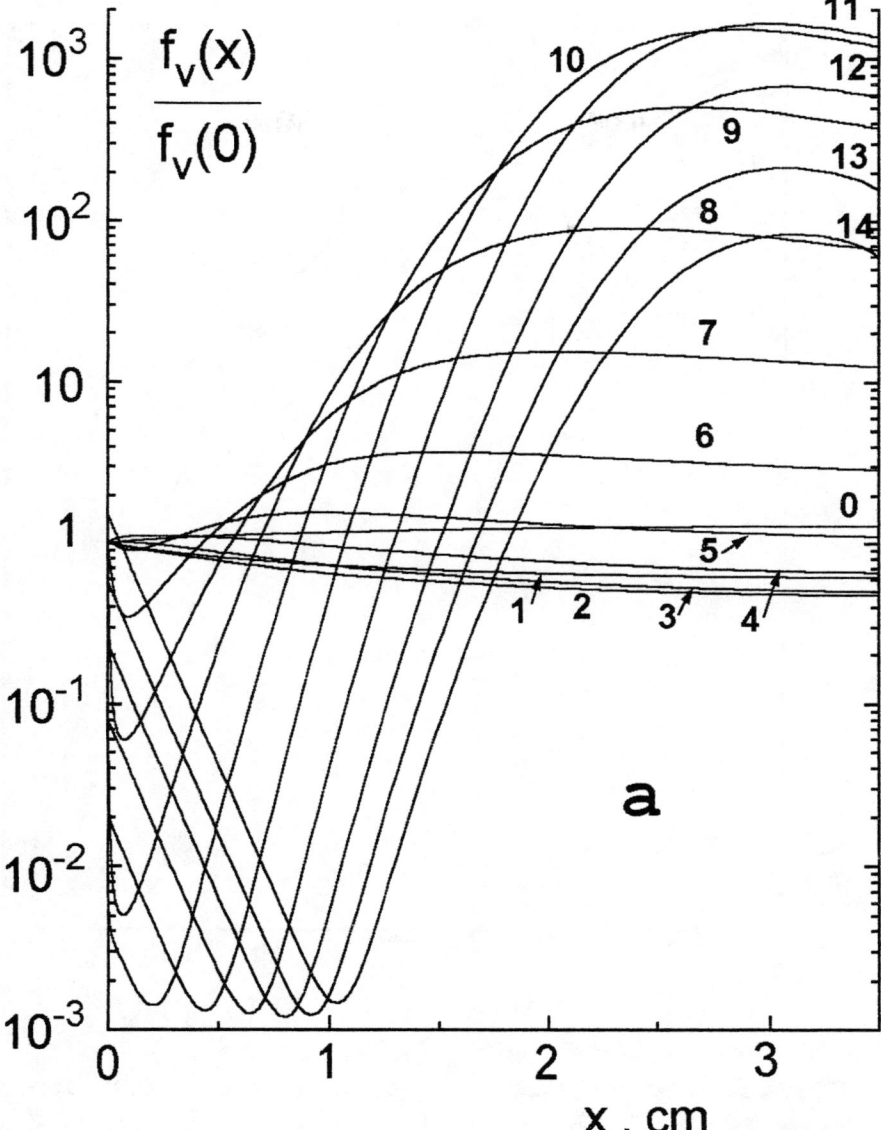

Fig.4a. Alteration of the vibrational distribution function along the length of the channel. The numbers of vibrational levels are shown near the curves.

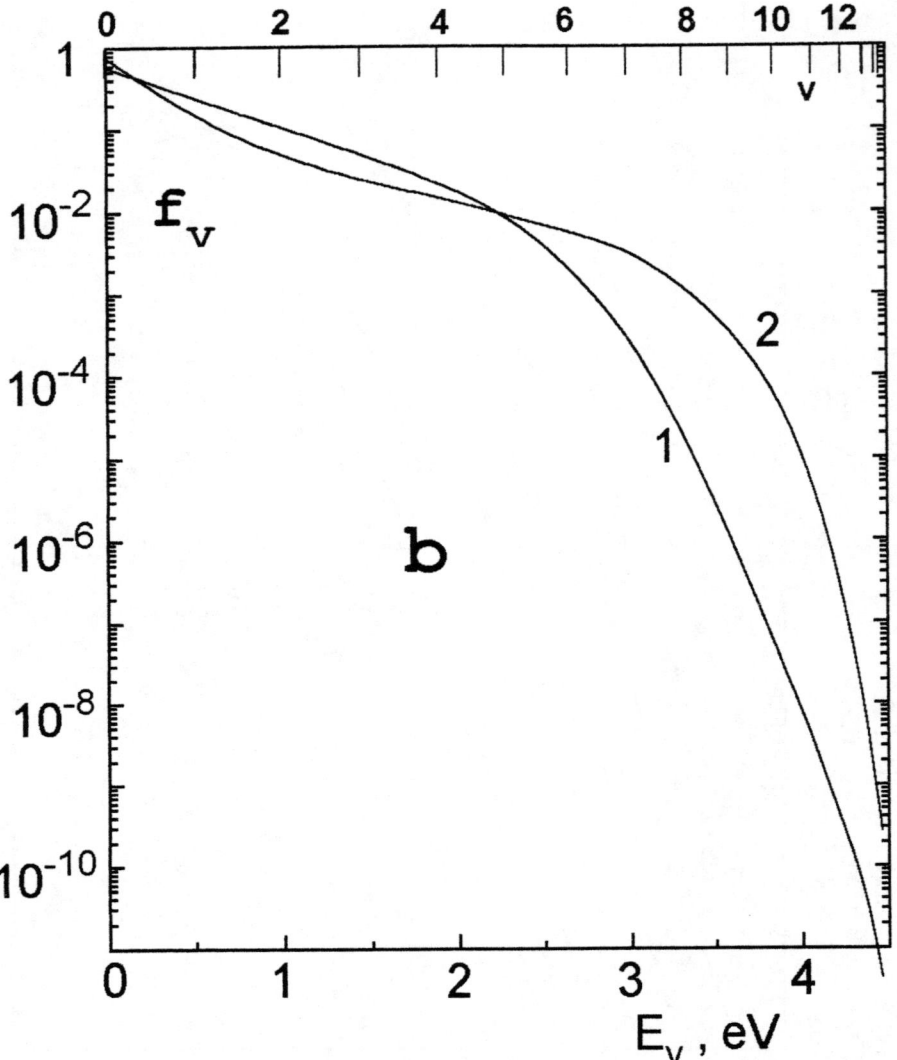

Fig.4b. The vibrational distribution functions in the inlet (x=0) and in the exit (x=h_2) of the channel. The length of the channel h_2 = 3.5 cm. 1 - $f_v(0)$, 2 - $f_v(h_2)$.

The gas discharge parameters and the gas temperature in the channel are the same as in Fig.1b.

In Fig. 4a-6a the ratio $f_v(x)/f_v(0)$ is shown. Here $f_v(x)$ is a VDF, which is created in the channel, $f_v(0) = f_v^{(0)}$ is the initial VDF. One can see that the process of the establishment of the VDF in the channel can be subdivided into two stages. At the first stage, a sharp diminution of high vibrational level populations takes place. This diminution is caused mainly by fast v-t relaxation due to the interaction between vibrationally excited H_2 molecules and H atoms. This is a well known consequence of the large values of corresponding rate constants. At the second stage, a very intensive (up to many orders of magnitude for the certain levels) increase of the vibrational level population occurs because of vibrational pumping due to v-v exchange in a cold gas. An increase of the populations N_v of vibrationally excited H_2 levels because of v-v exchange was observed previously in [25-27] in afterglows of the discharges, which plasma contains vibrationally excited hydrogen. In Fig. 5c the first stage of vibrational relaxation of H_2 molecules is shown on a large scale. Fig. 5c is related mainly to several upper levels. This figure illustrates an initial incrase (compare with 5a) of the populations of the upper levels, which are occupied due to atom into molecule recombination, which takes place because of gas temperature decrease at the inlet of the channel.

The results shown in Fig. 4 and 5 are obtained for the conditions, where surface desactivation of H_2 molecules is taken into account. In Fig. 6 the other results are shown. These results are obtained for the conditions where surface desactivation is neglected, i.e. $\gamma_v = 0$, the other parameters of the discharge and gas flux being the same as in Fig. 4 and 5. The comparison of 5a and 6a shows that, in the example considered here, the surface H_2 desactivation reduces the VDF $f_v(h_2)$ approximately by two orders of magnitude for the vibrational levels, which populations are very intensified in the channel. But in a number of other calculations, the difference between the results of computations, which take or take not into account a surface H_2 desactivation, is not so large. E.g. the calculations performed for the parameters, which correspond to Fig. 4, show that, in this case, the above mentioned difference is only of one order of magnitude.

In Fig. 4b-6b, the initial H_2 VDF $f_v(0) = f_v^{(0)}$ and the final ones $f_v(h_2)$ are shown. Moreover, in Fig. 5b and 6b, the values

$$\Gamma_v(h, T_e') = \frac{f_v(h) K_v(T_e')}{<K_{DA}(h, T_e')>}, \quad (7)$$

are shown additionally, where

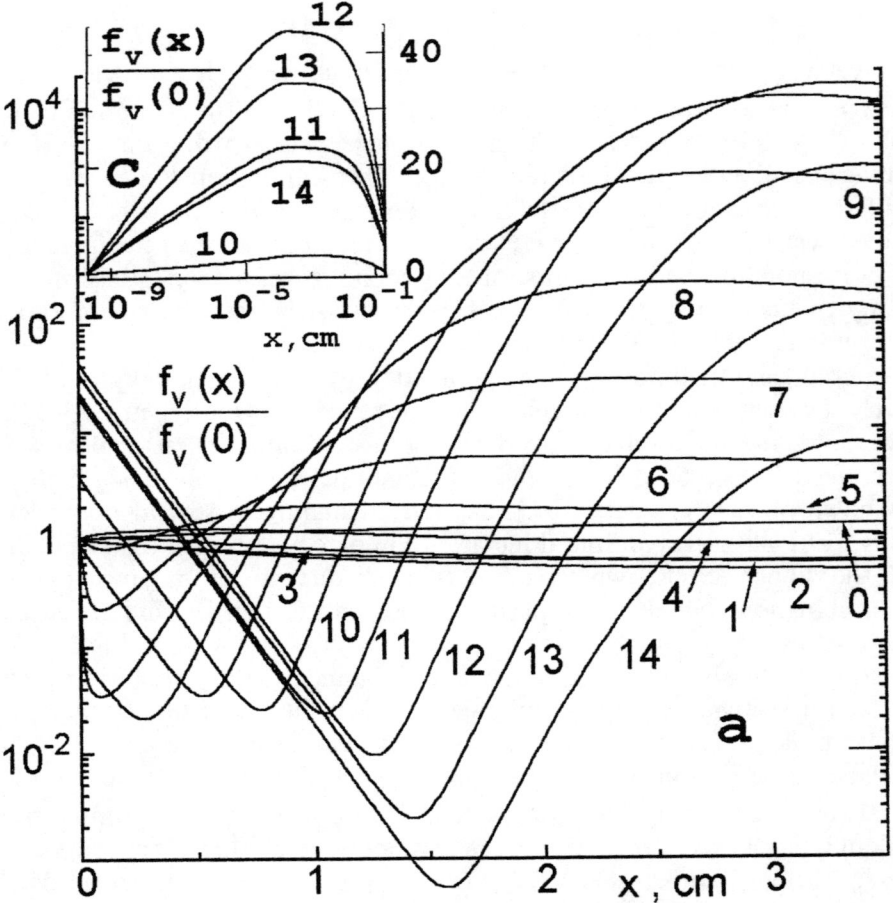

Fig.5a. Alteration of the vibrational distribution function along the length of the channel. The numbers of vibrational levels are shown near the curves.

Fig. 5c. The initial stage of the vibrational relaxation in the channel. The discharge plasma parameters are the same as in Fig. 2.

Gas temperature in the discharge - $T_0 = 0.06$ eV.

Gas temperature in the channel - $T = 0.03$ eV.

$j_s = 4.5$ A/cm^2, $j = 5$ A/cm^2, $U = 4.9$ V.

$\varphi_1 = 8.65$ V, $\varphi_2 = 0.75$ V.

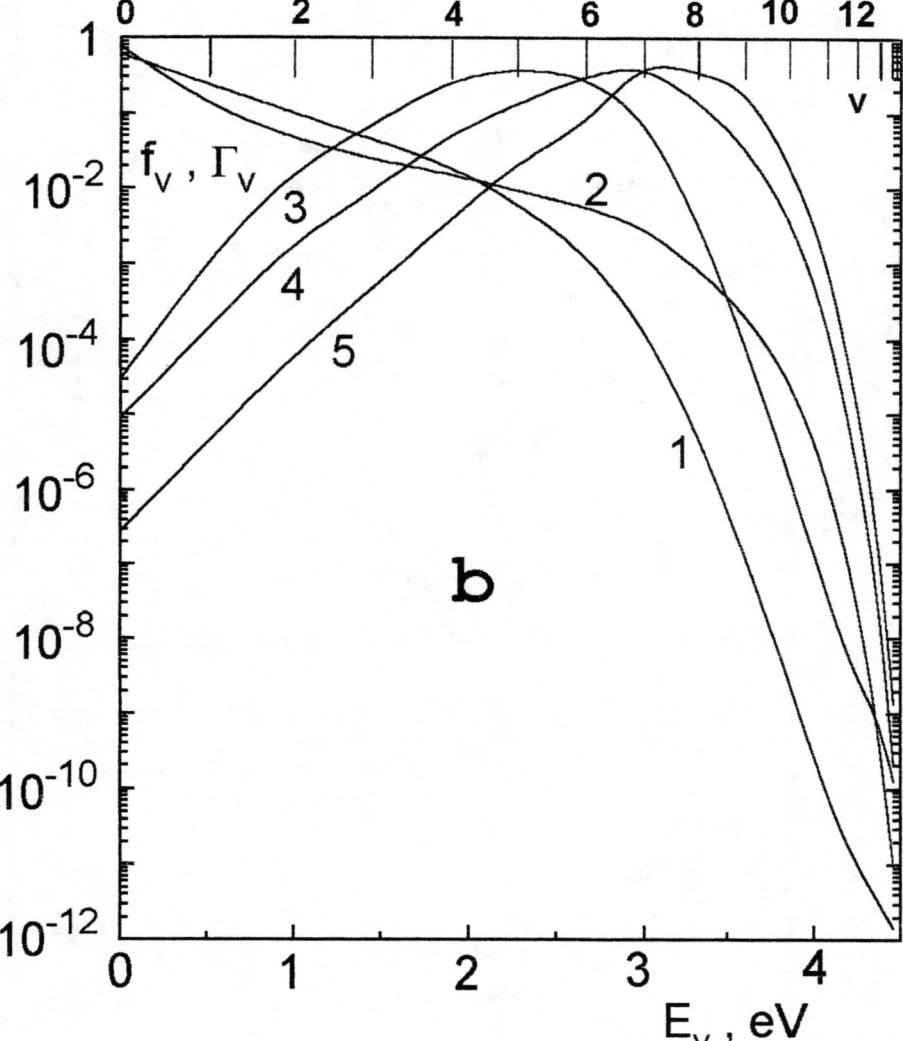

Fig.5b. Vibrational distribution functions and relative contribution of the various vibrational levels to the dissociative attachment in the inlet and in the exit of the channel.

1 - $f_v(0)$, 2 - $f_v(h_2)$, 3 - $\Gamma_v(0, 0.7\text{ eV})$, 4 - $\Gamma_v(h_2, 0.7\text{ eV})$, 5 - $\Gamma_v(h_2, 0.2\text{ eV})$.

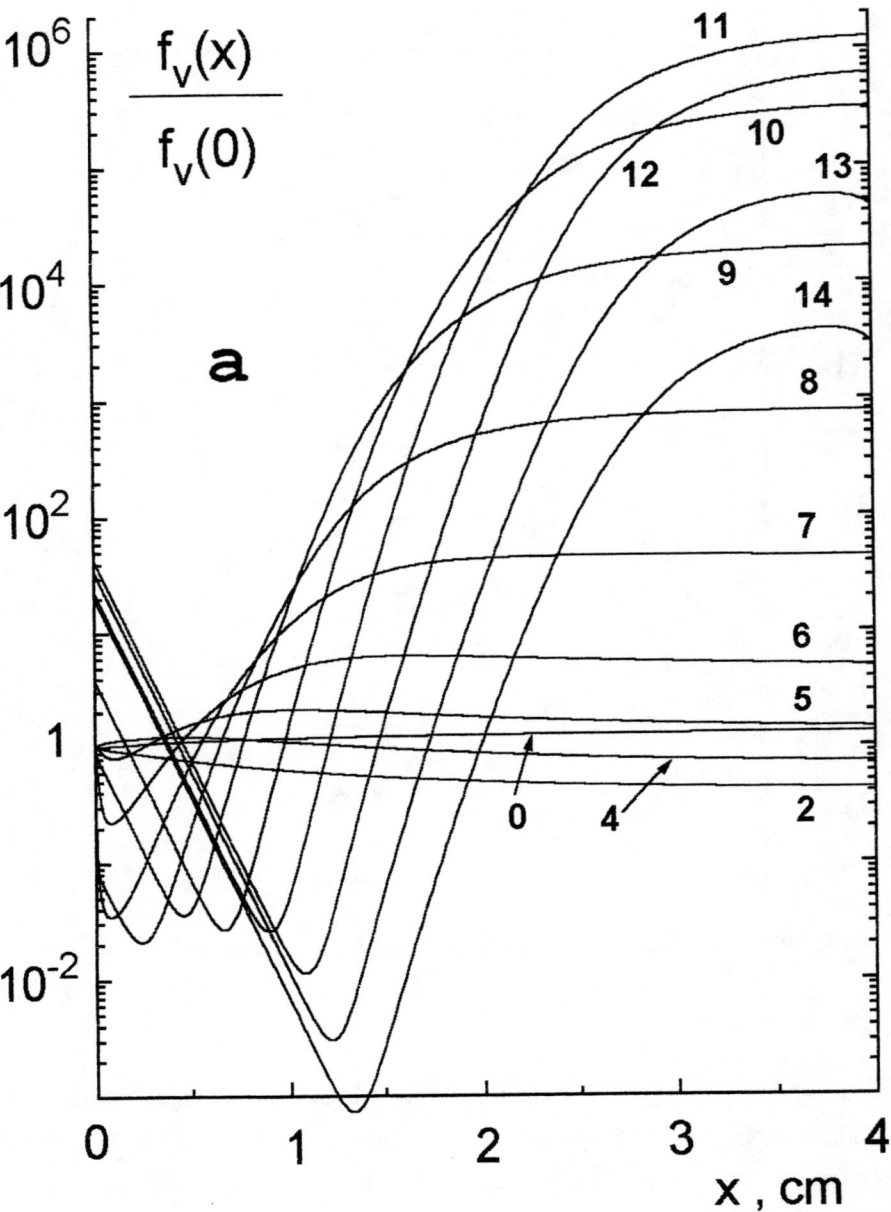

Fig.6a. The same as in Fig.5a but without taking into account the vibrational H_2 relaxation on the walls of the channel.

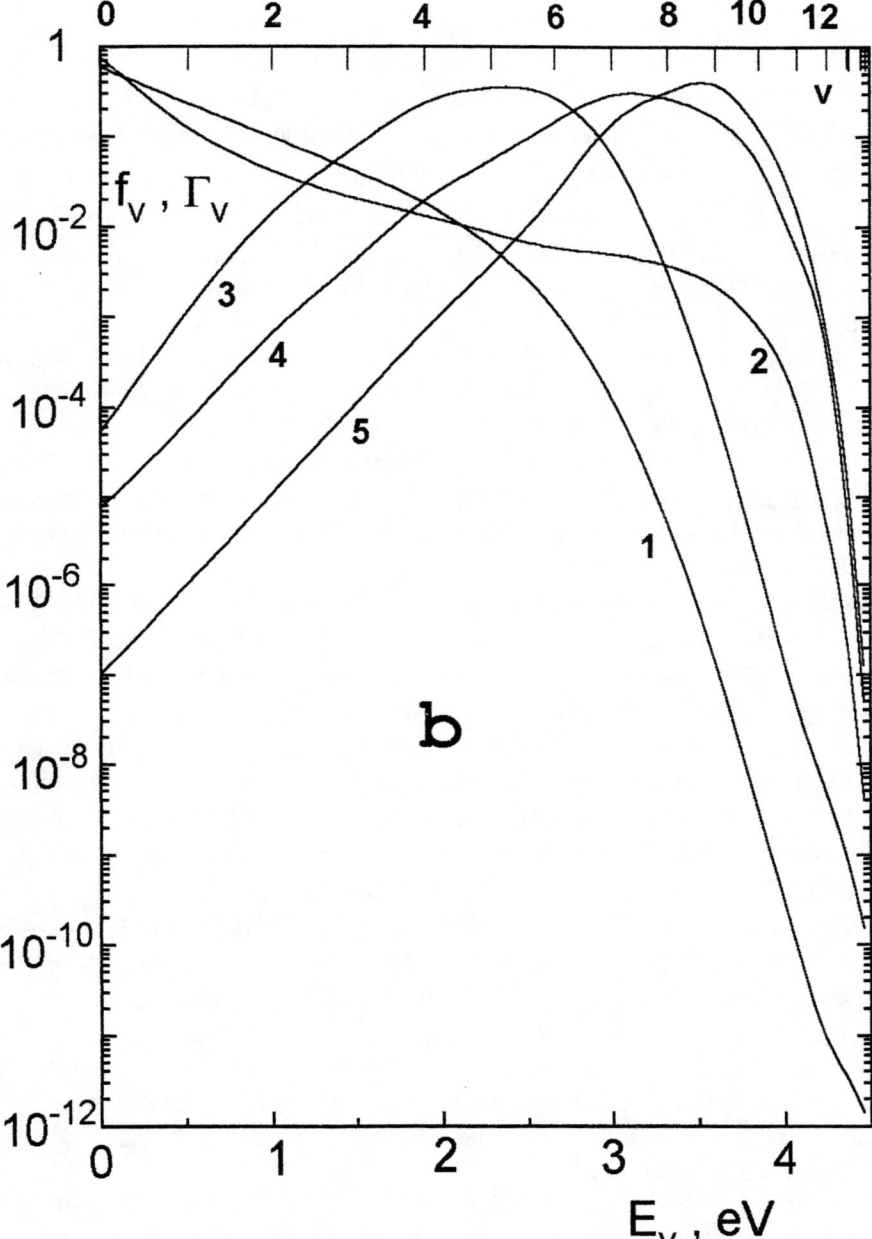

Fig.6b. The same as in Fig.5b but without taking into account the vibrational H_2 relaxation on the walls of the channel.

$$<K_{DA}(h,T_e')> = \sum_v f_v(h)K_v(T_e') \qquad (8)$$

is the effective rate constant, which is equal to the whole rate of H⁻ creation due to the DA of electrons to all vibrational levels. Γ_v is the probability of H⁻ creation due to the DA of electrons, which temperature is equal to the T_e', to H_2 molecules excited to v level. The VDF of these molecules is supposed to be $f_v(h)$. Γ_v is normalized to unit. In (7) and (8), $K_v(T_e)$ is the rate constant of the dissociative attachment to v level, and h is a length of the channel.

Γ_v describes a relative contribution of different vibrational levels to the process of DA. It is supposed that this process occurs in a chamber, in which vibrationally excited hydrogen flows and where the electrons have the Maxwellian distribution function with the temperature T_e'. One can see that, if the H_2 flux passed through the channel, then the maximum of Γ_v moves in the direction of large v numbers, i.e. in the direction of the greater K_v values [6].

Let us consider the dependence of Γ_v on T_e'. In Fig. 7 (compare with Fig. 8 in [4]) the effective rate constant of DA $<K_{DA}>$ is shown as the function of T_e' for several versions of calculations: without the intensification in the channel (curve 1) and after the intensification of the VDF in the channel (curves 2 and 3, which are calculated with and without the surface desactivation of H_2 molecules correspondingly). As it was mentioned above, the Γ_v maximums correspond to $T_e \cong 1$ eV. The most significant intensification of the value of $<K_{DA}>$ occurs at low temperature T_e', when the considerably high vibrational levels (v ≃ 7-9) give the main contributions to the DA, the populations N_v of these levels being intensified very significantly due to the vibrational pumping in the channel. But even if the whole volume, where DA occurs, is at the needed optimum T_e' value, the intensification is also significant (e.g. compare the curves 1 and 2 in Fig. 7). The most intensive increase of the vibrational pumping occurs at the levels v ≃ 10-11. Although these levels give comparatively small contribution to the value of $<K_{DA}>$, their vibrational pumping may be essential for plasmachemical applications.

Thus, it is shown that in the channel, where vibrationally pumped hydrogen flows, the significant intensification of the populations of vibrational states can be achieved in a sufficiently high share of the vibrational spectrum. The results were obtained by means of the consideration of electron-vibration kinetics in a low-voltage cesium-hydrogen discharge and in the channel, through which vibrationally pumped

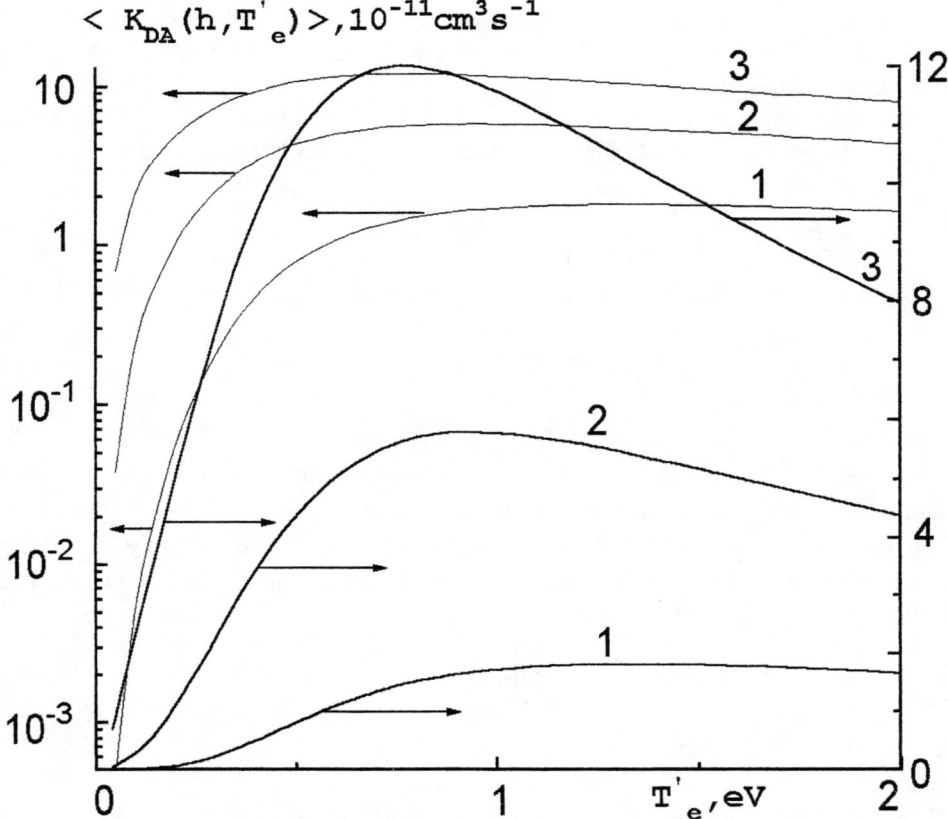

Fig.7. The dependence of the effective rate constant $< K_{DA}(h, T'_e) >$ of the dissociative attachment upon the electron temperature T'_e. The parameters of the discharge and of the gas flux in the channel are the same as in Fig.5 and Fig.6.

1 - $< K_{DA}(0, T'_e) >$

2 and 3 - $< K_{DA}(h_2, T'_e) >$

2 - The vibrational relaxation of H_2 molecules on the walls is included in the calculations.

3 - The vibrational relaxation of H_2 molecules on the walls is not included in the calculations.

hydrogen flows after passing the discharge section. Of course, the analogical results are applicable to the other modes of cesium-hydrogen and pure hydrogen discharges

Acknowledgements. We are grateful to Yu.Z. Ionih and S.M. Shkolnik for helpful discussion. This work is supported by INTAS (Grant № 94-316).

Appendix

Diffusion of hydrogen atoms in the channel is described by the equation

$$\vec{V} \cdot \nabla N_H - D_{12} \nabla^2 N_H = 0, \qquad (A.1)$$

where $N_H(x,y)$ is a hydrogen atom concentration in the channel (see Fig. 1a), $\vec{V} = \vec{V}(0)$ is a velocity of the flux in the section II, near the entrance of the channel. The recombination of atomic hydrogen into H_2 molecule is neglected in (A.1) (see. [28]) as well as the change of the velocity $\vec{V}$ of the flux at the initial small part of the channel. The boundary conditions for the equation (A.1) are written in the form:

$$N_H(x,0) = N_H(x,L) = 0, \qquad (A.2)$$

$$N_H(\infty,y) = 0. \qquad (A.3)$$

In addition to (A.2)-(A.3), the boundary conditions include the known $N_H(0,y)$ distribution of an atom concentration along the cross-section of the channel at $x = 0$.

Using the expansion of $N_H(x,y)$ in Fourier series

$$N_H(x,y) = \sum_{k=1}^{\infty} \tilde{N}_k(x) \cdot \sin(\pi k y / L) \qquad (A.4)$$

and substituting (A.4) into (A.1), we obtain

$$N_H(x,y) = \sum_{k=1}^{\infty} \tilde{N}_k(0) \cdot \exp\left\{-\left[\sqrt{1+(\delta k)^2} - 1\right] \cdot x/(\delta \Lambda)\right\} \sin(k y / \Lambda) \qquad (A.5)$$

where $\delta = 2\Lambda/(V \tau_d^{(II)})$, $\tau_d^{(H)} = \Lambda^2/D_{12}$. The mean value of $\langle N_H \rangle$, which is averaged over the cross-section, can be obtained from (A.5):

$$< N_H > = \sum_{k=1}^{\infty} \tilde{N}_k(0) \cdot \exp\left\{-\left[\sqrt{1+(\delta k)^2} - 1\right] \cdot x/(\delta \Lambda)\right\}\left[1+(-1)^{k+1}\right]/(\pi k) \quad (A.6)$$

In conditions, which are considered here, $\delta < 1$ ($\delta \cong 1/3 - 1/2$), and such values of x are essential, for which $x \sim V \tau_d^{(H)}$, i.e. $\Lambda/x \cong \delta/2$. Therefore, it is enough to consider only the item with $k = 1$ in (A.6). As a result, we obtain

$$< N_H > \cong (2\tilde{N}_1(0)/\pi)\exp\left[-x/(V\tau_d^{(H)})\right] \quad (A.7)$$

and

$$\frac{dI_H}{dx} = -\langle N_H \rangle / \tau_d^{(H)} \quad (A.8)$$

where $I_H = V<N_H>$ is the density of the hydrogen flux, which is averaged over the channel cross-section by the indicated way. One can see from (A.8) that above mentioned values Λ and $\tau_d^{(H)}$, indeed, are the effective diffusion length and the effective diffusion time for the diffusion of atomic hydrogen from the flux to the walls of the channel. If $2\Lambda < V \tau_d$, the equation of the type of (A.8) describes approximately the losses of atomic hydrogen, as well as the losses of the other impurities, which exist in the flux.

References

[1] Bacal M., Hamilton G.W. *Phys. Rev. Lett.*, **42**, 1538, (1979).
[2] Hiskes J.R. *J. Appl. Phys.*, **51**, 4592, (1980).
[3] Bacal M. *Nuclear Instruments and Methods in Physical Research.*, **B** 37/38, 28, (1989).
[4] Skinner D.A., Bruneteau A.M., Berlemont P., Courteille C., Leroy R., Bacal M. *Phys. Rev. E*, **48**, 2122, (1993).
[5] Bacal M., Skinner D.A. *Comments At. Mol. Phys.*, **23**, 283, (1990).
[6] Wadehra J.M. *Phys. Rev. A*, **29**, 106, (1984).
[7] Baksht F.G., Ivanov V.G. *Sov. Phys. Techn. Phys.* **37**, 223 (1992).
[8] Baksht F.G., Elizarov L.I., Ivanov V.G.. *Sov. J. Plasma Phys.* **16**, 479 (1990).
[9] Baksht F.G., Djuzhev G.A., Elizarov L.I., Ivanov V.G., Kostin A.A., Shkolnik S.M. *Plasma Sources, Sci., Technol.*, **3**, 88, (1994).
[10] Baksht F.G., Ivanov V.G.. *Pis'ma Zh. Tekn. Fiz.*, **23**, № 9, 59, (1997).
[11] Dushman S. *Scientific Foundation of Vacuum Technique*, NY-London, Wiley, 1962.
[12] Landau L.D., Lifshitz E.M. *Hydrodynamics* (in Russian), Moscow, "Nauka", (1986).

[13] Baksht F.G., Elizarov L.I., Ivanov V.G. et. al. *Sov. J. Plasma Phys.* **14**:56 (1988).
[14] Baksht F.G., Ivanov V.G., *Sov. J. Plasma Phys.* **12**:165 (1986).
[15] Baksht F.G., Ivanov V.G. *Zh. Tekn. Fiz.* **66**, № 9, 58, (1996).
[16] Schlichting H. *Grenzschicht-Theorie.* Verlag G. Braun, Karlsruhe, (1969).
[17]. Baksht F.G., Djuzhev G.A., Martsinovskiy A.M., Moizhes B.Ya., Pikus G.E., Sonin E.B., Yuriev V.G.. *"Thermionic converters and low-temperature plasma"*.(Russian edn. Moizhes B.Ya. and Pikus G.E.. Nauka, Moscow (1973). English edn. Hansen L.K., US Department of Energy, (1978))
[18] Ionih Yu. Z. *Optics and Spectroscopy* (in Russian), **51**, 76, (1981).
[19] Eletsky A.V., Palkina L.A., Smirnov B.M. *Transport Phenomena in Slightly Ionized Plasma.* (in Russian). Moskow. "Atomizdat", (1975).
[20] Blyth G. et al. *J. Chem. Soc. Faraday Trans.* I , **83**, 751, (1987).
[21] Madhavan P., Whitten J.L. *J. Chem. Phys.*, **77**, 2673, (1982).
[22] Harris J., Anderson S. *Phys. Rev. Lett.*, **55**,1583, (1985).
[23] Cacciatore M., Billing G.D. *Surf. Sci.* , **232**, 35, (1990).
[24] Reitner C.T., Auerbach D.J., Michelsen H.A. *Phys. Rev. Lett.*,**68**,1164, (1992).
[25] Gorse C., Capitelli M., Bacal M., Bretagne J., Lagana A. *Chem. Phys.* **117**,172, (1987).
[26] Baksht F.G., Elizarov L.I., Ivanov V.G., Nikitin A.G., Shkolnik S.M., *Pis'ma Zh. Tekn. Fiz.*, **19**, 39, (1993).
[27] Baksht F.G., Ivanov V.G., Nikitin A.G., Shkolnik S.M. *Sov. Techn. Phys . Lett.* **20**, 927 (1994).
[28] Baksht F.G. *Zh. Tekn. Fiz .*, **52**, 3, (1982).

Negative Ion Yields in Hydrogen Scattering from Graphite Surfaces.

M. A. Gleeson[#,*], W. R. Koppers[#], K. Tsumori[†] and A. W. Kleyn[#].

\# Association EURATOM-FOM, FOM Institute for Atomic and Molecular Physics,
Kruislaan 407, 1098 SJ Amsterdam, the Netherlands.
* Physics Department, Dublin City University, Glasnevin, Dublin 9, Ireland.
† National Institute for Fusion Science, Furo, Chikusa, Nagoya 464-01, Japan.

Abstract. We compare the negative ion fraction obtained for scattering of hydrogenous ions (H_x^+, x=1-3) from highly oriented pyrolytic graphite (HOPG), with that obtained for scattering from a polycrystalline graphite surface. In contrast to the HOPG surface, which has a negative ion yield of the order of 1-2%, the polycrystalline sample has yields of up to 30%.

INTRODUCTION

The study of the neutralisation and formation of ions (both positive and negative) is important in gaining a proper understanding of plasma-wall interactions, divertor physics and negative ion sources. We have studied the ion yields resulting from scattering of hydrogenous ions (H^+, H_2^+ & H_3^+) at graphite surfaces. Appreciable negative ion yields (20-30%) have already been observed for scattering from a polycrystalline graphite (Le Carbonne Lorraine) sample disk (1). In this work we compare those results with similar measurements taken from a HOPG sample. In the case of HOPG, the ion yield is of the order of only 1-2%. Possible reasons for the difference in the ion yield between the two graphite surfaces are suggested.

EXPERIMENTAL

The experimental set-up has been described in detail elsewhere (2,3). It consists of two vacuum chambers, one for sample preparation and characterisation and one for surface scattering experiments. The sample is mounted on a 2-axis goniometer in the scattering chamber, which allows the orientation of the sample with respect to the incident ion beam to be varied both polarly and azimuthally. Two detectors are available whose position with respect to the sample can be varied both in and out of the plane of scattering. An electrostatic energy analyzer (ESA) is used to measure the energy and angular distributions of the scattered negative and positive ions, while the scattered ion fraction (energy-integrated) can be measured with a fraction detector (1).

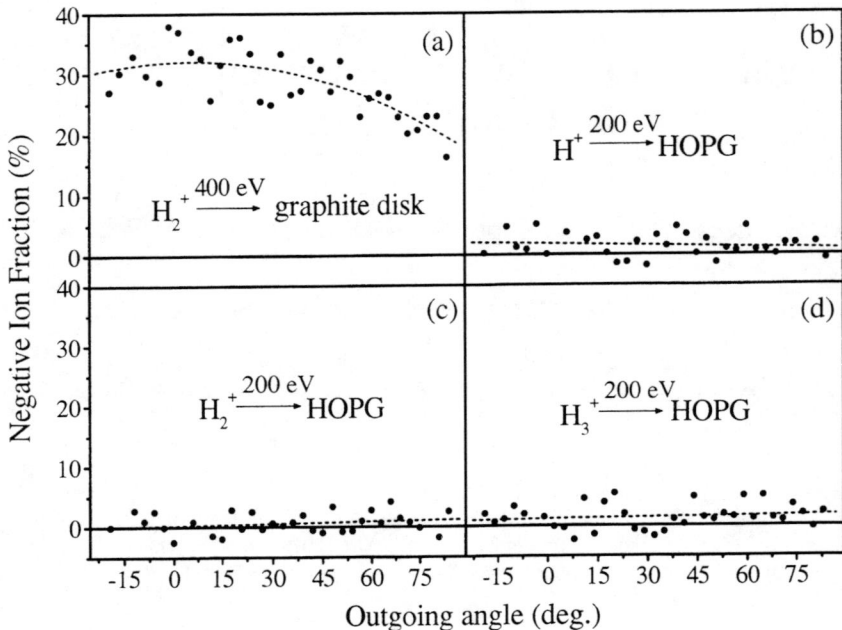

FIGURE 1. Angular distribution of the negative ion fraction measured for H_2^+ incident on the graphite disk at 400 eV, (a), and for H^+, H_2^+ and H_3^+ incident on the HOPG sample at 200 eV, (b), (c) & (d), respectively. In all cases the angle of incidence was 70°. All angles are defined with respect to the surface normal. The dashed lines are fits intended to guide the eye. The zero-fraction line is also shown.

The preparation of the graphite sample disk has been described previously (1). After cleaning the surface cleanliness was checked by XPS, which showed a weak oxygen signal and a molybdenum signal of ~3% based upon the cross-section corrected C 1s peak intensity. The HOPG surface was prepared by removal of the outermost layers using adhesive tape immediately prior to insertion into the preparation chamber. Once under vacuum, the sample was annealed to 800 K. No contaminants were detectable by XPS on the sample.

RESULTS and DISCUSSION

Figure 1 shows a comparison of the negative ion fraction obtained for H_2^+ (400 eV, $\theta_i = 70°$) incident on the graphite disk with that obtained for H^+, H_2^+ and H_3^+ (200 eV, $\theta_i = 70°$) incident on the HOPG sample. The negative fraction is defined as

$$\frac{I_{H^-}}{\left(I_{H^-} + I_{H^0} + I_{H_x^+}\right)}, \quad x = 1, 2, 3$$

where I is the measured angle-resolved flux of the various scattered particles. Clearly there is a significant difference between the charged fraction measured from the two graphite surfaces. In the case of H_2^+ incident on the graphite disk, negative ion yields of the order of 30 % are observed. Comparable ion yields were also obtained for H^+ and H_3^+ incident on this sample at 400 eV. In the case of HOPG, the negative ion yield is of the order of only a few percent and cannot be accurately determined due to the scatter in the data points. Energy distributions of the scattered ions (positive and negative) were measured for H_x^+ ions incident on both graphite samples in the 100-500 eV incident energy range. In both cases the scattering was consistent with dissociation of the molecular ions prior to collision with the surface. In all cases, peaks that were consistent with single hydrogen atoms having undergone binary elastic collision with C atoms dominated the energy analysis spectra. The fraction of scattered positive ions measured from the graphite disk was comparable to that of the negative ions (1), while for the HOPG sample it was again of the order of 1-2 %.

Negative ion yields comparable to those observed from the graphite disk have been observed for low-work function metal surfaces (e.g. Cs and Ba covered surfaces) (2,4-6). However, the high work function of graphite (5 eV, (7)) is incompatible with the conventional theories used to explain charge transfer at metal surfaces (6). Studies on hydrogen and deuterium interaction with non-metallic surfaces in the early eighties showed negative ion yields exceeding 5% for particles backscattered with energies of a few KeV (8). However, at backscattered energies of around 200 eV the observed yield was very small. Wurz et al have observed H^- ion yields of 3.0-5.5% for H_2^+ incident on polycrystalline diamond surfaces at energies ranging from 300 to 800 eV (9). This observed yield was stable and reproducible over long time periods. These results are intermediate between the yields we observe from the HOPG and the graphite disk.

The reason for the large ion yields that were observed from the graphite disk compared with the small yields from the HOPG sample is as yet unclear. However a number of possible suggestions for the difference between the two samples can be made. Differences in the structure of the samples and the roughness of the surfaces may play a role. The crystallites in the HOPG sample are all oriented with respect to the surface normal (the crystallite layers are oriented parallel to the surface plane), leading to a large anisotropy between the crystal properties parallel to the surface plane and those perpendicular to this plane. In contrast, the crystallites in the graphite disk are randomly oriented leading to effectively uniform bulk properties along all sample directions. In addition, the surface of the graphite disk, which was subjected to cycles of Ar^+ bombardment, is undoubtedly far rougher than that of the HOPG sample, which received no sputter treatment. The roughness of the graphite disk may explain the high negative ion yield, via a mechanism similar to that proposed by Rechtien et al (10,11). In their studies of O_2 scattered from Si {001}, they suggested that the high O_2^- yield observed might arise due to electron pick-up from surface dangling bonds. Re-neutralisation is inhibited because the hole created is localized on the dangling bond. A high degree of surface roughness will result in a large number of dangling bonds existing on the surface of graphite disk, facilitating negative ion formation via such a mechanism.

The simultaneous observation of high positive ion yields from the graphite disk (1) means that another process of ion formation must exist. The graphite disk contained low levels of impurities, such as Mo, which are most likely introduced at the production stage. The presence of low levels of impurities may be sufficient to alter the electronic structure and density of states around the Fermi level of the graphite sample, resulting in the observed ion yields. One possibility is that oxidation at the grain boundaries in the graphite disk may result in the crystallites in the sample acting as a collection of "micro-insulators". The polycrystalline nature of the sample will give rise to a large number of grain boundaries within the sample, with the open structure of graphite allowing significant penetration of hydrogen atoms. Hence, the negative ions observed might arise from hydrogen scattering from the surface of the disk, whereas the majority of the positive ions may originate from scattering at defect sites and grain boundaries deeper in the sample structure.

Clearly further work is required to identify the source of the enhanced ion formation from the graphite disk. Further experiments are planned to this end. Initially, the effect of roughening the HOPG surface will be studied. It is also planned to study the ion yields obtained from other graphite surfaces, and also from diamond samples.

ACKNOWLEDGEMENTS

F. G. Giskes is gratefully acknowledged for help and technical support throughout this work. M. A. G. thanks the EURATOM-DCU association for the provision of mobility funding. The work described here was performed as a part of an association agreement between the Stichting voor Fundamenteel Onderzoek der Materie (FOM) and EURATOM, and was made possible through the financial support of the Nederlandse organisatie voor Wetenschappelijk Onderzoek (NWO) and EURATOM.

REFERENCES

1. Tsumori, K., Koppers, W. R., Heeren, R. M. A., Kadodwala, M. F., Beijersbergen, J. H. M., and Kleyn, A. W., *J Appl. Phys.*, **81**, 6390-6396 (1997).
2. van Slooten, U., Teodoro, O. M. N. D., Kleyn, A. W., Los, J., Teillet-Billy, D. and Gauyacq, J.P., *Chem. Phys.*, **179**, 227-240, (1994).
3. Koppers, W. R., *Ph. D. Thesis*, FOM-Institute for Atomic and Molecular Physics, Kruislaan 407, 1098 SJ, Amsterdam, the Netherlands (1997).
4. Los, J. and Geerlings, J. J. C., *Phys. Rep.*, **190**, 133-190 (1990).
5. van Wunnik, J. N. M. and Los, J., *Phys. Scr.*, **T6**, 27-34 (1983).
6. van Os, C. F. A., van Amersfoort, P. W. and Los, J., *J. Appl. Phys.*, **64**, 3863-3873 (1988).
7. Robrieux, B., Faure, R. and Dussaulcy, J. P., *C. R. Acad. Sci. Ser B.*, **278**, 659-662 (1974).
8. Verbeek, H., Eckstein, W. and Bhattacharya, R. S., *Surf. Sci.*, **95**, 380-390 (1980).
9. Wurz, P., Schletti, R. and Aellig, M. R., *Surf. Sci.*, **373**, 56-66 (1997).
10. Rechtien, J.H., Imke, U., Snowdon, K. J., Reijnen, P. H. F., van den Hoek, P. J., Kleyn, A. W. and Namiki, A., *Surf. Sci.*, **227**, 35-42 (1990).
11. Rechtien, J.H., Imke, U., Snowdon, K. J., Reijnen, P. H. F., van den Hoek, P. J., Kleyn, A. W. and Namiki, A., *Nucl. Instrum.Methods Phys. Res. B*, **48**, 339-343 (1990).

FIRST RESULTS FROM A DOUBLE VLASOV MODEL FOR NEGATIVE ION EXRACTION FROM VOLUME SOURCES —
The Possibility of an Enhanced Transverse Space Charge Limit

J. H. Whealton, D. K. Olsen, and R. J. Raridon

Oak Ridge National Laboratory

Abstract. A new negative ion source extraction model has been formulated and implemented which explicitly considers the motion of positive ions and the volume generation of negative ions. It is found that (1) for high-beam currents, the beam current is limited by a transverse space charge limit, meaning that an increase in negative ion density at the extraction sheath will result in a lower beam current (this result is universally observed at high beam currents); (2) there is a saddle point with a potential barrier preventing most volume produced negative ions from being extracted (the combination of 1 and 2 indicates that most of the negative ions being created do not find their way into the beam); (3) introduction of cesium may cause, most importantly, an increase in the transverse space charge limit (there is an abundance of experimental data supporting this effect); (4) cesium may also result in an increase in the fraction of volume produced negative ions which are extracted; (5) cesium may also result in a reduction of extracted electrons by dint of a less negative bias on the plasma electrode with respect to the adjacent plasma, thus allowing the transverse space charge limit budget to be taken up wholly by the ions. (The combination of 3–5 represents the way an actual increase in the beam current can be achieved); (6) a strong ion time scale sheath instability due to violation of Bohm criteria produces an anomalous ion temperature that increases with beam current routinely seen in experiments; and (7) introduction of cesium may result in a reduction in this instability. These insights may lead to improvements in volume negative ion sources, and the most important finding of an increased

---

Research sponsored by the LDRD Program of Oak Ridge National Laboratory managed by Lockheed Martin Energy Research Corp. for the U.S. Department of Energy under Contract No. DE-AC05-96OR22464.

Differences between Negative Ion and Positive Ion Extraction

Several phenomena surrounding the production of extractable negative ions have apparently not yet been explained in a self-consistent manner. For example, why does cesium addition add to the negative ion output and avoid the negative current saturation as a function of plasma density that is observed so often without cesium? If the extracted negative ion and electron current is transverse space charge limited, as supported by a multitude of experimental evidence, then why should an increase in negative ion production, as cesium injection is expected to produce, increase the extracted beam current above and beyond merely replacing the space charge of the reduced extracted electron flux? Why does the RMS emittance of the extracted beam not appear to increase when cesium is added? These questions form an apparent contradiction. We approach these problems by considering a more accurate physics model for negative ion extraction, which is cognizant of most of the five major asymmetries between positive ion extraction that the analysis is well developed, and negative ion extraction, where it has been less so [1].

$$\nabla^2 \phi\left(\vec{r},t\right) = \int f\left(\vec{r},\vec{v},t\right) d\vec{v} - \exp\left[-\phi\left(\vec{r},t\right)\right] \tag{1}$$

$$\frac{\partial f\left(\vec{r},\vec{v},t\right)}{\partial t} + \vec{v} \bullet \nabla f\left(\vec{r},\vec{v},t\right) + \left\{\vec{v} \times \vec{B} - \nabla \phi\right\} \bullet \nabla_v f\left(\vec{r},\vec{v},t\right) = 0 \tag{2}$$

The first asymmetry is that the electrons cancel charge imbalance excursions in positive ion sources much better than positive ions do in negative ion sources. This is explicitly addressed in the present model by considering the positive ions as a coupled Vlasov equation added to the conventional system of Poisson-Vlasov equations for a self-consistent treatment of the negative ions and the electromagnetic fields.

$$\frac{\partial f_+}{\partial t} + \vec{v}\bullet\nabla f_+ + \left\{\vec{v}\times\vec{B}-\nabla\phi\right\}\bullet\nabla_v f_+ = \delta\left(S_1\right)$$

(3)

$$\frac{\partial f_-}{\partial t} + \vec{v}\bullet\nabla f_- - \left\{\vec{v}\times\vec{B}-\nabla\phi\right\}\bullet\nabla_v f_- = G\left(\vec{r}\right)$$

(4)

$$\frac{\partial f_e}{\partial t} + \vec{v}_e\bullet\nabla f_e - \frac{M}{m_e}\left\{\vec{v}\times\vec{B}-\nabla\phi\right\}\bullet\nabla_v f_e = \delta\left(S_2\right)$$

(5)

$$\nabla^2\phi = \int\{f_+(v) - f_-(v) - f_e(v)\}\,dv - e^{-\phi}$$

(6)

The second asymmetry is the observation that the plasma electrode is biased negative to the local plasma potential for negative ion sources just like in positive ion sources and for the same reason—to contain the plasma electrons electrostatically. Therefore, in a positive ion source, there is a continuous monotonic downhill run for the positive ions generated deep in the plasma from formation to extraction. For negative ions, there is a saddle point formed within the plasma since the plasma potential lies intermediate between that of the plasma electrode and the acceleration electrode. This observation was made ten years ago [1, 2] but has attracted scant attention. Negative ions formed from one side of the ridge, formed from the saddle point to the plasma electrode, go toward the source plasma instead of the extraction aperture. In these cases, the location and control of such ridges could be important.

The third asymmetry is that negative ions, due to their short mean free extinction path, must be born in a volume close to the extraction aperture in order to be extracted. For positive ions, it is conventional to assume that they are transported from deep within the plasma. A fourth asymmetry is that electrons are extracted along with the negative ions, unlike the case for positive ion sources. This effect will not be considered in this paper. A fifth asymmetry is that the Bohm sheath stability criteria could be violated more extensively in certain regions for negative ion sources than in the case for positive ion sources, at least in the absence of cesium. This is because any process that increases the negative space charge in the pre-extraction region causes the curvature of the potential to become positive (as evidenced from the structure of the Poisson Equation), which is precisely the condition for the Bohm instability. Analysis capable of considering sheath produced ion-acoustic waves from the Bohm instability has been considered for positive ions [2] and will be a necessary future step in the modeling of negative ion sources.

Experimental evidence for the existence of unstable sheaths in negative ion sources, takes the form of extracted beam temperatures (as interpreted from emittance measurements), which are as much as an order of magnitude higher than the temperature of negative ions in the plasma. We will average over these instabilities as a temporary expedient, which will deny us the possibility of explaining some features of the ion beam emittance.

A standard positive ion extraction result is shown in Fig. 1a, showing a monatomic downhill run from center of plasma to extraction, with an attraction toward the plasma electrode greeting the ions not extracted. In order to elucidate the phenomena of negative ion extraction, we will consider separately a low- and high-density regime. At low densities positive ions falling downhill from the center of the plasma as shown in Fig. 2a. They are accelerated until they reach the saddle ridge. Then the positive ions are repelled by the accelerator fields and are attracted into the plasma electrode as shown. This is in contrast to positive ions shown in Fig. 1. For volume produced negative ions, the trajectories are shown in Fig. 2b. Here we see that only a small fraction of the negative ions produced are extracted, and the rest are attracted to the center of the plasma since they are repulsed by the plasma electrode. The densities and sheath properties are not unlike the case considered in [3] (shown in Fig. 3a); however, in [3] a trivial variation of a positive ion extraction model was used with an ad hoc representation of only the first asymmetric property as mentioned above.

The situation becomes more interesting at higher densities. For illustration purposes, Fig. 4 shows a higher plasma density. In many cases, such as those shown in [3] (see Fig.3b herein), the beam (negative ions and electrons) is transverse space charge limited. The very nature of a transverse space charge limit means, if the generation rate is increased (beyond any decrease in electron space charge), the beam current actually decreases. This is because the excess generation not only gets intercepted by the electrodes but causes some of the formerly transmitted ions to now be intercepted by the accelerator structure. Therefore, the question of how the cesium addition could increase the beam current beyond the decrease in electron space charge assumes greater importance. It has been found that a cesium surface coating very near the extraction apertures is especially beneficial, sometimes adding substantially to beam current beyond the decrease in extracted electron space charge [3, 4]. Since a cesium surface coating is known to produce negative ions when bombarded by positive ions, this would seem, at first sight, to be a possible explanation for the beneficial effects of cesium. However, since the beam current is usually transverse space charge limited, an increase in production current will not result in an increase in beam current (beyond the decrease in electron space charge). In fact, it may make it lower. The explanation of the effect of cesium must be from another source. In those cases where the transverse space charge limit appears to be increased with the presence of cesium, the extraction sheath somehow must become more concave since all other things are the same. This would not usually happen, i.e., compare Fig. 3a with Fig. 3b, which shows the result of increasing density. However, this extra concavity could be obtained if there were an additional source of positive space charge due to the cesium. Since surface sputtering of neutral cesium by the impinging positive ion flux is inevitable, the ionization (Cs^+) mean free path is short, and the cesium ions are heavy and slow the opportunity for an excess of positive charge is present. An example of the effect of positive charge is shown in Fig. 5 which, except for an abundance of positive charge, is the same as Fig. 4. Not only can the sheath be seen to be further toward the plasma in Fig. 5 than that of Fig. 4 (thus allowing in some cases a higher transverse space charge limit), but the fraction of negative ions that get extracted in a specific case increases from 21 percent to 37 percent, almost doubling. This by itself does not result in an increased beam but means that less arc power is required. The fraction of volume produced negative ions being extracted as a function of plasma density is shown in Fig. 7 for a case where the initial velocity is low.

Another feature of negative ion extraction from volume sources is the presence of ion acoustic instabilities in the sheath due to violation of the Bohm sheath criteria [1]. This violation occurs whenever the negative space charge in the presheath region exceeds the positive space charge (see the simulation of Fig. 1b for positive ion extraction). This will generally occur in some presheath regions in volume negative ion sources. The instability will take the form shown in Fig. 6, with a relatively stable region slightly dominated by positive ions followed by an unstable region (on ion time scales). Negative ions born in the unstable region will have an enhanced "temperature" due to their bouncing on the electric potential waves. In the instance of a potential barrier near the extraction sheath, these instabilities may actually increase negative ion extraction, although we do not so far have much evidence for this. Injection of positive space charge as may be possible with cesium will tend the presheath toward stability, thus lowering the ion temperature (Fig. 4 vs Fig. 5).

As a natural consequence of negative ions and electrons being ejected from the plasma electrode, and returning Cs^+ ions intercepting it (those that escape the sheath adjacent to the plasma electrodes are sputtered off as neutrals), an isolated plasma electrode will be biased less negative with respect to the plasma than would otherwise be expected. Therefore, more plasma electrons will intercept the plasma electrode than would otherwise be the case, and fewer electrons will make it to the extraction region. This is an explanation for the dearth of extraction of electrons from the plasma when cesium is injected.

In summary, a new model has been described that takes into account four differences between positive- and negative-ion extraction. The new model appears to explain such important observations as reduced electron extraction and increased beam current when cesium is added to negative ion volume sources.

REFERENCES

1. J. H. Whealton, M. A. Bell,., R. J. Raridon, , K. E. Rothe,, and P. M. Ryan, "Computer modelling of negative ion beam formation," *J. Appl. Phys.*, **64,** 6210, 1988.

2. J. H. Whealton, "Review of Computer Modelling of Negative Ion Beam Formation," 4th International Symposium on the Production and Neutralization of Negative Ions and Beams (1986), AIP Conf. Proc. **158**, 432, 1987.

3. J. H. Whealton, P. S. Meszaros, R. J. Raridon, K. E. Rothe, M. Bacal, J. Brunetean, and P. Devynek,, "Validation and Comparison of Nonlinear Negative Ion Extraction Theory for two Experimental Configurations," *Revue Phys. Appl.*, **24**, 1989.

4. M. F. Bacal, El Balglutic-Suba, L. I. Elizarov, and A. Y. Tontegode "Effect of Cesium Seeding on Hydrogen Negative Ion Volume Production," ICIS '97 (1997), paper P-F.01, p.180, Taormina, Italy, 1997

5. K-N Leung et al., NSNS ion source review, 1997.

6. P. Allison, H-V Smith, J. D. Sherman, "H–Ion Source Research at Los Alamos, Proc. 2nd Symposium on the Production and Neutralization of Negative Ions and Beams, BNL, **51304**, 1997.

7. A. Holmes (86 BNL/305)

ACKNOWLEDGEMENTS

The authors wish to thank M. A. Akerman, M. Bacal, and J. B. Green for their technical advice and P. J. Abbott, J. M. Shover, and E. D. Stratman for their assistance.

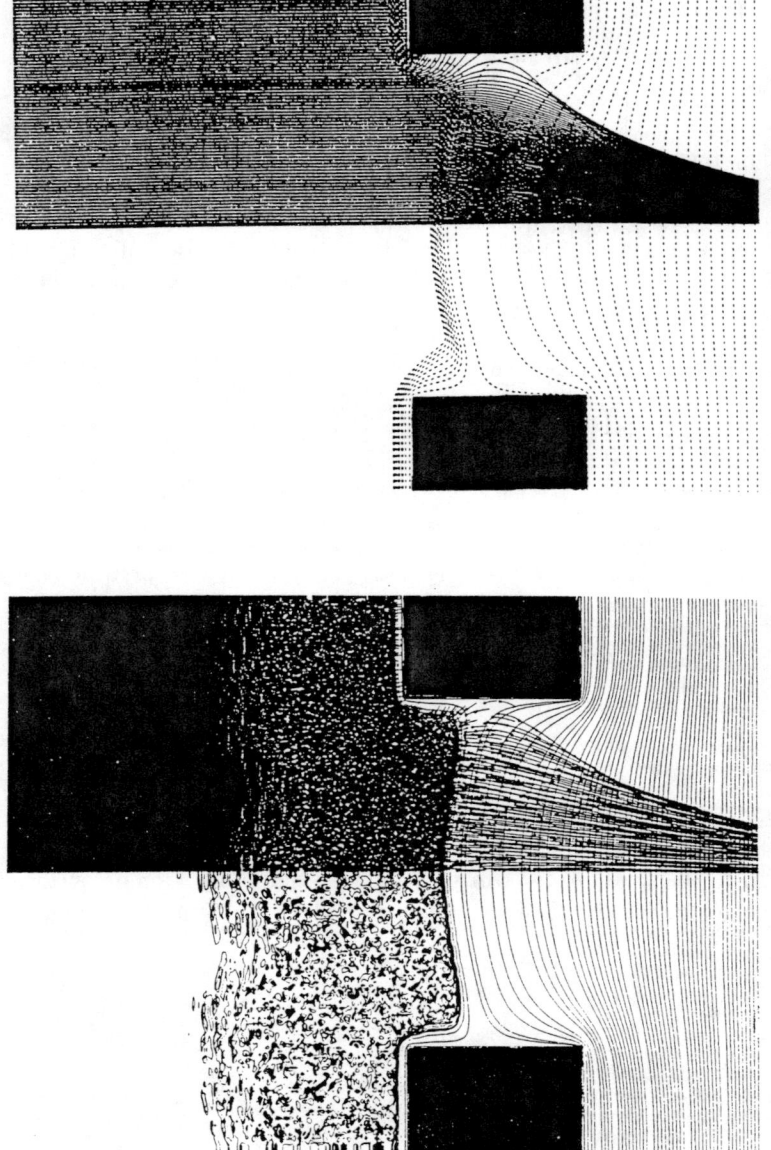

Figure 1a. A typical positive ion sheath and plasma electrode structure near optium perveance showing a downhill run of the positive ions from the source plasma to extraction---no saddle point, no potential barriers.

Figure 1b. Time dependent solution to the Vlasov-Poisson equations for positive ion extraction (electrons are Boltzmann) in a region where the Bohm sheath stability criteria is not satisfied resulting in ion-acoustic waves in the presheath and concomitant RMS emittance growth.

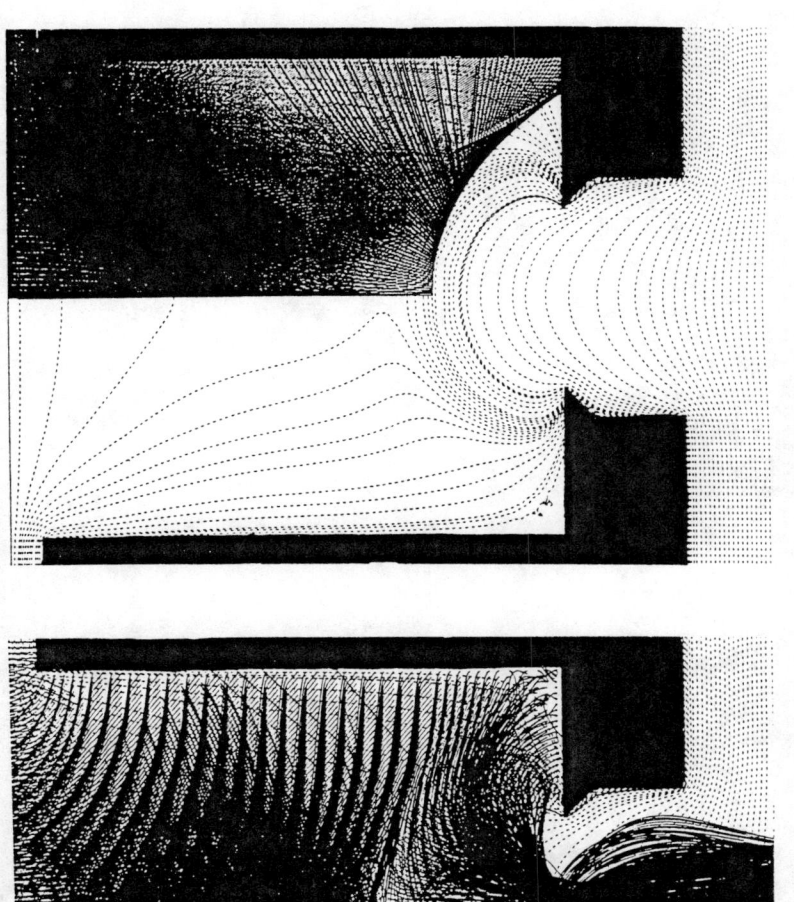
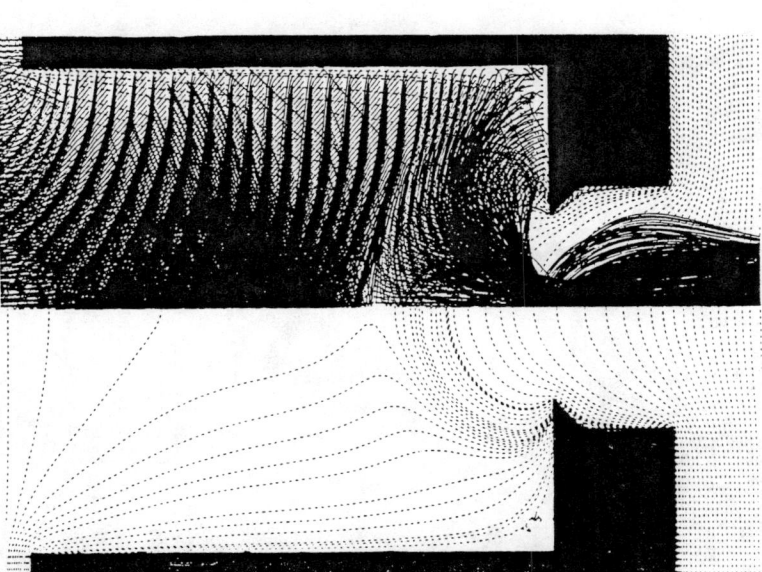

Figure 2. Positive ion trajectories (Fig. 2a) indicated by solid lines from the source plasma arriving toward the extraction sheath (low density) and being repelled by the accelerator fields. A saddle point in the electrostatic potential is formed as shown by the dashed contours; electrons are represented by a modified Boltzmann distribution. Negative ions formed on the extraction side of the saddle point ridge will be inclined to be extracted; negative ions formed on the plasma side will tend to not get extracted. Electrons are shown in Fig. 2b. The plasma electrode is typical for volume negative ion sources.

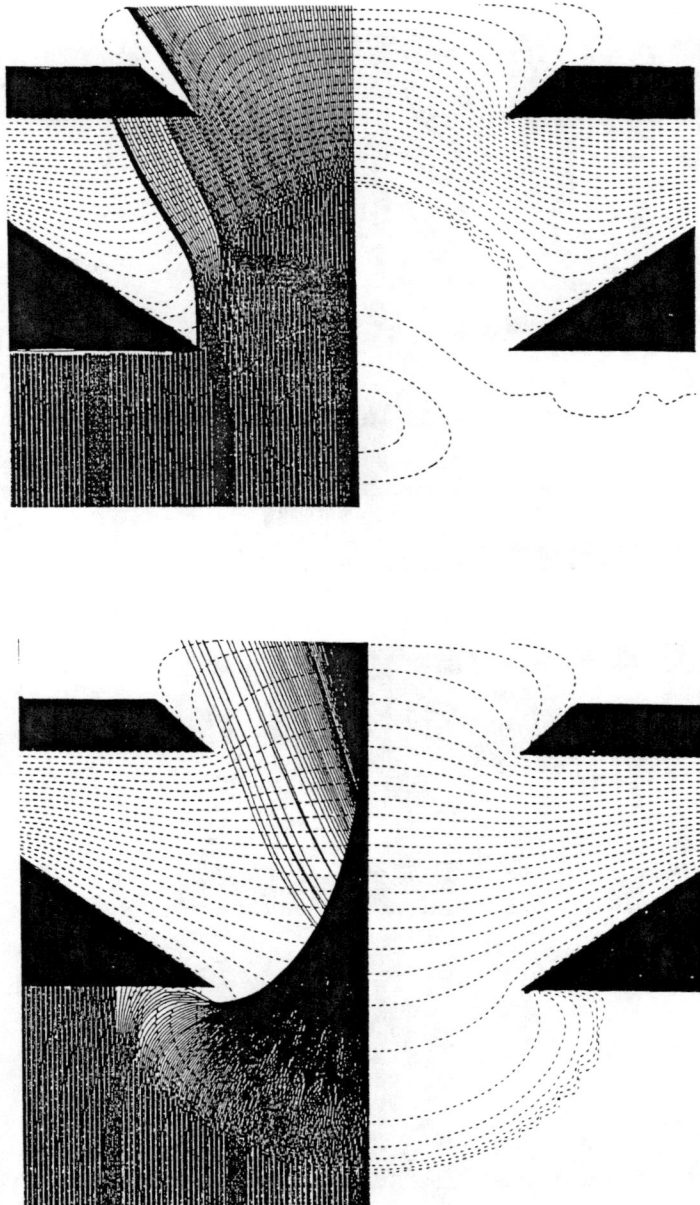

Figure 3a. Results of heuristicly modified positive ion model for very low perveance indicating amplification of the ion beam current due to accelerator electric field penetration. This result is especially significant because of the apparent agreement with experiment.

Figure 3b. Results of heuristic modified positive ion model for very high perveance, beyond the transverse space charge limit indicating suppression of the ion beam current due to accelerator electrode interception. This result also is apparently in agreement with experiment.

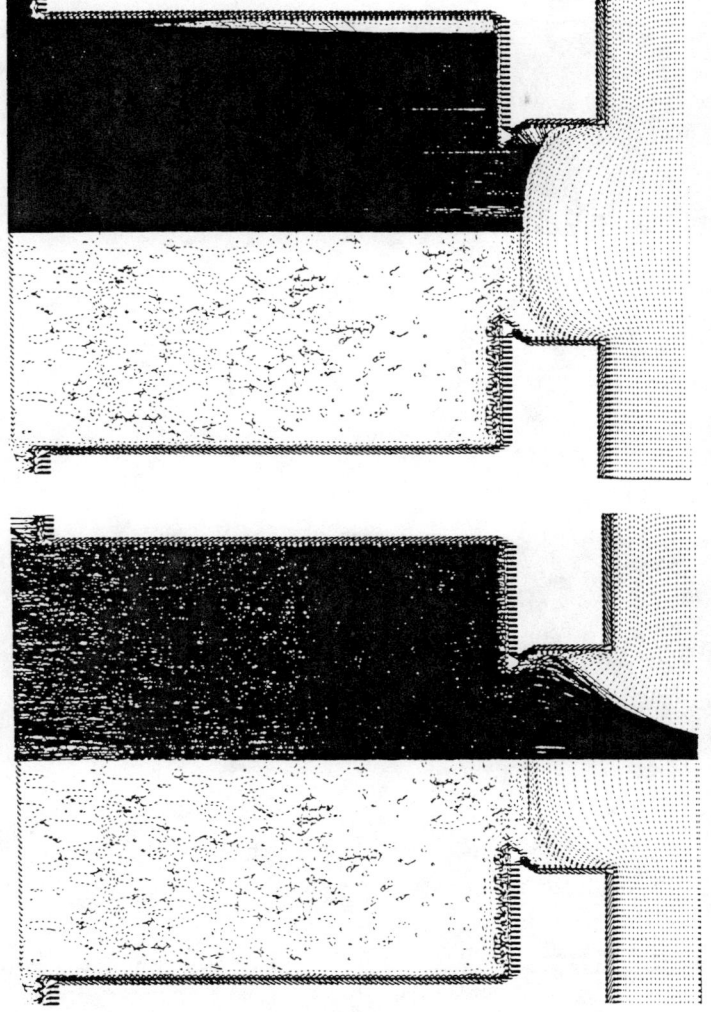

Figure 4. Positive ion trajectories in a negative ion source at moderately high densities showing thin ridge compared to Fig. 1. Instabilities have been suppressed by space charge under-relaxation.

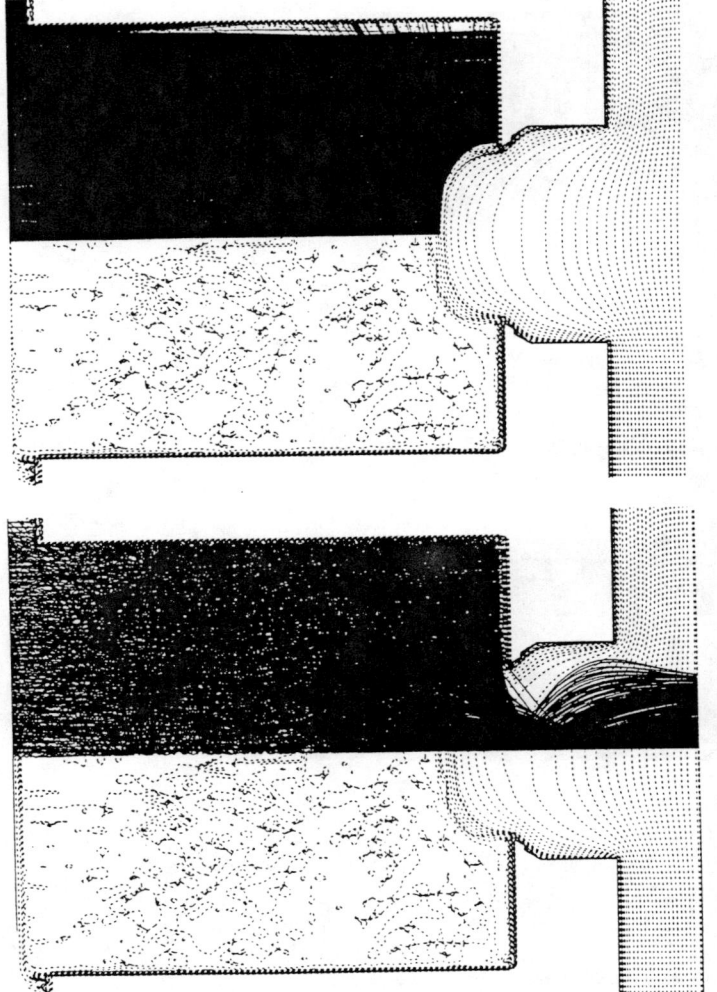

Figure 5. Similar to Fig. 4, but positive space charge added in the presheath region.

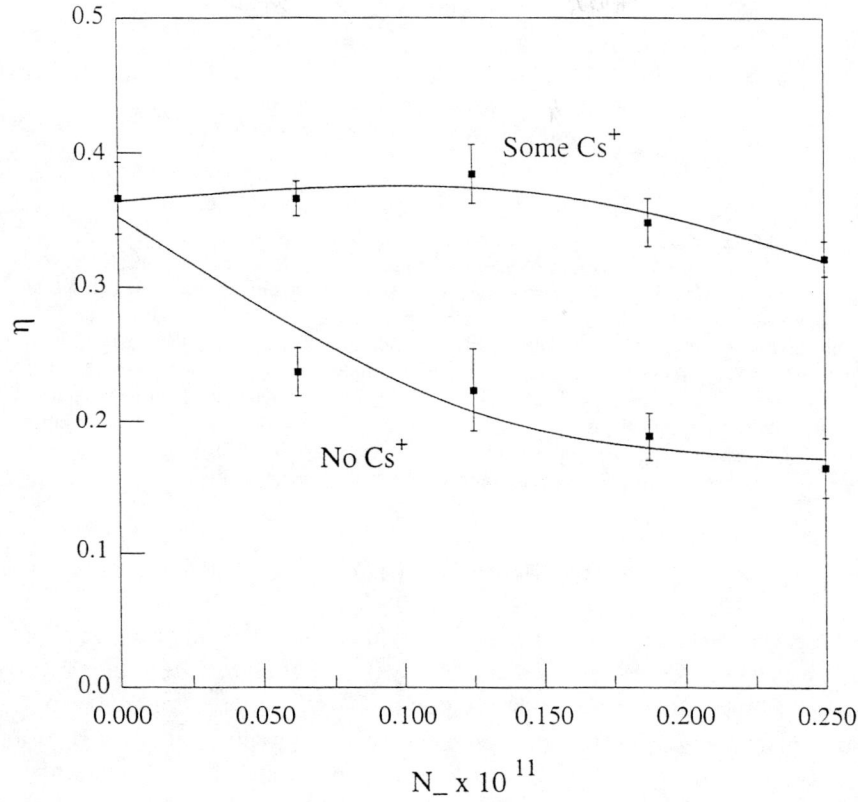

Figure 6. Fraction of volume produced negative ions extracted as a function of negative ion density—the potential barrier prevents the rest from getting out. One of the effects of the introduction of Cs^+ (simulated by addition of positive charge in the presheath) is that the extracted fraction remains at a relatively high value as compared to the case labeled no Cs^+ where no positive charge was added to the presheath. The error bars refer to the fluctuations of the results due to physical instabilities.

Modeling Negative Ion Production in Volume Sources

Osamu Fukumasa* and Kyogo Yoshino

*Department of Electrical and Electronic Engineering,
Faculty of Engineering, Yamaguchi University, Ube 755, Japan*

Abstract. Effects of cesium vapor injection on H⁻ production in a tandem volume source are studied numerically as a function of plasma parameters. Model calculation is performed by solving a set of particle balance equations for steady-state hydrogen discharge plasmas. Here, the results with a focus on electron temperature and gas pressure dependence on H⁻ volume production are presented and discussed. Considering H⁻ surface production due to H atoms and positive hydrogen ions, enhancement of H⁻ production and pressure dependence of H⁻ production observed experimentally are qualitatively well reproduced in the model calculation, where stripping loss in the extraction and acceleration regions is taken into account. For enhancement of H⁻ production, so-called electron cooling is not very effective if plasma parameters are initially optimized with the use of a magnetic filter.

INTRODUCTION

Recent experimental investigations have revealed that the addition of cesium (Cs) or barium to a hydrogen discharge can enhance the H⁻ output current by a several factor and cause a substantial reduction in the electron-to-H⁻ ratio in the extracted beam.[1,2] It has also been reported [3,4] that the optimum pressure, p_{opt}, giving the highest H⁻ current for a certain arc current, I_a, is almost independent of I_a and that the value of p_{opt} decreases to 0.8-1.0 Pa when Cs vapor is seeded into a plasma source. Although these effects have been observed by many researchers, the mechanism of Cs catalysis in H⁻ production remains to be clarified.

To date, we have studied source modeling [5,6] and Cs effects on enhancement of H⁻ yield.[7] Based on some experimental results (for example, correlation between the H⁻ current and the work function of the plasma grid [9] and dependence of H⁻ current on barium washer voltage [10]), we assume that the dominant process of enhancement is surface production, where the surface has a low work function because of the Cs coverage. In this paper, to elucidate further the Cs effects, we will discuss enhancement of H⁻ yield as functions of some plasma parameters, i.e. electron density n_e, hydrogen gas pressure p and T_e. Taking into account H⁻ surface production, the model calculation well reproduces the characteristic features of enhancement of H⁻ production observed experimentally.[3,4]

SIMULATION MODEL

To study H⁻ production in a tandem two chamber system, we used the simulation model [5,6] shown in Fig.1. Two chambers of volume $L \times L \times L_1$ (the first) and $L \times L \times L_2$ (the second) are in contact with each other in the region of magnetic filter, where $L = L_1 + L_2 = 30$ cm. We assume that fast electrons, e_f, are present only in the 1st chamber because the magnetic filter prevents e_f from entering the 2nd chamber. We consider four ion species (H^-, H^+, H_2^+ and H_3^+), two electron species (e and e_f) and three species of neutral particles [H, $H_2(v'')$ and H_2]. Particles other than e and e_f are assumed to move freely between the two chambers without being influenced by the filter. The number of particles passing through the filter is represented by flux nv, where n and v are the particle density and velocity, respectively.

In the present model, two kinds of reaction process at the wall surface are included. One is H⁻ surface production caused by Cs injection. The following four processes are considered [7,11] : $H_n^+ + \text{wall} \rightarrow nH^-$ (n = 1,2,3), and $H + \text{wall} \rightarrow H^-$. The other is effect of $H_2(v'')$ surface production due to wall recombination of H and neutralization of positive ions. [12,13]

In modeling, surface production rates of negative ions are estimated as follows : The term representing wall loss of H atoms is expressed as $-(\gamma_1 + P_{csH})N_H/\tau_H$, where γ_1 is a wall recombination coefficient, P_{csH} indicates the probability of H⁻ formation at the wall, N_H is H density, and τ_H is a confinement time of H. Then, the H⁻ production rate at the wall surface is expressed as $P_{csH}N_H/\tau_H$. We also assume that recombination of H to H_2 at the wall is written as $\gamma_1 N_H/(2\tau_H)$. Therefore, production of $H_2(v'')$ is expresses as $P_0 \gamma_1 N_H/(2\tau_H)$, where P_0 is the probability of finding $H_2(v'')$ in H_2 formed at the wall. In the same way, the rates of production of H⁻ and $H_2(v'')$ from positive ions at the wall are also estimated.

For each chamber, 19 rate equations for H, $H_2(v''$ = 1-14), H^-, H^+, H_2^+ and H_3^+ are derived by taking into account the above-mentioned reaction processes, other collisional reaction processes occurring in hydrogen plasma and the interaction between the two chambers. Besides these rate equations, there are two constraints for each chamber : i.e., the charge neutrality and particle number conservation.

NUMERICAL RESULTS AND DISCUSSION

The procedure for numerical simulation is as follows : To determine the electron density dependence of the H⁻ production, calculations are performed for various electron densities, $n_e(1)$, in the 1st chamber on the assumption that other plasma parameters are kept constant: i.e., for example, the electron density ratio between two chambers $n_e(2)/n_e(1) = 0.2$, density of e_f in the 1st chamber $n_{fe}(1)/n_e(1) = 0.05$, p = 0.67 Pa(5 mTorr), T_e in the 1st chamber $T_e(1) = 5$ eV and T_e in the 2nd chamber $T_e(2) = 1$ eV, and the filter position $L_1 : L_2 = 28 : 2$ cm. According to previously obtained

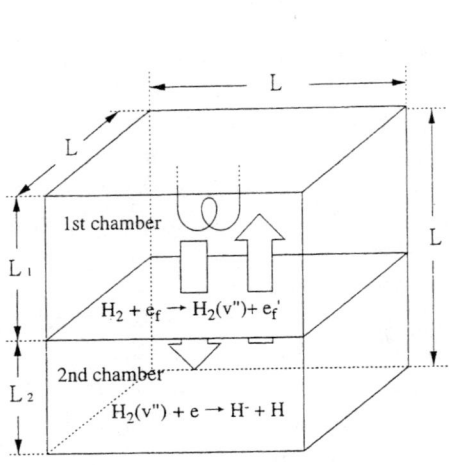

Fig.1 Simulation model for the tandem two-chamber system.

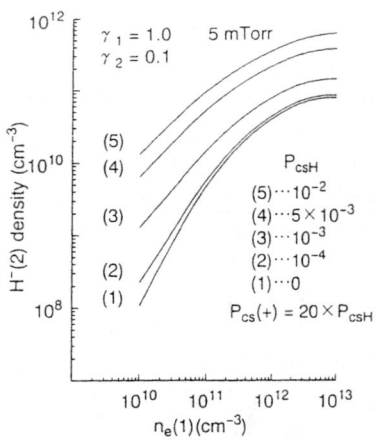

Fig.2 Effects of the surface production due to both H atoms and positive ions on enhancement of H⁻ production : H⁻ density, H⁻(2), in the second chamber versus electron density in the first chamber $n_e(1)$. Parameter is the probability of H⁻ formation at the wall, P_{cs}.

results,[5,14] these plasma conditions were chosen to optimize H⁻ pure-volume production in the 2nd chamber. They are also used in the present study.

Figure 2 shows the H⁻ densities, H⁻(2), in the 2nd chambers for various different values of P_{cs}. Wall conditions are as follows. For positive ions, $P_{cs1} = P_{cs2} = P_{cs3} = P_{cs}(+)$, and $P_{cs}(+) = 20\ P_{csH}$. $P_0 = P_1 = 0.01$, and $P_2 = P_3 = 0.3$. When $P_{csH} = 0$, i.e. curve (1), H⁻ ions are produced by the so-called two-step pure volume process. With increasing P_{csH}, as was shown previously,[7,11] H⁻(2) increases markedly. In a high density region, i.e. at $n_e(1) = 5 \times 10^{12}$ cm⁻³, H⁻(2) is enhanced by a factor of four for curve (4) and by a factor of seven for curve (5). Since the same factor of enhancement was obtained when the effect of surface H⁻ production in only the 2nd chamber was taken into account, one can conclude that the 2nd chamber is the most effective area for surface enhancement of H⁻ production. In the present simulation, as the magnetic filter is set at $L_1 = 28$cm, the total area of the wall surface in the 2nd chamber is nearly equal to the area of the end wall which corresponds to the plasma grid of the tandem volume sources.[3,4] Thus, the numerical results above mentioned agree well qualitatively with the experimental results.[3,4,9,10] As determined experimentally, the H⁻ yield rises with plasma grid temperature peaking at around 250 ℃ [3] or 300 ℃ [4]. This spatial or localized dependence of H⁻ production enhancement is caused not by a volume process but by a surface process.

Theoretically, the probability, β^-, of incoming H atoms being converted to H⁻ ions on the wall is given as $\beta^- = (2/\pi) \exp[-\pi(\phi - A)/2av]$,[15] where ϕ is the work function, A is the electron affinity, a is the decay constant and v is the normal velocity of the incident particle. The effect of Cs is expressed through the value of ϕ. For example, ϕ is 1.45 eV for the surface covered with half of a monolayer of Cs and 2.1 eV for the surface covered with a monolayer of Cs. If we take the temperature of H atoms to be 0.5 eV, $\phi = 1.8$ eV [9] and $a = 3.08 \times 10^{-5}$ eV·sec/m, [15,16] we can estimate $\beta^- = 4.87 \times 10^{-3}$ for H atoms. Impinging positive ions will be accelerated by sheath potential. Namely, positive ions have rather large v compared with H. For protons with energy of 1eV, β^- is 2.05×10^{-2}. Therefore, the probability of incoming positive ions being converted to H$^-$ ions on the wall could be much higher than that of thermal H atoms.[17] Then, in the present calculation, $P_{CS}(+) = 20\ P_{CSH}$. Although P_{CSH} is treated as a numerical parameter, values of P_{CSH} are quite reasonable except that $P_{CSH} = 10^{-2}$. To discuss Cs effects quantitatively, we must estimate precisely the relationship between P_{CS} in our simulation and the theoretical value β^- for the corresponding experiment.

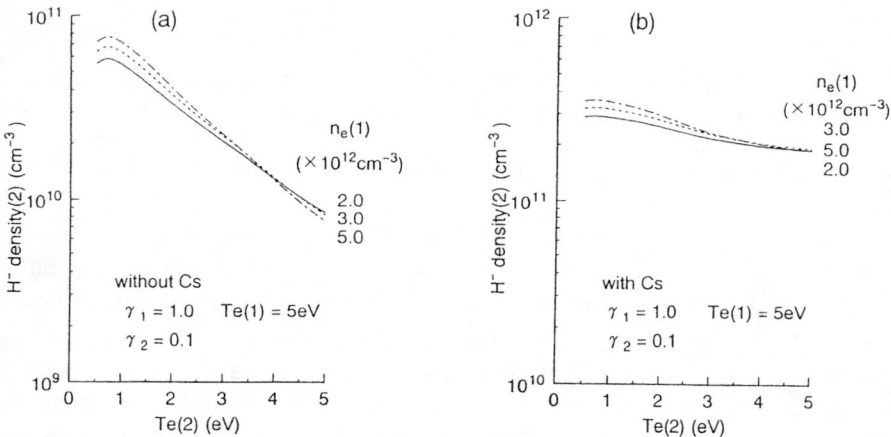

Fig.3 Dependence of H$^-$ production on electron temperature : H$^-$(2) versus $T_e(2)$ in the second chamber, (a) without Cs, and (b) with Cs. Parameter is $n_e(1)$.

For discussion of the effect of electron cooling due to Cs injection, model calculation has been performed as a function of T_e of the 2nd chamber, $T_e(2)$. Here, wall condition is described by curve (4) in Fig.2, and the parameter is $n_e(1)$. A typical example of numerical results is shown in Fig.3. For the pure volume case, in Fig.3(a), H$^-$(2) depends strongly on $T_e(2)$. With decreasing $T_e(2)$, H$^-$(2) increases and then reaches the maximum value at about 1 eV.[14] On the other hand, in the presence of Cs, in Fig.3(b), the enhancement of H$^-$ production due to surface processes is so marked that the increase in H$^-$ density due to a small reduction of T_e due to Cs injection could be masked. Therefore, electron cooling is not so effective for enhancing H$^-$ yield, if plasma parameters including T_e are well optimized with the use of the magnetic filter. In experiment, however, electron cooling seems to be one of the reasons for the reduction in the electron-to-H$^-$ ratio in the extracted beam.

Figure 4 shows H$^-$(2) as a function of p for various $n_e(1)$. In this calculation, the

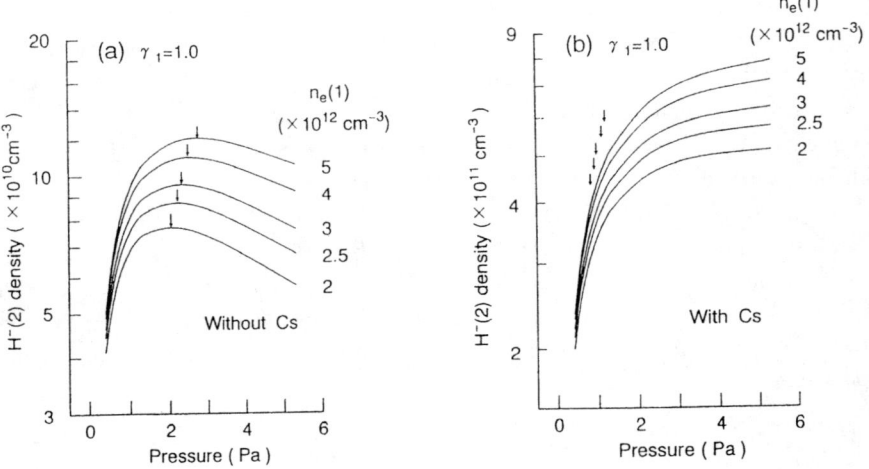

Fig.4 Effects of hydrogen gas pressure on H⁻ production : H⁻(2) versus gas pressure p, (a) without Cs, and (b) with Cs. In (a), the arrows show the point where H⁻ density is maximum. In (b), the arrows show the point where H⁻ density corresponds to four times the maximum H⁻ density in the absence of Cs. Parameter is $n_e(1)$.

wall conditions are described by curve (4) in Fig.2. In Fig.4(a), i.e. in the absence of Cs, H⁻ ions are produced by a pure volume process. Apparently, there is an optimum pressure p_{opt} for each $n_e(1)$, and the value of p_{opt} increases with $n_e(1)$ (see the arrows in the figure). In experiment,[3,18] this is a typical tendency in a multicusp volume source where the parameter is not n_e but arc current I_a. On the other hand, in Fig.4(b), i.e. in the presence of Cs, the pressure dependence of H⁻(2) for each $n_e(1)$ has almost the same pattern. As in the low-pressure region, enhancement of H⁻ production also appears clearly, the arrows which show the points where H⁻(2) corresponds to four times the value for H⁻(2) in the pure volume case shift to the low-pressure region. In this case, however, no optimum pressures are observed clearly.

Figure 5 shows the production rate and destruction probability of H⁻(2) corresponding to the result in Fig.4(b), where $n_e = 5 \times 10^{12}$ cm⁻³. The predominant production process is surface production due to H. At $p = 1$ Pa, for example, both surface production due to H⁺ and volume production VP have next dominant contribution, and the contribution of H_3^+ is the third. At $p = 1$ Pa, the density distribution of H⁺ : H_2^+ : H_3^+ in the 2nd chamber is 49 : 24 : 27. Probability of H⁻ surface production from H_3^+ is effectively twofold larger than that from H_2^+. For destruction of H⁻ ions, electron detachments caused by H and H_2 are predominant for all regions. In the low-pressure region ($p \leq 1$Pa), however, both electron detachments caused by H⁺ and loss flux of negative ions $\Gamma_{2\to1}$, i.e. flow of H⁻(2) across the filter to the 1st chamber, are predominant.

Experimental results, except those in ref. 18, show the extracted H⁻ current as a function of p. In refs.3 and 4, the H⁻ current is enhanced severalfold, and p_{opt} is

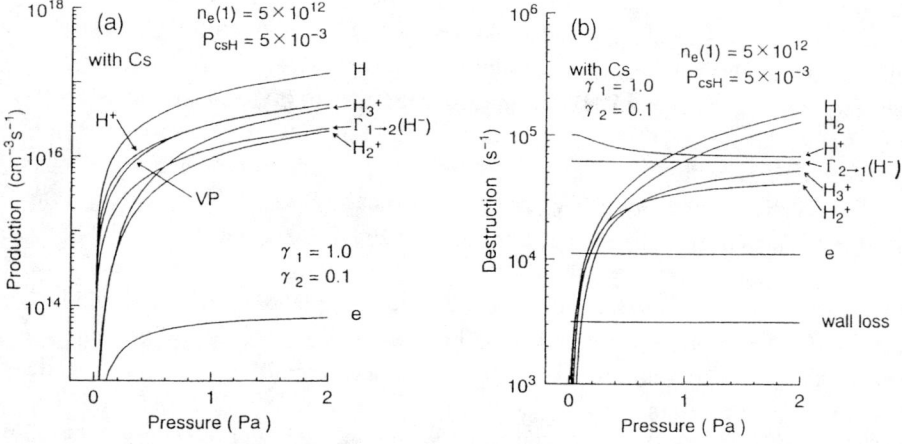

Fig.5 Production rates (a), and destruction probabilities (b) for H$^-$(2) density versus p, corresponding to the results shown in Fig.4(b), where $n_e = 5 \times 10^{12}$ cm^{-3}.

reduced to 0.5-0.8 Pa and is almost constant irrespective of I_a or arc power P_{arc}. As described in ref.19, p_{opt} is reduced again to 0.3-0.4 Pa although p_{opt} increases gradually with P_{arc}. On the other hand, Fig.4 shows not the extracted H$^-$ current but the H$^-$ density in the 2nd chamber. Therefore, strictly speaking, we could not directly compare the numerical results shown in Fig.4 with the experimental ones. Because, the pressure dependence of H$^-$ current depends strongly on stripping loss of H$^-$ ions along the beam axis.[4,19]

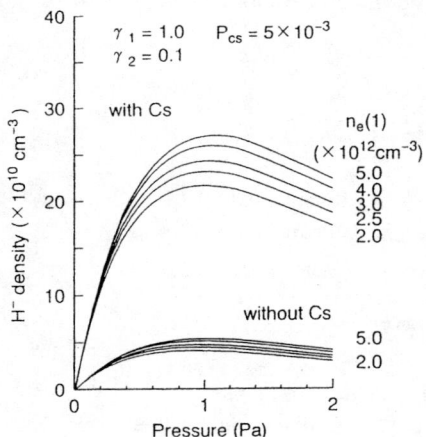

Fig.6 Pressure dependence of the extracted H$^-$ ions : H$^-$ density versus p, corresponding to the results shown in Fig.4, where stripping loss of H$^-$ ions in the acceleration grid region is included.

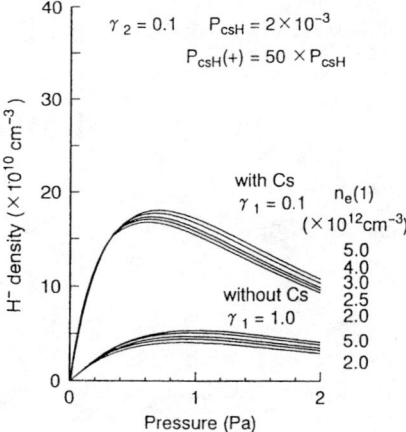

Fig.7 Pressure dependence of the extracted H$^-$ ions : H$^-$ density versus p. In this simulation, wall parameter γ_1 is varied from 1.0 (without Cs) to 0.1 (with Cs). Parameter is $n_e(1)$.

In order to discuss pressure dependence of the extracted H^- current, stripping loss of H^- ions in the acceleration grid region is considered. According to gas pressure distribution along the beam axis estimated by the Monte Carlo simulation,[19] we calculate the survival factor F against the stripping loss of H^- ions, i.e. $H^- + H_2 \rightarrow H + H_2 + e$ and $H^- + H \rightarrow 2H + e$. F is a decreasing function of pressure. In Fig.6, H^- densities corresponding to the extracted H^- ions, product of $H^-(2)$ in Fig.4 with F, with and without Cs are given as a function of pressure with varying $n_e(1)$. With Cs, H^- density increases and p_{opt} becomes clear. But it does not shift to low-pressure region as in experiment.

Figure 7 shows another example of pressure dependence of the extracted H^- ions. In Cs case, the value of γ_1 is reduced to 0.1 and p_{opt} shifts to low-pressure region. This agrees with experiments. As γ_1 is a parameter controlling H density, H density increases with decreasing γ_1. Therefore we speculate that H density in hydrogen discharge with Cs becomes high compared with pure hydrogen discharge. Details are now under study.

CONCLUSIONS

We have theoretically studied cesium effects on enhancement of H^- yield and plasma parameter dependence of H^- density in negative-ion volume sources. The characteristic features of the experimental results obtained in the presence of cesium are well reproduced in the model. H^- surface production due to H and positive ions (H^+ and H_3^+) contribute predominantly to H^- enhancement. For destruction of H^-, H and H_2 contribute predominantly. Detailed discussion including another wall effect, of γ_1 will be reported in a forthcoming paper.

ACKNOWLEDGMENTS

This work was supported by a Grant-in-Aid for Scientific Research from the Japanese Ministry of Education, Science, Sports and Culture.

REFERENCES

1) S.R.Walther, K.N.Leung and W.B.Kunkel : J.Appl.Phys. **64**, 3424 (1988).
2) K.N.Leung, C.A.Hauck, W.B.Kunkel and S.R.Walther : Rev.Sci.Instrum. **60**, 531 (1989).
3) Y.Okumura, N.Hanada, T.Inoue, H.Kojima, Y.Matsuda, Y.Ohara, M.Seki and K.Watanabe : Proc.5th Int.Symp.Production and Neutralization of Negative Ions and Beams (AIP Press, NY, 1990) p. 169.
4) A.Ando, T.Tsumori, Y.Oka, O.Kaneko, Y.Takeiri, E.Asano, T,Kawamoto, R.Akiyama and T.Kuroda : Phys. Plasmas 1, 2813 (1994).

5) O.Fukumasa and S. Ohashi : J.Phys. **D 22**, 1931 (1989).
6) O.Fukumasa : J.Appl. Phys. **71**, 3193 (1992).
7) O.Fukumasa, T.Tanabe and H.Naitou : Rev.Sci. Instrum. **65**, 1213 (1994).
8) J.R.Peterson : Proc.4th Int.Symp.Production and Neutralization of Negative Ions and Beams (AIP Press, NY, 1989) p. 113.
9) Y.Okumura, Y.Fujiwara, T.Inoue, K.Miyamoto, N.Miyamoto, A.Nagase, Y.Ohara and K.Watanabe : Rev.Sci.Instrum. **67**, 1092 (1996).
10) K.N.Leung, C.F.A.van Os and W.B.Kunkel : Appl.Phys.Lett. **58**, 1467 (1991).
11) O.Fukumasa and E.Niitani : 7th Int.Symp.Production and Neutralization of Negative Ions and Beams (AIP Press, NY, 1995).
12) J.R.Hiskes and A.M.Karo : J.Appl.Phy. **67**, 6621 (1990).
13) O.Fukumasa, K.Mutou and H.Naitou : Rev.Sci. Instrum. **63**, 2693 (1992).
14) O.Fukumasa : J.Phys. **D 22**, 1668 (1989).
15) B.Rasser, J.N.M.Van Wunnik and J.Los : Surf. Sci. **118**, 697 (1982).
16) T.Okayama : private communication (1992).
17) P.Berlemont, D.A.Skinner and M.Bacal : Chem. Phys.Lett. **183**, 397 (1991).
18) C.Courteille, A.M.Bruneteau and M.Bacal : Rev.Sci.Instrum. **66**, 2533 (1995).
19) Y.Takeiri, A.Ando, O.Kaneko, Y.Oka, K.Tsumori, R.Akiyama, E.Asano, T.Kawamoto, T.Kuroda, M.Tanaka and H.Kawakami : Rev.Sci.Instrum. **66**, 2541 (1995).

Modeling of Negative Ion Transport in a Plasma Source

David Riz[1,2], Jérôme Paméla[2].

[1] *Laboratoire de Physique des Milieux Ionisés, Laboratoire du CNRS, Ecole Polytechnique, 91128 Palaiseau, France.*

[2] *Département de Recherches sur la Fusion Contrôlée CE Cadarache, 13108 St Paul lez Durance, France.*

A code called NIETZSCHE has been developed to simulate the negative ion transport in a plasma source, from their birth place to the extraction holes. The ion trajectory is calculated by numerically solving the 3-D motion equation, while the atomic processes of destruction, of elastic collision H^-/H^+ and of charge exchange $H^-/H^°$ are handled at each time step by a Monte-Carlo procedure. This code can be used to calculate the extraction probability of a negative ion produced at any location inside the source. Calculations performed with NIETZSCHE have allowed to explain, either quantitatively or qualitatively, several phenomena observed in negative ion sources, such as the isotopic H^-/D^- effect, and the influence of the plasma grid bias or of the magnetic filter on the negative ion extraction. The code has also shown that, in the type of sources contemplated for ITER, which operate at large arc power densities (> 1 W.cm^{-3}), negative ions can reach the extraction region provided if they are produced at a distance lower than 2 cm from the plasma grid in the case of « volume production » (dissociative attachment processes), or if they are produced at the plasma grid surface, in the vicinity of the extraction holes.

INTRODUCTION

Negative ion based neutral beam injection is one of the candidates for heating and/or current drive on the next magnetic fusion device ITER (International Thermonuclear Experimental Reactor). The ITER neutral beam injection system should be based on 3 injectors which contain each an electrostatic 1 MeV accelerator and a negative ion source delivering more than 40 A of D$^-$ (20 mA.cm^{-2}) [1].

Study of negative ion sources has started in the 60's. Considerable progress has been achieved since and the first negative ion based neutral beam injection into a tokamak has been performed in 1996 in Japan on JT-60U [2]. Today, a total current of 13.5 A [3] and current densities up to 20 mA.cm^{-2} have been obtained for D$^-$. However, despite these good results, the building of a 40 A D$^-$ source is still a difficult challenge.

The extrapolation of the prototype sources performances to a large source is still uncertain. It is then important to understand all the processes which occur in a negative ion source. Many 0-D codes have been developed[4, 5, 6, 7, 8], but the negative ion transport is not taken into account. Consequently, a numerical code called NIETZSCHE (*Negative Ion Extraction and Transport ZSimulation Code for HydrogEn species*) has been developed to perform a 3-D simulation of the negative ion transport in a plasma source[9]. It could be a useful tool to understand the problem of extrapolation of the prototype source results to a large source.

In the sec. II, a brief history of negative ion source development is presented. Then, the negative ion transport and the NIEZSTCHE code are described in the sec. III. The simulation of negative ion transport has been performed for both volume produced (sec. IV) and surface produced (sec. V) ions.

HISTORY OF NEGATIVE ION SOURCE DEVELOPMENT

Surface sources

In 1972, it has been shown that a flux of hydrogen (H^+ or H^0) may be converted to H^- [10] after hitting a metal surface covered by cesium to reduce the work function and then improve the electron transfer from the metal to the hydrogen[11]. In this first surface source, a potential difference between the plasma and the cathode covered by cesium (few hundreds of Volts), accelerates the positive ions toward the cathode and pushes the negative ions produced on its surface toward the extraction hole, leading to a total H^- current of a few hundreds of mA (3.7 A.cm^{-2}). Unfortunately, this type of sources has two major drawbacks. First, the high negative ion energy in the plasma (> 100 eV) leads to a high transverse energy and a too high divergence (~ 40 mrad[12]). Secondly, the cesium could be introduced in the tokamak and pollute the walls. This is why another process has been studied : the volume production.

Volume sources

By the end of the 70's, it was shown that a large amount of negative ions could be produced in the volume of a plasma by dissociative attachment of an electron on vibrational excited molecules[13]:

$$e + H_2^*(v'') \rightarrow H_2^- \rightarrow H^- + H \tag{1}$$

Consequently, a source optimized for this process was developed, where the plasma is divided in two regions by a magnetic filter (driver and production zone)[14]. The molecules are excited by primary electrons in the filaments zone called driver and diffuse freely across the magnetic field toward the second zone where the electron temperature around 1 eV is favorable for the negative ion production. This source has produced 2.7 mA (38 mA.cm^{-2}) of H$^-$ [15] but the good current densities obtained on small prototypes have never been obtained in large sources developed for fusion application. Indeed, in Japan at JAERI[16], a 11.3 liters source has delivered 1.26 A (9.5 mA.cm^{-2}) of H$^-$. Consequently, cesium was introduced in the discharge, leading to the so called hybrid production sources.

Hybrid production sources

In 1989, it has been shown that it was possible to increase strongly the negative ion current extracted from a volume source after injection of cesium vapor into the discharge[17]. It is generally admitted that the surface produced negative ions are added to the volume produced negative ions. Other authors suggest that cesium ionized in the discharge decreases the electron temperature of the plasma and then increases the volume production rate of negative ions[18, 19]. However, recent spectroscopic studies of the cesium lines in a large source seems to show that the extracted negative ions are mainly produced by a surface process[20]. Today, many sources produce 20 mA.cm^{-2} of D$^-$ and a large source, at JAERI, has delivered 13.5 A of D$^-$ [3].

PRESENTATION OF THE NIETZSCHE CODE

Fundamental processes for the negative ion transport

When a negative ion is created, it moves inside the source until destruction or extraction. Its trajectory is first governed by electric and magnetic field, respectively $\vec{E}$ and $\vec{B}$, through the motion equation

$$M\frac{d\vec{v}}{dt} = -e(\vec{E} + \vec{v} \times \vec{B}) \qquad (2)$$

where M is the ion mass, $\vec{v}$ its velocity, t the time and e the electron charge. Its trajectory depends also on all "collision" processes. The main non destructive collision reactions are elastic collisions (H$^-$/H$^+$)[21] and charge exchange (H$^-$/H^0)[22]. The three most important destruction processes are[23] :

Electronic Detachment (ED) : $\quad e + H^- \rightarrow 2e + H$ (3)
Mutual Neutralization (MN) : $\quad H^+ + H^- \rightarrow 2H$ (4)
Associative Detachment (AD) : $\quad H^- + H \rightarrow e + H_2(v")$ (5)

Structure of the code

The goal of the NIETZSCHE code is to calculate the extraction probability of a negative ion produced at any location inside the source. Production mechanisms are not simulated here.

The 3-D geometry of the modelled source and its magnetic configuration have to be specified first. The magnetic cartography is calculated by another code, developed by Ciric at JET[24]. The plasma characteristics come from experimental data or modeling. Only electrons, H^+/D^+ positive ions and H^o/D^o atoms are considered. A 1-D profile is used for n_e, T_e, V_p (plasma potential) and n_H, T_{H^+} and T_H are assumed to be uniform.

For volume produced negative ions, the initial energy is fixed around 0.3 eV and the ions are launched isotropically in all directions. In the surface production case, the ion created on the wall is accelerated by the plasma sheath potential to a few eV toward the plasma, so it is launched in a direction perpendicular to the surface.

The code calculates a negative ion trajectory until this ion is destroyed or extracted. The motion equation (2) is numerically solved by an order 4 Runge-Kutta method. When an ion hits a wall, it is destroyed if its energy is greater than the potential difference between the plasma and the wall, and reflected if its energy is lower. At each time step of the motion equation solving, a Monte-Carlo procedure handles the atomic processes. If the ion is not destroyed, it may undergo an elastic collision or a charge exchange. For an elastic collision with a H^+/D^+ ion, the H^-/D^- will exchange energy with it and have a direction perpendicular to its initial direction. For charge exchange with H^0/D^0, the negative ion is replaced by another ion which can go in any direction with the H^0/D^0 energy.

The code calculates successively the transport of a large amount of negative ions in the plasma source. The extraction probability p of a negative ion is then obtained by dividing the number of extracted particles by the number of launched particles N, with the following relative error bar :

$$\frac{dp}{p} = \frac{2\sqrt{Np(1-p)}}{pN} \qquad (6)$$

In our calculation, we launch typically between 1000 and 10000 particles to obtain an error bar lower than 10 %.

VOLUME PRODUCTION

Many experimental measurements about volume production have been performed at Cadarache on the Dragon source[25, 26, 27]. This large source, built at Culham[28], has a rectangular cross section (22.7 cm long x 50 cm wide x 75 cm high). In order to understand experimental results, a simulation of the volume produced negative ions transport has been performed. The plasma characteristics (n_e, T_e) used in the calculations, come from Langmuir probes. We consider a uniform V_p in the driver and the production zone and we fix T_H and T_{H^+} to 0.5 eV in all the source. Negative ions are assumed to be produced with a 0.3 eV initial energy.

Effective production zone

In the Dragon source, the production zone is very large (more than half of the source volume) because of a special "tent" magnetic filter. However, because of the destruction mean free path, the negative ions produced far from the plasma grid will not be able to be extracted. The limit of the effective production zone is determined using NIETZSCHE.

The figure 1 shows the probability for an H⁻ to reach the grid versus its birth place. These calculations have been preformed for a 50 kW arc power and two different atomic densities ($n_H = 10^{13}$ cm^{-3} and $n_H = 10^{14}$ cm^{-3}). In the two cases considered, the probability becomes negligible for H⁻ created at distances of more than 2.5 cm into the discharge. This probability decreases when the atomic density is increased because of the destruction rate increase. The effective production zone represents only 5 % of the total source volume. Moreover, the H$_2$(v") produced in the driver too far from the effective production zone might be destroyed before creating an H⁻. The production zone is consequently too deep and relatively inefficient with this type of filter. A magnetic filter closer to the plasma grid would reduce the distance traveled by H$_2$(v") outside the driver and thereby enhance volume production.

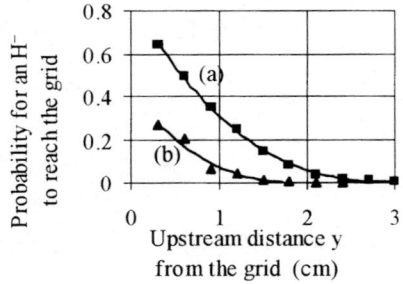

FIGURE 1 Probability for an H⁻ to reach the grid versus its birth location. $P_{arc} = 50$ kW. (a) $n_H = 10^{13}$ cm^{-3} (b) $n_H = 10^{14}$ cm^{-3}

Isotopic effect

In the Dragon experiment, an isotopic effect between H/D has been observed on the plasma and the negative ion current. The electron density in the plasma grid vicinity is higher with D_2 than with H_2 and I_{D^-} is lower than I_{H^-}. An evolution of the isotopic effect on current with arc power has also been observed. Consequently, the probability to reach the grid for H^- and D^- versus the arc power has been calculated, and is found to be lower in the case of D^-. This effect is due to two phenomena. First, the destruction rate is higher for D^- because of its lower velocity and of the higher electron density and secondly, the D^- trajectory is also changed because of the different reaction rates for elastic collisions and charge exchange.

The isotopic effect on the extracted current is the result of a competition between a higher production and destruction of D^- in comparison with H^-. It is then necessary to take the production mechanism into account to estimate the negative ion flux J_{H^-/D^-} falling on the grid. The ratio J_{H^-}/J_{D^-} can be written as follows :

$$\frac{J_{H^-}}{J_{D^-}} = \frac{\int n_{e(H_2)} n_{H_2(v'')} <\sigma v>_{DA(H_2)} P_{H_2} dy}{\int n_{e(D_2)} n_{D_2(v'')} <\sigma v>_{DA(D_2)} P_{D_2} dy} \qquad (7)$$

where $n_{v''}$ is the vibrational excited molecules density, $<\sigma v>_{DA}$ the dissociative attachment reaction rate, P the calculated probability given by NIETZSCHE for H_2 and D_2, and y the negative ion birth location from the plasma grid. We assume that the plasma parameters and the extraction probability are uniform in the two other directions of the source. The $n_{v''}$ profile is unknown but however it is possible, with some approximations, to estimate the ratio J_{H^-}/J_{D^-}. We have seen previously that the effective production zone is very short, so the product $n_{v''}<\sigma v>_{DA}$ which is only important in this zone may be assumed to be constant. Calculations performed by Skinner and al.[29] show that $n_{v''}$ is the same for a hydrogen or a deuterium discharge and $<\sigma v>_{DA}$ is also identical. Consequently, these terms may be removed from the equation (7).

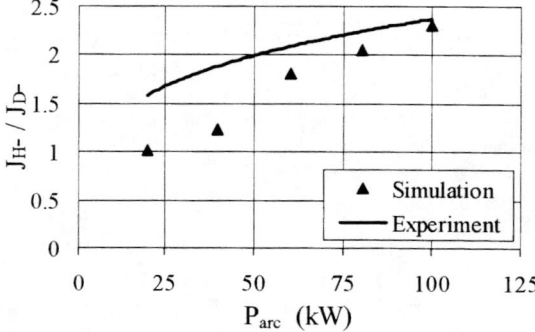

FIGURE 2 Ratio between J_{H^-} and J_{D^-} versus the arc power.

Figure 2 compares the ratio calculated from equation (7) with experimental data versus the arc power. The simulation underestimates the isotopic effect at low arc power but the J_{H^-}/J_{D^-} evolution is reproduced. The difference may come from the various hypothesis on the plasma parameters.

Effect of plasma grid bias

The experiment on the Dragon source has shown that the extracted negative ion current decreases when the plasma grid is polarized to a few volts. This phenomenon has been already observed in other sources running at high arc power density[16] whereas in some experiments at low arc power density[30,31], the negative ion current increases up to a maximum value before decreasing for more positive bias. The negative ion transport has been analysed to understand this difference.

The experiment shows that above a bias $V_{grid} = 3$ Volts, the plasma potential becomes lower than V_{grid}. The bias induces two transport modes : one where negative ions are pushed by the grid ($V_{grid} < 3$ V) and one where negative ions are attracted by the grid ($V_{grid} > 3$ V). The extraction probability has then been calculated for two cases ($V_{grid} = 0$ V and $V_{grid} = 5$ V) for a 50 kW arc power and negative ions produced at 0.5 cm from the grid at various position on an axis parallel to two extraction apertures (see figure 3).

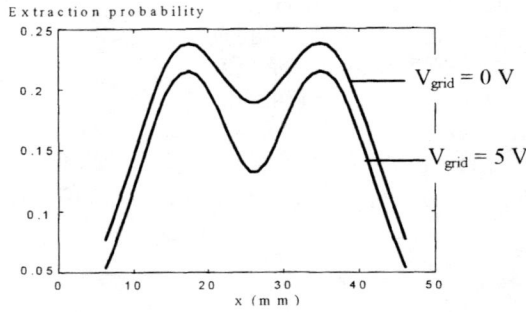

FIGURE 3 Extraction probability of an H⁻ produced at 0.5 cm from the plasma grid on an axis parallel to the extraction apertures. $P_{arc} = 50$ kW.

It was not possible to obtain a good probe measurement with the biased grid, so the plasma parameters (n_e, T_e) have been kept constant with V_{grid}. The two bumps on figure 3 correspond to negative ions launched directly in front of the apertures.. The figure shows that the extraction probability is lower for $V_{grid} = 5$ V. The destruction of negative ions on the plasma grid is then an important process. When the grid is not polarized, a fraction of the negative ions are extracted after a few reflections on the grid. The comparison between the two curves shows that a large amount of the negative ions launched between two apertures are extracted thanks to these reflections. These calculations only concern the direct effect of V_{grid} on the negative ion transport because we have not taken into account the change of (n_e, T_e) which also plays also a role on ions trajectory and destruction. However, we can analyse the difference between sources running at low and high arc power. For the low arc power sources, it is generally believed that the plasma grid bias creates a decrease of the electron density close to the grid, which induces a flux of negative ions coming from further in the source and then an increase of the extracted current. This effect has been experimentally observed by laser photodetachment measurements[32]. But in sources running at high arc power, the plasma density is almost 100 times larger and as it was shown in IV.A, negative ions coming from the center of the discharge are destroyed before reaching the grid. Consequently, a negative ion current drop is observed because of the plasma density decrease and the additional destruction on the grid.

SURFACE PRODUCTION

Experiments with cesium injection have been performed at Cadarache on the Dragon[27, 33, 34] and Kamaboko[20] sources. The Kamaboko source, which has been developed at JAERI[35, 36], is a semi cylindrical chamber (34 cm in diameter and 34 cm in length). These experiments have shown that the negative ions produced in these sources come mainly from a surface process, so we will study the transport of surface produced negative ions to analyse the experimental results. It was not possible to obtain good measurements with the Langmuir probes polluted by cesium so plasma characteristics measured before injection are used.

Effective production surface

Negative ions may be produced on the source walls or on the plasma grid surface. According to the code calculations it appears that only the negative ions produced on the grid, in the vicinity of the extraction holes, stand a significant chance to be extracted. In large sources as Dragon or Kamaboko, the walls are too far and the extraction probability is equal to zero, even under the most favorable hypothesis. One should notice that the negative ions created on the grid are accelerated into the discharge by the plasma sheath potential and are reflected inside the source thanks to the elastic collisions and charge exchange processes. The extraction probability is consequently relatively low (~ 20 %).

Isotopic effect

It has been shown on Dragon and Kamaboko that cesium injection decreases the isotopic effect [34, 20]. The ratio J_{D^-}/J_{H^-} is equal to 0.7 and stays constant with the arc power. This occurrence is typical with cesium and has already been observed in other experiments [37, 38]. The extraction probability P has been calculated for H^- and D^- produced on various position (x, z) of the plasma grid. In order to analyse the experimental results, the extracted ion current density has been estimated as follows :

$$J_{neg} = \frac{e}{S_{extr}} \iint_{grid} (k_0 J_0 + k_+ J_+) P(x,z) dx dz \qquad (8)$$

where S_{extr} is the extraction surface, J_0 and J_+ are respectively the H^0/D^0 and H^+/D^+ flux on the grid, k_0 and k_+ the negative ion production rate respectively by H^0/D^0 and H^+/D^+.

J_{neg} has been estimated in arbitrary units, by assuming that the production rates and the incoming particle flux are uniform over the grid surface, for two extreme hypothesis. We consider that negative ions are only produced by H^+/D^+ in the first case ($k_+ J_+ \gg k_0 J_0$) and by H^0/D^0 in the second case ($k_+ J_+ \ll k_0 J_0$). In the two cases, we consider the same production rate for hydrogen and deuterium. The calculation predicts an isotopic effect opposite to the experimental results in the first hypothesis because the D^+ flux is three times larger than the H^+ flux. On the other hand, for the second hypothesis, the calculation gives the result presented on figure 4. The good qualitative accordance between simulation and experimental results shows the major role that atoms seem to play in negative ion sources with cesium.

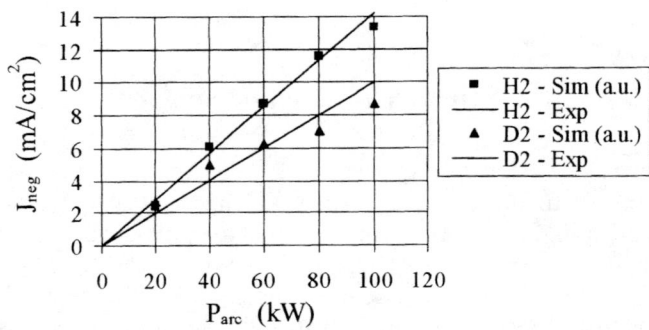

FIGURE 4 Evolution of negative ion current density with arc power. Comparison between experimental results and simulation in the hypothesis that negative ions are mainly produced by atoms. Initial E_{H^-} = 1 eV.

Influence of the magnetic filter

The Kamaboko source studied at Cadarache with a 800 gauss.cm original magnetic filter, has delivered 1.4 A (20 mA/cm^2) of D$^-$ with an extracted electron current I_e about 14 A intercepted by the second grid of the extraction system [20]. This electron current has to be strongly decreased to fill the ITER requirement (I_e/I_{D^-} < 1). One way of reducing I_e is to enhance the magnetic filter strength to prevent as much as possible the extraction of the electrons from the plasma. The filter decreases the H$^+$/D$^+$ flux on the plasma grid but should not change the H^0/D^0 flux. Consequently, if the negative ion surface production is mainly due to atoms, it should be possible to reduce I_e without any alteration of I_{D^-}. At Cadarache, different magnetic filters (1100, 1400 and 1800 gauss.cm) have then been tested on Kamaboko[20]. The increase of the filter strength up to 1800 gauss.cm has induced a drop of I_e by a factor 20 without any change on the D$^-$ current, so that the Kamaboko source fills now the ITER requirement (I_e/I_{D^-} = 0.5).

The extraction probability of D$^-$ produced on the plasma grid surface has been calculated for all the tested filters. Langmuir probe measurements have been performed in order to find the plasma parameters which are used in the transport simulation. These parameters can not be measured close to the grid because of the high magnetic field and some extrapolation has been made. For example, the electron density at the grid level is between 10^{11} and 10^{12} cm^{-3} for the 1100 gauss.cm filter case. The calculations have been performed for these two limits with $n_D = 10^{13}$ cm^{-3}. The mean extraction probability k_{extr}, presented on figure 5, has been calculated as follows :

$$k_{extr} = \frac{1}{S_{extr}} \iint_{grid} P(x,z) dx dz \quad (9)$$

In all the considered cases, we see that the extraction probability does not depend on the magnetic filter. Taking into account the experimental results (I_{D^-} independent of the filter), we conclude that the flux of negative ions produced on the plasma grid does not depend on the filter either. Consequently, since the D$^+$ flux which hits the grid depends on the filter strength, it seems that atoms are the main responsible for the surface production of negative ions. This result confirms the conclusion of the isotopic effect study.

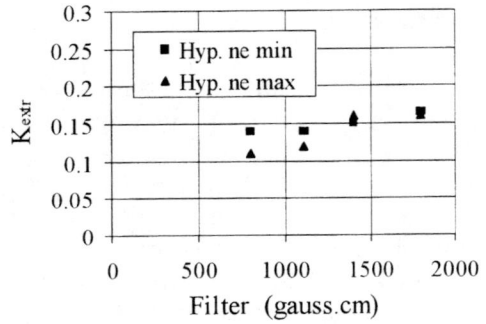

FIGURE 5 Evolution of the mean extraction probability with the magnetic filter. Initial E_{H^-} = 1 eV.

CONCLUSION

Calculations with the NIETZCHE code have shown that in the sources contemplated for ITER, the negative ions are extracted if they are produced at a distance lower than 2 cm from the plasma grid in the case of volume production or if they are produced on the plasma grid surface, in the vicinity of the extraction holes, after being accelerated by the plasma sheath potential toward the discharge and reflected by the collision processes. These sources, originally optimized for volume production, have a very short effective production surface. This may explain why it is necessary to introduce 2.5 $W.cm^{-3}$ of arc power to extract 20 $mA.cm^{-2}$ of D^- whereas some similar sources directly optimized for surface production[39] have produced 13.3 $mA.cm^{-2}$ of H^- for only 0.2 $W.cm^{-3}$, but with a high divergence. Consequently, a compromise between these two concepts would be perhaps a better solution for ITER.

REFERENCES

[1] ITER, Design Description Document, Chap 5.3 (1995)
[2] M.Kuriyama and al., Symp. On Fusion Technol., (1996)
[3] K.Watanabe and al., Int. Symp. on Production and Neutralization of Negative Hydrogen Ions and Beams, NY, 309 (1996)
[4] C. Corse, M. Capitelli, J. Bretagne, M. Bacal, Chem. Phys. **93**, 1 (1985)
[5] P. Berlemont, D.A. Skinner, M. Bacal, Proc. Int. Symp. on Production and Neutralization of Negative Hydrogen Ions and Beams, NY, 77 (1992)
[6] O. Fukumasa, J. Phys. D: Appl. Phys., **22**, 1668 (1989)
[7] M. Ogasawara et al., Proc. 13th Symp. on ISIAT, Tokyo, 83 (1990)
[8] A.J.T. Holmes, Plasma Sources Sci. Technol. (1996) 453
[9] D. Riz, J. Paméla, Proc. Int. Symp. on Production and Neutralization of Negative Hydrogen Ions and Beams, NY, 3 (1996)
[10] Y. Belchenko, G.I. Dimov, V.G. Dudnikov, Nucl. Fusion, **14**, 113 (1974)

[11] L.W. Swanson, R.W. Strayer, J. Chem. Phys., **48**, 2421 (1968)
[12] J.W. Kwan and al., Rev. Sci. Instrum., **57**, 831 (1986)
[13] M. Bacal and al., Proc. Int. Symp. on Production and Neutralization of Negative Hydrogen Ions and Beams, NY, 95 (1980)
[14] K.N. Leung, K.W. Ehlers, M. Bacal, Rev. Sci. Instrum.,. **54**, 56 (1983)
[15] R.L. York and al., Rev. Sci. Instrum., **55**, 681 (1984)
[16] Y. Okumura and al., Proc. Int. Symp. On Production and Neutralization of Negative Hydrogen Ions and Beams, NY, 309 (1986)
[17] K.N. Leung and al., Rev. Sci. Instrum.,. **60**, 531 (1989)
[18] S.J. Cox and al.,, Proc. Int. Symp. on Production and Neutralization of Negative Hydrogen Ions and Beams, NY, 397 (1992)
[19] M. Bacal and al., Proc. Int. Symp. on Production and Neutralization of Negative Hydrogen Ions and Beams, NY, (1996)
[20] R. Trainham and al., 7th Int. Conf. on Ion Sources (1997)
[21] D.L. Book, NRL Plasma Formulary, Naval Research Laboratory N°177-4405, 31-35 (1990)
[22] D.G. Hummer and al., Phys. Rev., **119**,668 (1960)
[23] R.K. Janev and al., Elementary Processes in Hydrogen-Helium Plasmas, Springer-Verlag, chap 7 (1987)
[24] D. Ciric, private communication
[25] C. Jacquot, R. Hemsworth, D. Riz, A. Holmes, 5th European Conf. on the Production and Application of Light Negative Ions, 26, (1994)
[26] C. Jacquot, D. Riz, R. Hemsworth, euratom-cea report N° eur-cea-fc-1543
[27] C. Jacquot and al., Proc. Int. Symp. On Production and Neutralization of Negative Hydrogen Ions and Beams, NY, 214 (1996)
[28] A.J.T. Holmes and al., AEA Fusion report, 185 (1991)
[29] D.A. Skinner et al., Phys. Rev. E, **48**, 2122 (1993)
[30] K.N. Leung, K.W. Ehlers, M. Bacal, Rev. Sci. Instrum., **54**, 56 (1983)
[31] M. Bacal, J. Bruneteau, P. Devynck, Rev. Sci. Instrum., **59**, 2152 (1988)
[32] A.A. Ivanov and al., Phys. Rev. E, **55**, 956 (1997)
[33] Y. Belchenko, C. Jacquot, J. Paméla, D. Riz, Rev. Sci. Instrum., **67**, 1033 (1996)
[34] C. Jacquot, J. Paméla, Y. Belchenko, D. Riz, Rev. Sci. Instrum. **67**, 1036 (1996)
[35] M. Hanada and al., Fusion Technology, Elsevier Science B.V, 617 (1994)
[36] N. Miyamoto and al., Proc. Int. Symp. On Production and Neutralization of Negative Hydrogen Ions and Beams, NY 300 (1996)
[37] Y. Takeiri and al., Rev. Sci. Instrum., **66**, 2541 (1995)
[38] K. Watanabe and al., Proc. Int. Symp. On Production and Neutralization of Negative Hydrogen Ions and Beams, NY, 326 (1992)
[39] K.N. Leung, K.W. Ehlers, Rev. Sci. Instrum.,. **53**, 803 (1982)

DIAGNOSTICS

Measurement of the Temporal Behaviour of the H⁻-Density in a Pulsed Hydrogen Multipole Discharge using the Photodetachment Technique

T. Mosbach, H.M. Katsch and H.F. Döbele

Institut für Laser- und Plasmaphysik, Universität - GH Essen, Universitätsstr. 5, 45117 Essen, Germany

Abstract. An increase of the H⁻-density in a pulsed low-pressure hydrogen multicusp discharge is observed after the current pulse, i.e. in the "post-discharge". The H⁻-density is determined by laser induced photodetachment at a Langmuir probe. A semiempirical model is used to infer the total density of vibrationally excited molecules in the electronic ground state from a comparison with the measured shape of the transient H⁻-density in the post-discharge. The total density of vibrationally excited molecules is found to be independent of the discharge current up to 20 amps. The density evolution of vibrationally excited molecules during the discharge is obtained from measurements of the maximum H⁻-density in the post-discharge as a function of the duration of a single discharge pulse. After 200 µs the density is saturated. The lifetime of vibrationally excited molecules in the post-discharge is determined by varying the time intervall between discharge pulses of short duration. The lifetime is found to be approximately 1 ms.

INTRODUCTION

Negative hydrogen ions are generated in multipole devices mainly by dissociative attachment of electrons to highly vibrationally excited hydrogen molecules. The main destruction mechanism in these devices is by collisions with fast electrons for pressures below ~ 1 Pa and medium plasma densities. In this experiment the so called "time filter" is applied to eliminate the fast electrons within approximately 2 µs which consists in the fast switch-off of the discharge. With the destructive fast electrons being absent in the post-discharge the H⁻-density increases while the plasma density decreases.

A simple semiempirical model allows to describe the evolution of the transient H⁻-density in the post-discharge. A comparison with the measured temporal behaviour of the H⁻-density in the post-discharge with this model allows to determine the concentration of vibrationally excited molecules and represents therefore a possibility to test theoretical models which are being controversially discussed (1).

MODELLING OF H⁻ GENERATION IN THE POST-DISCHARGE

The following assumptions - partly supported by measurements - are made to evaluate the transient H⁻-density in the post-discharge:

1.) the electron temperature $T_e(t)$ in the post-discharge exhibits a fast drop after current shut-off (supported by Langmuir probe measurements)

2.) the total density of vibrationally excited H_2-molecules in the post-discharge is constant during 100 μs after switching off the discharge

3.) the H⁻-density production rate Δ^+ in the post-discharge is given by the product of the electron density, the density of vibrationally excited H_2 and the electron temperature dependent attachment rate

4.) the H⁻-density loss rate in the post-discharge is determined by ambipolar diffusion to the walls, Δ_D, and by mutual neutralization ($H^+ + H^-$) in the post-discharge, Δ_{MN}.

The time dependence of the H⁻-density is therefore given by:

$$\frac{dn_-(t)}{dt} = \Delta^+ - (\Delta_D + \Delta_{MN})$$

The H⁻-density in the post-discharge can be compared with the results of the model calculation. In this calculation the ratios of the excited state populations are chosen as proposed by Hiskes (2). The total population serves as a free parameter in the comparison. It is therefore possible to deduce the unknown density of the excited molecules by measurements of the plasma parameters electron density, electron temperature and H⁻-density. (More details are given in (3)).

EXPERIMENTAL SET-UP AND DIAGNOSTICS

The experiment is a typical one-chamber discharge. Two heated tungsten wires act as cathodes. The grounded metallic cylindrical vacuum vessel (diameter 100 mm, length 250 mm), surrounded by permanent magnets of alternating polarity, acts as anode. No magnets are mounted at the end flanges of the vessel (thus giving only incomplete confinement). The discharge voltage can be switched with adjustable on- and off-durations determining the pulse and pause durations. Typical plasma parameters are: electron density $n_e = 10^{16} - 10^{17}$ m⁻³, electron temperature $T_e = 1 - 5$ eV, gas pressure $p_n = 0.1 - 2.5$ Pa, discharge current $I_d < 20$ A, discharge voltage $U_d = 90$ V (for more details on the experimental set-up see (3)). The plasma parameters (electron density and temperature) are measured with an electrostatic cylindrical probe. The ion species present are identified by extraction and mass analysis. Further more, laser induced photodetachment is applied to obtain the temporal behaviour of the H⁻-density in the post-discharge. A ruby laser (1J / pulse, pulse duration 60ns and wavelength 694.3nm) provides photodetachment of H⁻ ions. The detached electrons are detected by an electrostatic cylindrical probe located coaxially with the laser beam. This diagnostic technique was developed by Bacal and Hamilton and is described in detail in (4). The techniques mentioned allow time-resolved measurements.

MEASUREMENTS AND RESULTS

Formation and relaxation of vibrationally excited hydrogen molecules

With the discharge configuration outlined above the end value of the plasma density is reached after about 50 µs whereas the maximum H⁻-density in the post-discharge reaches its saturation value later (see fig. 2). The total density of the vibrationally excited molecules reaches its value correspondingly at a later time as inferred from the saturation of the maximum H⁻-density in the post-discharge. Figure 1 represents the maximum H⁻-density as a function of the pulse duration of a single discharge. For 1 amp the maximum H⁻-density saturates at 200 µs; for 4 amp the saturation is reached at 100 µs.

In order to determine the lifetime of vibrationally excited molecules in the post-discharge the discharge pulse duration has to be short enough so that a stationary density of the vibrationally excited molecules is not reached. If the lifetime of vibrationally excited molecules in the post-discharge lasts longer than the time at which the next discharge pulse starts, this pulse will further enhance the density of vibrationally excited molecules. The density of vibrationally excited molecules at the end of the following discharge pulse is therefore higher than for the preceding pulse. If the maximum H⁻-density in the post-discharge is indicative for the total density of vibrationally excited molecules, we expect an increasing H⁻ signal with decreasing discharge pause duration. We expect alternatively that with increasing duration of the discharge pause the H⁻ signal becomes independent of the duration of the discharge pause, if the lifetime of vibrationally excited molecules is reached. Figure 2 demonstrates this

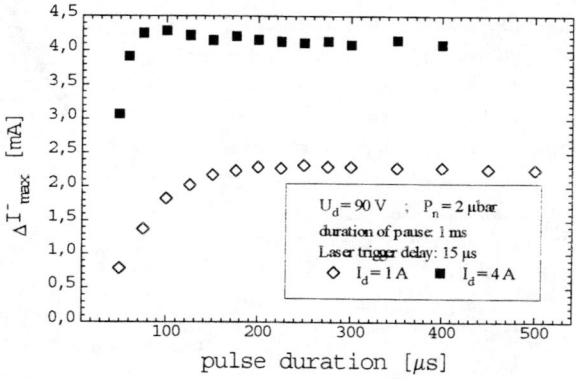

Fig. 1: Maximum H⁻-density in the post-discharge as function of the pulse duration for discharge currents of 1 and 4 amps.

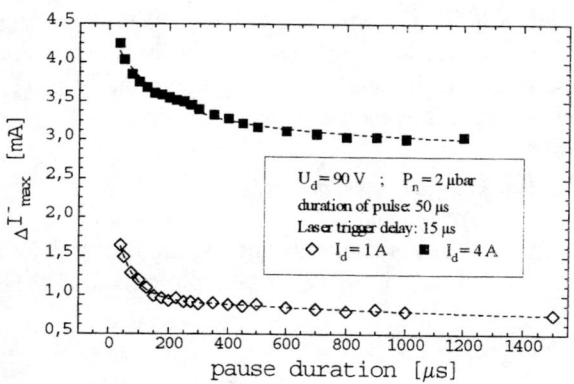

Fig. 2: Maximum H⁻-density in the post-discharge as function of the pause duration for discharge currents of 1 and 4 amps.

behaviour. We therefore infer a lifetime of vibrationally excited molecules in the post-discharge in our case of approximately 1 ms.

Relative H⁻-densities in the post-discharge for different values of the discharge current

The measurements discussed in the following refer to discharge pulse durations long enough so that the vibrational population has reached the saturation value. Figure 3 shows the relative H⁻-density (the ratio of the H⁻ to the electron density) as a function of time in the post-discharge for different values of the discharge current. This quantity turns out to be independent of the discharge current, although the plasma density increases in proportion with the discharge current. The comparison with the model outlined yields the result that the total density of vibrationally excited molecules does not increase with plasma density. This results is in agreement with results of VUV laser absorption spectroscopy measurements by Stutzin up to v"=8 (5). This result is in contradiction, however, with selfconsistent model calculations based on the EV-process being mainly responsible for the formation of vibrationally excited molecules. With this process dominating one would expect an increase of vibrationally excited molecules with increasing plasma density.

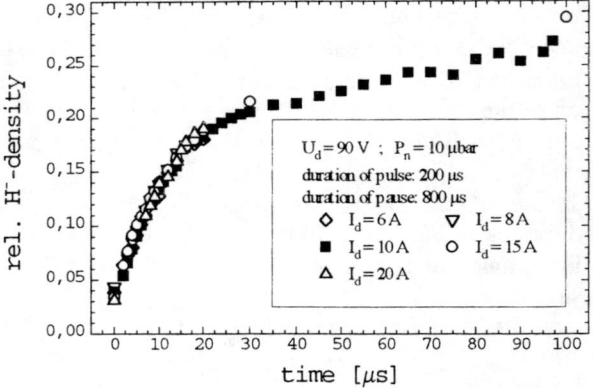

Fig. 4: Temporal behaviour of the relative H⁻-density in the post-discharge for different values of the discharge current.

ACKNOWLEDGMENTS

This work was supported by the Deutsche Forschungsgemeinschaft in the frame of the Sonderforschungsbereich 191 "Grundlagen des Niedertemperaturplasmas".

REFERENCES

1. Graham, W.G. ; *Plasma Sources Sci. Technol.* **4**, 281 (1995)

2. Hiskes, J.R. ; Karo, A.M. ; *Appl. Phys. Lett.* **54**, 508 (1989)

3. Katsch, H.M. ; Quandt, E. ; Köster, A. ; *J. Phys. D* **28**, 493 (1995)

4. Bacal, M. ; *Plasma Sources, Science and Technology* **2**, 190 (19939

5. Stutzin, G.C. ; Young, A.T. ; Döbele, H.F. ; Schlachter, A.S. ; Leung, K.N. ; Kunkel, W.B. ; *Rev. Sci. Instrum.* **61**, 619 (1990)

Measurement of negative ion beam emittance

C. Michaut, J. Bucalossi[†] and D. Riz

*Laboratoire de Physique des Milieux Ionisés, Laboratoire du C.N.R.S.,
Ecole Polytechnique, 91128 Palaiseau Cedex, France*

Abstract. The purpose of this experiment is to improve the optics of an intense H⁻ negative ion beam, to be used in future tokamaks. The emittance diagnostic has been installed in the experimental set-up INCA, to perform an optimization of a negative-ion injector pre-accelerator. The first stage of this work consists to test the operation of the electric-sweep scanner, to measure the beam current in the 2D-phase space and to investigate the diagnostic performances. In addition to the measurement of H⁻ angles, we can scan the angles of the fast H⁺ created along the beam in the transport region and test the efficiency of the electron magnetic trap installed in the second accelerator grid (extraction electrode). We also observe the presence of a neutral component. This analyzer not only provides a diagnostic of the beam divergence but also improves of the knowledge of the beam components. We studied two accelerator configurations. The first one permitted to obtain an optimal beam optics under 15 keV. Then the accelerator geometry was modified to ensure a higher optimum high-voltage. We will show several experimental results performed with two accelerator configurations.

INTRODUCTION

In view of the transport on long distances of energetic neutral beams produced by the neutralization of the negative beam, a good understanding of the beam formation is essential. In fact, only the intense beams formed with sufficiently good optics could be accelerated to the energy required for the tokamak plasma heating. Therefore we began

---

[†] Permanent address: Département de Recherches sur la Fusion Contrôlée, CE Cadarache, 13108 Saint-Paul Lez Durance

the study of the H⁻ ion beam emittance on our experimental test-stand INCA[1]. The emittance diagnostic has been built in the laboratory using the design described by a Los Alamos team[2]. INCA is composed of an H⁻ volume multicusp source and a single-aperture three-grid accelerator. In its initial configuration the beam optics is optimum for an acceleration voltage around 15 kV[3], although the maximum voltage of the power supply is 60 kV.

The analyzer used for the emittance diagnostic is also named electric-sweep scanner (ESS). The scanner consists of two slits, separated by a pair of deflector plates driven by a ramp voltage. Behind the second slit, there are in the listed order: an electron suppressor plate and a grounded collector. The analyzer is driven by a motorized manipulator which permits to scan a beam transverse section. Considering the diagnostic geometry, the angular resolution given by the designers is ±0.5 mrad, the maximum analyzable angle is ±91 mrad.

A complete characterization of the beam requires emittance measurements in two transverse directions. If the studied system has a cylindrical symmetry, investigating only one transverse direction is sufficient. This is the case in our experimental set-up.

THE EXPERIMENTAL FACILITY INCA

The experimental facility INCA allows the study of mechanisms occurring in the negative-ion beam formation and during the beam transport. Previously, INCA was completely described[1] in relation with studies of the beam acceleration[3] and of the effect of the pressure[4] on the beam transport. This experimental set-up consists in a negative-ion volume source, a single-aperture three-grid accelerator, a one meter transport region. Figure 1 gives a general view of the experimental device INCA. Let us note that the ion source is on the top of the tank, so the beam is propagating vertically towards the bottom.

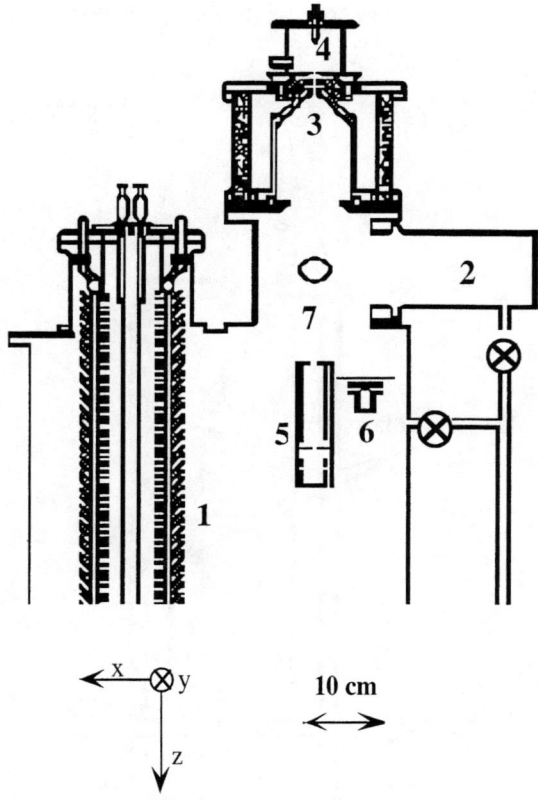

FIGURE 1. Experimental set-up INCA which comprising a large speed Titanium pump (1), a turbomolecular pump (3), a three-grid accelerator (3), an H- ion volume source (4), an electric-sweep scanner (5), a Faraday cup (6) situated in the beam transport region (7).

The charged particle beam current is measured using a Faraday cup (FC), while the incident angles of the charged particles are determined using the electric-sweep scanner. These diagnostics are mounted on independent manipulators and are retractable. When the electric-sweep scanner is outside the beam, the FC collects the beam, which can be thus measured before each emittance measurement.

The emittance diagnostic

The ESS is driven across the negative-ion beam using a motorized manipulator, located at around 90 cm from the source extraction aperture, i.e. 86 cm from the accelerator exit. Figure 2 represents a schematic longitudinal section of the analyzer. In front of the first slit, a thick copper plate ensures that no thermal overheating of the analyzer occurs.

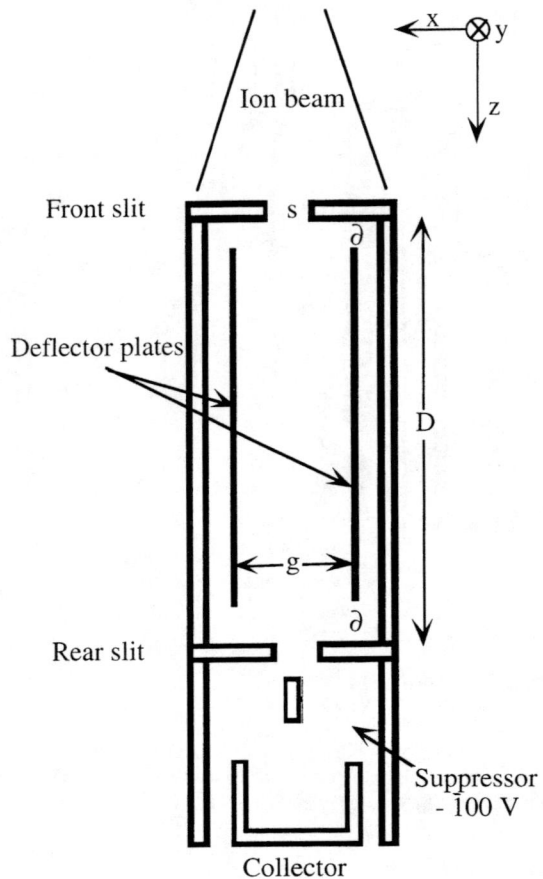

FIGURE 2. Schematic view of the electric-sweep scanner. It is composed by two slits of width s (0.1 mm) the distance between the slits is D (98 mm); two deflector plates of length D-2∂ (with ∂=4 mm) separated by a gap g (5 mm); a secondary electron suppressor and a collector.

The scanner linear motion is following the X-axis, while the slits are perpendicular, along the Y-axis. At each manipulator position, a beam slice goes through the front slit. Between the two deflector plates, a sweeping uniform electric field is applied deflecting each entered particle using a ramp voltage connected to one of the two plates, while the other one is grounded. Of course the assumption of a monoenergetic beam is made. A unique electric field permits a particle entered with an initial angle X' to pass through the rear slit to reach the collector. A simple relation relies the initial angle to the applied voltage V.

$$X' = \frac{V}{\Phi} \frac{(D-2\partial)}{4g} \qquad (1)$$

The amplified collector current and the deflection voltage are simultaneously measured by a two-channel digitizer, which is provided with a synchronous trigger to ensure the precision of each experimental data. As the ramp voltage is swept, we collect the current as a function of initial angle, I(X'). The motorized manipulator takes one constant step forward across a beam section, then it stops during the ramp voltage time. Finally the beam current I(X,X') is recorded in the 2D-phase space. Emittance is given by the area of the beam in the phase space. Figure 3 shows the electronic design driving the emittance diagnostic.

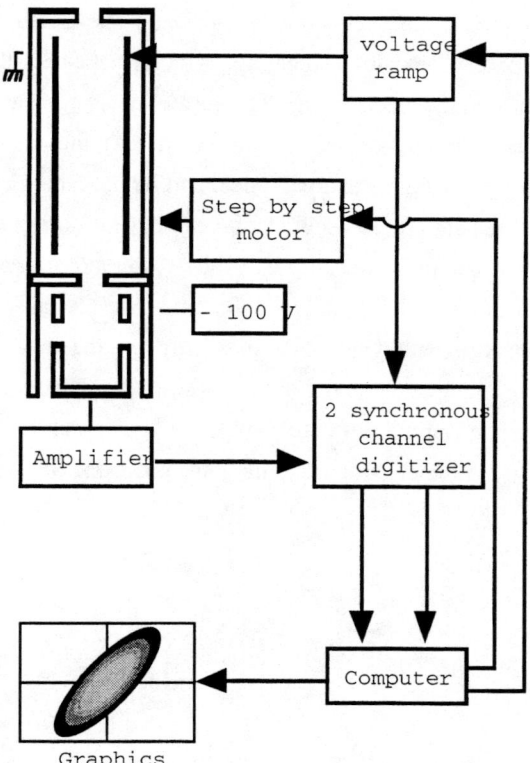

FIGURE 3. Electric set-up permitting to drive the analyzer, the transverse electric field and the measurement acquisition.

EXPERIMENTAL PROGRESS

The beam line from the volume source to the current diagnostics situated in the transport region is entirely of cylindrical symmetry. Therefore a particle reaching the analyzer with no angle (X'=0) should have a trajectory exactly on the revolution symmetry axis (X=0). Effectively, when the accelerator alignment is correct the beam current trace is centered on the axis intersection on the emittance diagram. When the accelerator alignment is not correct, the beam trace does not appear at the diagram center.

Check of magnetic trap

As the negative ions are extracted from the source by a positive voltage, the source plasma electrons are also extracted. In the second electrode of the accelerator, there is a magnetic field trap to remove the electrons from the beam. The electrons are collected by the extraction electrode and lead to its heating, which can cause the damage of the permanent magnets used in the electron trap. The current measurement by a Faraday cup is not adequate to detect an electron leak in the beam. Of course, the electron trap is previously verified by measuring its magnetic field and by producing an Argon plasma in the source. But the permanent magnet degradation can appear during experiments. Thus in few experiments, we obtained two ellipses of the same sign current in the emittance diagram. The second ellipse is shifted from the first one, because the electrons are deflected by the magnetic field when they pass through the defective electron trap. Figure 4 shows the current level contours in the phase space. The second trace under the H⁻ ellipse is due to the bad trapping of electrons. Thus the emittance diagnostic can also determine the efficiency of the electron trapping in the second accelerator stage.

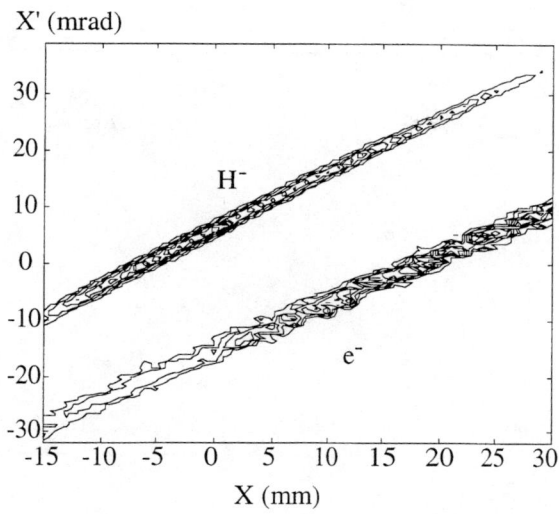

FIGURE 4. Electron trace on the emittance diagram. Plasma electrode aperture diameter $\Phi EP = 0.45$ cm

Pressure effect

The collisions of the H⁻ beam with a hydrogen gas target lead to the formation of fast H⁰ and H⁺ depending on the target pressure. Here the transport region acts as a gas target. The effects of the transport region pressure on the beam have been widely discussed in previous works[5,6]. Direct measurement of the beam current, as made by a Faraday cup, gives the difference between negative and positive charge contributions and it is not possible to identify separately each of these currents. As emittance measurements are effected by a deflection electric field, the positive and negative charges can not be selected with the same electric field. Since fast H⁺ are created in the transport region when the beam particles have their full energy, their trajectories follow the H⁻ trajectories. Thus, they reach the front slit of the analyzer with the same angle. As their charges are opposite, the deflection voltage permitting an H⁺ ion to pass through the rear slit will be the same one, with an opposite sign, than for H⁻. Thus we observed when the pressure in the transport region is sufficiently high, the positive and negative traces on the same emittance diagram. Figure 5 shows how the ellipses from H⁻ and H⁺ cross at the zero initial angle.

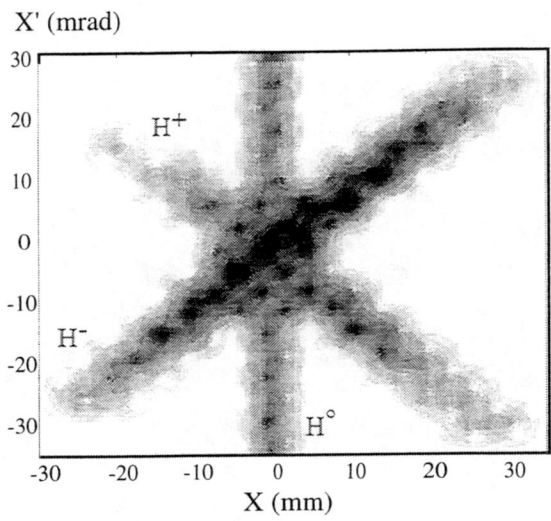

FIGURE 5. Logarithmic representation of the absolute value of the beam current in the 2D-phase space. We observe the presence of several types of beam particles: negative, positive and neutral. Emittance diagram obtained with the new grid set. Pressure: P_{source} = 6 mTorr, P_{trans} = 2.4 10^{-4} Torr.

In Fig. 5, X'-axis represents the initial angles computed from Eq. 1. But in this last relation, the particle charge sign has been simplified, taking into account only negative particles. Thus the obtained angles are directly proportional to the applied deflection voltage. It is interesting to remark that the electric-sweep scanner is more than an emittance diagnostic. In fact the beam components can be known with this device. Only the section where the crossing between the two traces occurs is not quantitatively exploitable, for the same reason as current measurement using Faraday cup. However an important effort on the emittance diagram analysis should be done to lead to the beam component proportion.

New grid set

The goal of installing the emittance diagnostic is to determine the best configurations of single-aperture accelerator in view of optimizing beam optics. A new three-grid set has been designed and built taking into account 3D-simulation results performed at CE Cadarache. As it has been reported[3, 7] previously, for a given extraction voltage the negative ion beam can be focused for the optimum acceleration voltage. Therefore the optimum running conditions are searched with the new accelerator installed on INCA.

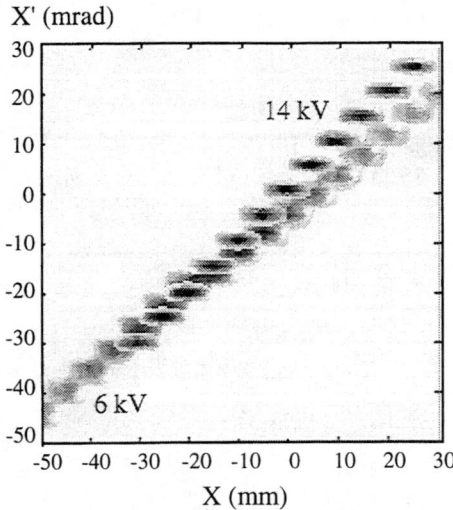

FIGURE 6. Emittance diagram of H⁻ beam for two different acceleration voltages (6 kV and 14 kV), obtained with the new grid set. Plasma Electrode aperture diameter $\Phi_{EP} = 1.4$ cm. Pressures: $P_{source} = 5$ mTorr, $P_{trans} = 3.1\ 10^{-4}$ Torr; Extraction electrode parameters: $V_{extr} = 2$ kV, $I_{extr} = 42$ mA.

Figure 6 compares the emittance diagrams recorded with two different acceleration voltages. All the other experiment parameters are unchanged. We see how the ellipse turns left when the acceleration voltage is increased, therefore the beam becomes more narrow. Moreover, in the same time the incident angle range also decreases with the beam energy, the beam particle trajectories get closer to the beam propagation direction. This is the proof that the beam focusing depends on the ratio between the acceleration and extraction voltages.

ACKNOWLEDGMENTS

This work is a part of the Collaboration of CE and CNRS on controlled fusion. We are extremely grateful to L. Travers and P. Van Coillie for technical support.

REFERENCES

[1] P. Devynck, C. Michaut, M. Bacal, Z. Sledziewski, F.P.G. Valckx, J.H. Whealton and R.J. Raridon, Proc. Fourth European Workshop on the Production and Application of Light Negative Ions, March 1991, Belfast (N. Ireland). Editor: W.G. Graham (p. 109).

[2] P.W. Allison, J.D. Sherman and D.B. Holtkamp, IEEE Trans. Nucl. Sci. NS-30, 220004 (1983).

[3] C. Michaut, P. Devynck, M. Bacal, Z. Sledziewski, F.P.G. Valckx, J.H. Whealton, R.J. Raridon, Rev. Sci. Instrum. 63, 2774 (1992).

[4] C. Michaut, M. Bacal, Rev. Sci. Instrum. 65, 1424 (1994).

[5] M. Bacal, C. Michaut, L.I. Elizarov, F. El Balghiti, Rev. Sci. Instrum. 67, 1138 (1996).

[6] C. Michaut, M. Bacal, Proc. XIIth ESCAMPIG, Eindhoven, The Netherlands, 23-26 Août 1994, p. 352.

[7] A.J.T. Holmes and M.P.S. Nightingale, Rev. Sci. Instrum, 57, 2402 (1986).

SOURCES/H$^-$, D$^-$

Development of Large Scale Negative Ion Source for LHD-NBI

K. Tsumori, T. Takanashi, *S. Asano, M. Osakabe, Y. Takeiri, Y. Oka,
E. Asano, T. Kawamoto, R. Akiyama *T. Okuyama, *Y. Suzuki
and O. Kaneko.

National Institute for Fusion Science, Oroshi 322-56, Toki, Gifu 509-52, Japan

TOSHIBA CO. Ukishima-chou 4-1, Kawasaki, Kanagawa, Japan

Abstract. In this paper we report the first results on a prototype negative ion source for LHD-NBI. The ion source with a multi-cusp configuration consists of the external magnetic filter and has an inner volume of 140 cm in height x 35 cm in width x 25 cm in depth. Two cesium lines are installed to seed cesium vapor into arc plasma for enhancement of H$^-$ current. The H$^-$ ion beam was extracted from an area of 120 cm x 25 cm and was accelerated via single-stage accelerator. An H$^-$ current of about 4 A with the energy of 73 keV was obtained at the arc power of 250 kW for a duration of 0.3 sec.

INTRODUCTION

Negative-ion based neutral injector (N-NBI) becomes indispensable as an important plasma-heating device for plasma confinement devices. That is because when the plasma confinement device becomes larger, it is necessary to obtain a long penetration length of a neutral beam by increasing the injection energy, and H$^-$ beam has higher neutralization efficiency than H$^+$ beam. The high-power negative ion sources have been researched and developed in several laboratories (1 - 4). In a project of Large Helical Device(LHD) two neutral beam lines with negative ion sources are planned to be equipped for plasma heating (5, 6). The beam line consists of two ion sources and extracted H$^-$ beams from those sources will be merged into one neutral beam via a pair of gas neutralizers (5). The beam lines were designed to produce 180 kV-40 A of H$^-$ beam and 15 MW of injection power per a beam line with the beam-divergence angle of less than 10 mrad. The current corresponds to 40 mA / cm2 of the current density. To obtain this high current density, cesium, Cs, vapor is seeded to enhance the H$^-$ current. We have successively investigated the 1/6 and 1/3 scaled ion sources, and successfully obtained H$^-$ ion beams more than 16 A with the current density of more than 31 mA / cm^2 (2, 7). Recently, an H$^-$ beam with 21 mA / cm^2 of high current density was extracted for a duration of 10 sec (8).

In the following sections, we report the structure of the prototype negative ion source and its peripheral devices, the plasma unformity in the longitudinal direction, and the beam acceleration characteristics.

EXPERIMENTAL APPARATUS

The ion source is composed of two parts, which are an arc chamber and a beam accelerator. A schematic diagram of the source is shown in Fig. 1, and its inner volume is 140 cm (H) x 35 cm (W) x 25 cm (D). The chamber is a multi-cusp source with an external magnetic filter, and water-cooling channels are imbedded in the chamber walls which are made of oxygen-free copper. In order to make the magnetic field uniform in the longitudinal direction, the cusp lines are arranged straight at a back and side plates of the chamber and the bending parts of the cusp magnets are concentrated at the top and the bottom plates of the chamber. The strength of the cusp magnet is 1.4 kGauss on the inner chamber wall. An external magnetic filter is adopted to improve arc efficiency of H⁻ extraction. A dipole filter field is produced by a pair of external magnets installed between the arc chamber and a plasma grid as shown in Fig. 1. The field separates arc chamber into a plasma generator and a beam extraction region near the plasma grid. A maximum field strength at the center line of the chamber was set at 50 Gauss and the magnetic thickness was 500 Gauss•cm.

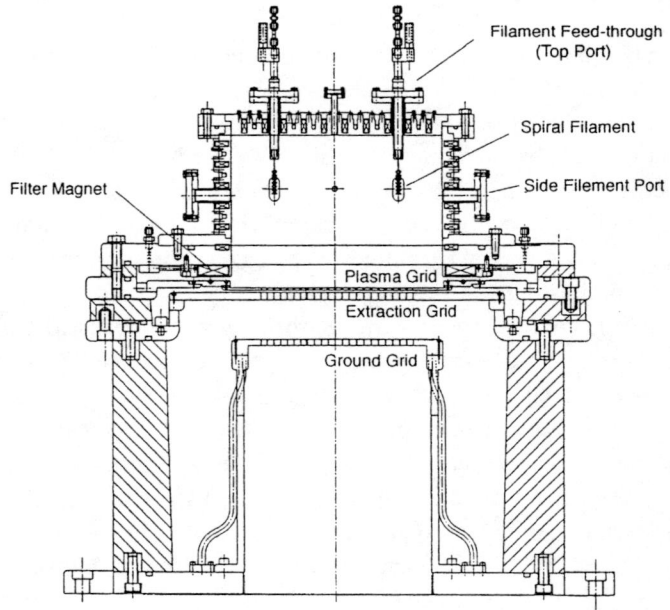

FIGURE 1. A cross-sectional view of a prototype negative ion source for LHD NBI.

The arc chamber equips 48 filament ports; 24 ports are attached from a back plate and the other ports are attached from side plates. Spiral type filaments are set from the back plate. This type of filament was expected to obtain a high electron-emission rate, and it possibility to reduce the total number of the filaments. In order to obtain the H⁻ current density of 40 mA/cm^2, cesium vapor is seed into an arc plasma. Two Cs lines with valves were attached through back plate. The oven temperature was kept below 200 °C, and the other parts of the line were kept from 270 °C to 300 °C to avoid Cs condensation. The Cs valves were left to open until the H⁻ current increased and saturated. After the saturation the oven temperature was adjusted not to decrease the H⁻ current. The arc chamber equips several ports for probe measurement. Plasma parameters, such as electron density and its temperature, were measured make use of Langmuir probes.

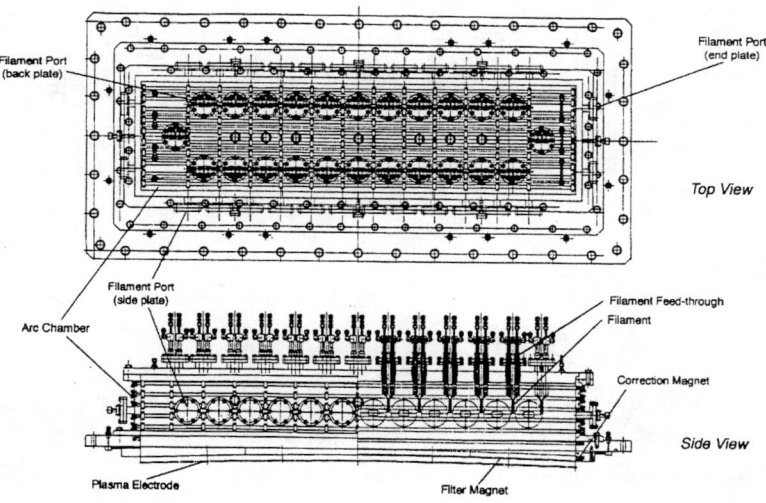

FIGURE 2. Top and side view of a prototype negative ion source for LHD NBI with inside structure.

The acceleration system consists of three electrodes, which are a plasma grid, an extraction grid and a grounded grid. The total beam extraction area is 120 cm (H) x 25 cm (W), and the grids are separated vertically into five segments. These segments are aligned to focus the H⁻ beam vertically at a distance about 13 m from the grounded grid. Each segment contains 22 (H) x 16 (V) apertures. A structure and electric connection of the grids are schematically shown in Fig. 3. The plasma grid is made of molybdenum, and the grid is not cooled directly to enhance the surface effect on H⁻ yield by increasing the temperature in an atmosphere of Cs vapor. Size of an aperture in the plasma grid is 9.0 mm in diameter, and the beam transparency is 38 %. The aperture is shaped like a

JAERI-type plasma grid to fix a beam- extraction interface. The extraction grid is made of oxygen- free copper. Water-cooling channels is aligned in the grid to remove heat due to extracted electron and to cool electron-deflection magnets installed inside the grid. The grid is 12 mm thick and aperture size is 10.5 mm in diameter. The cooling channel is placed on a side of plasma grid. There is no electron trap in an aperture to increase the beam transparency. The water-cooled grounded grid is also made of oxygen-free copper. The aperture size is as the same as that of the extraction grids. The aperture axis is displaced in the horizontal direction to compensate the H⁻ beam trajectories bent by the electron-deflection magnets. The aperture axes are also arranged to converge all the beamlets at the point 13 m apart from grounded grid.

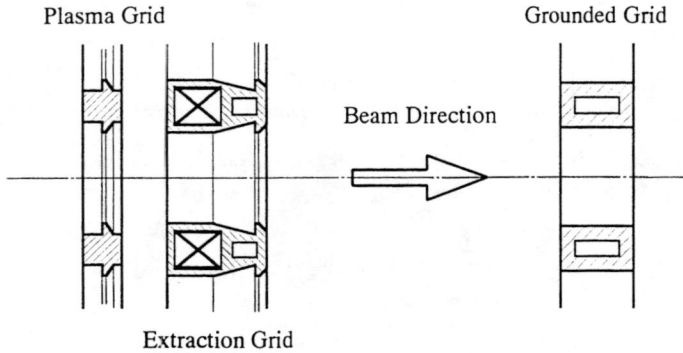

FIGURE 3. A schematic diagram of beam acceleration system.

Beam acceleration system composed three electrodes and they are separated into five units to focus H⁻ beam vertically. In this experiment, H⁻ ion beams are extracted from a central 1/5 segment and the rest is masked by plates in this initial experiment. Total H⁻ current and its profile of were measured by movable calorimeter array having a cross shape. The array in a test chamber is located at 4.1 m apart from the grounded grid. Electron currents were measured electrically and heat loads onto the extraction and grounded grids, which are brought by charged particles and stripped H⁻, were measured by means of water calorimetry.

RESULTS AND DISCUSSIONS

Plasma Uniformity

The arc chamber has an elongated box shape, whose longest side is along to a longitudinal direction. One of the key subject in this experiment was to adjust an arc plasma uniform in the longitudinal direction, because non uniformity of arc plasma seems not only to have an influence to the profile of H⁻ ion beam but also to induce an unstable arc discharge. In order to adjust the discharge balance, discharge currents flowing to filaments were controlled. A filament power supply is divided into twelve insulated supplies, and a pair of filaments connected to each supply. A variable resistor is inserted between every insulated filament supply and arc power supply as shown in Fig. 4. Discharge current flowing to every filament was monitored using current transformers.

In the case of low input arc power, for instance an arc power at less than 80 kW, the plasma uniformity was measured by Langmiur probes. In a higher arc power, the uniformity was checked by comparing each arc current flows to individual filament.

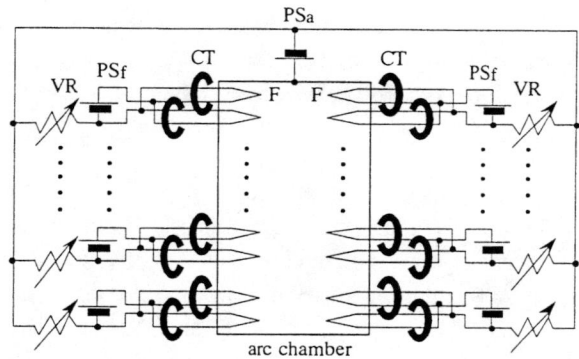

FIGURE 4. Electric diagram of an arc and filament power supply PSf, which is provided twelve isolated electric power supply. An external variable resistor VR, is inserted each isolated power supply and arc power supply, PSa. Each current flowing to a filament F, is monitored by current transformer, CT.

Figure 5a shows a longitudinal distribution of electron and ion densities measured by Langmuir probes after an adjustment of the external resistors; the densities were measured under the condition of the arc power at 60 kW and the pressure at 3 mTorr. Before the adjustment arc plasma shifted extremely downward, and plasma density was very low around the top plate. Arc current flowing to each filament is shown in the same figure, open square open diamond. The notations of (R) and (L) indicate right and left row of filament array on the back. The current profile coincides with the distributions of electron and ion densities. Figure 5b shows electron density as a function of the arc current through each filament. It is noticeable that the electron density has a linear correlation with arc current through filament in this power range. Assuming that the coincidence between a profile of plasma density and that of arc current to filaments can be extensive in a high arc-power case, the current adjustment is a useful method to make plasma uniformity.

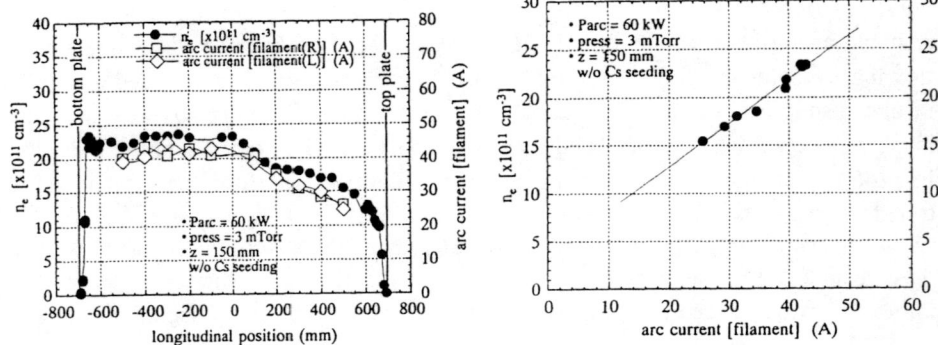

FIGURE 5. (a) A distribution of plasma density in a vertical direction of the arc chamber; ne and ni indicate electron and ion densities, respectively. The probe measurement has been done under a condition of an arc power of 60 kW and an operational pressure of 3 mTorr. (b) Electron density coincides with the arc current.

Figure 6a and 6b show distributions of arc currents to individual filaments before and after current adjustment in a high arc power range. Before adjusting the current, it is shown that the current to the top filament was very low and the profile distributed downward extremely. The current distribution became considerably symmetric after balancing the arc current. The non uniformity in longitudinal direction of arc chamber was reported by Okumura et. al (9). The direction of the plasma drift coincides to that of grad B of positive ions. There was no change in the drift direction when we changed the current directions to filaments.

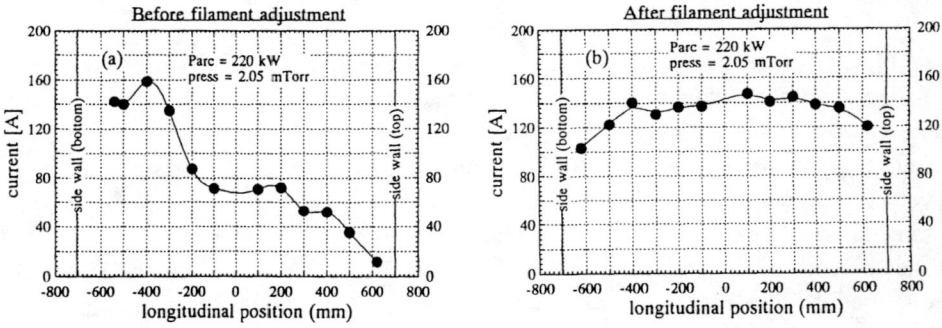

FIGURE 6. (a) A longitudinal current distribution before adjusting arc current to filaments by external variable resistor. (b) The distribution after adjustment.

Plasma uniformity was checked using a filament of different shape. Twenty-four hair-pin filaments were attached from the side plates. In this case a non-uniformity of the arc plasma was much smaller than that observed in the plasma generated by spiral filaments installed from back-plate port. When the hair-pin filaments installed from side plates, the tips are imbedded in a region caused by external magnetic filter. In this case generated charged particles are expected to be bounded by transversal magnetic field caused by external magnetic filter.

Beam Acceleration

Arc power dependence of total H⁻ current is shown in Fig. 7a. In this case, Cs is seeded in arc plasma and an operational gas pressure was set at 4 mTorr. The H⁻ current goes up linearly with increasing the input arc power and attains about 4 A at an arc power of 250 kW. The beam energy was 73 kV and duration was 0.3 sec. This current is a half value of an expected value of 8 A / segment. Around 250 kW the H⁻ current starts to saturate due to insufficiency of extraction field. Maximum arc power was limited at about 250 kW by following reasons. The first limit was abnormal arcings, which seem to be induced by unbalance of arc plasma in vertical direction of the arc chamber. The second was high heat load onto grounded grid. The grounded grid, indeed, was melted by an irregular concentration of accelerated beam onto the grid. In Fig. 7b extraction and acceleration current is shown as functions of input arc power. These currents increase linearly as well as H⁻ current.

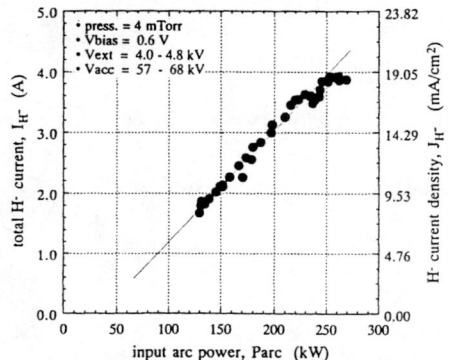
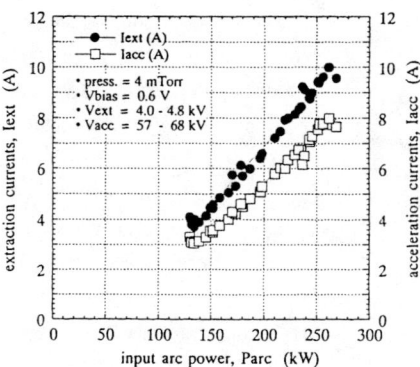

FIGURE 7. Arc power dependence of (a) total H⁻ current, (b) extraction current, close circle, and acceleration current, open square, at an operational pressure of 4 mTorr.

Comparing an efficiency of H⁻ current density per arc power per unit volume of arc chamber, this ion source is 9.8 $(mA/cm^2)/(W/cm^3)$ and that of previous 1/3 scaled source with rod filter was 6.5 $(mA/cm^2)/(W/cm^3)$ (2). The efficiency of this source is as 1.5 times high as that of the 1/3 scaled source. On the other hand, the efficiency is lower than that of the 1/3 scaled source with an external filter(10), whose efficiency was 12.6 $(mA/cm^2)/(W/cm^3)$. That is considered due to the weakness of the cusp magnetic field of the source in this experiment; the cusp field strength of this ion source is 1400 Gauss on the wall and the 1/3 source with external filter was 1800 Gauss.

Pressure dependence of extraction current, acceleration current and H⁻ current is plotted in Fig. 8a; the arc power is fixed at 230 kW. In this figure H⁻ current has a hump around a pressure of 4.5 mTorr and extraction current has a tendency to increase with increasing the pressure.

By measuring electric currents and thermal load currents calculated from thermal loads onto the grids, currents blocked by the grids and current passing through the grids can be roughly separated in a scheme of the zeroth order approximation. These currents link as following simple equations:

$$Q_{eg} = V_{ext} \cdot I_{eg} \quad [1]$$
$$Q_{gg} = V_{acc} \cdot I_{gg} \quad [2]$$
$$I_{ext} = I_{eg} + I_{acc} \quad [3]$$
$$I_{acc} = I_{gg} + I_{H^-} + I_p \quad [4]$$

where Q_{eg} : Thermal load onto extraction grid,
       Q_{gg} : Thermal load onto grounded grid,
       V_{ext} : extraction voltage,
       V_{acc} : acceleration voltage,
       I_{eg} : current converted form thermal load onto extraction grid, That is Q_{eg}/V_{ext},
       I_{gg} : current converted form thermal load onto grounded grid, that is Q_{gg}/V_{acc},
       I_{ext} : extraction current,
       I_{acc} : acceleration current,
       I_{H^-} : H⁻ current,
       I_p : current passing through grounded grid, that is $I_{acc} - I_{gg} - I_{H^-}$.

Here, thermal load currents of I_{eg} and I_{gg} are equivalent to currents blocked by extraction grid and grounded grid, respectively. Figure 8b shows separated blocked and passing currents calculated by using the equations [1]-[4] as functions of filling gas pressure. In this figure, main constitution of each current except for H⁻ current is considered to be electron. A current blocked by extraction grid, Ieg, goes up with decrease the pressure and following this both of the passing current, Ip, and the current blocked by grounded grid increase slightly. This tendency suggests that electron beam leaks to acceleration region via extraction apertures. On high pressure side, passing current, Ip, and current blocked by grounded grid, Igg, goes up with increase of H⁻ current. In this pressure range, H⁻ ions are stripped in acceleration region, and the increase of the passing current and the current blocked by grounded grid seems to come from stripping electron. The blocked current includes a contribution of Ho beam neutralized from H⁻ in acceleration region.

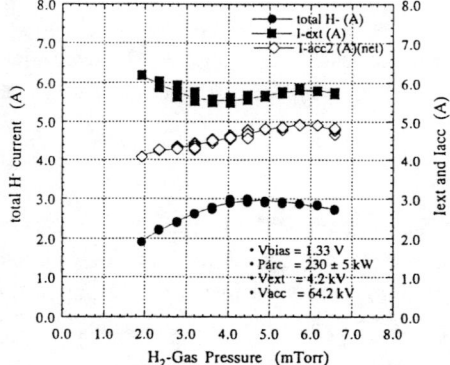 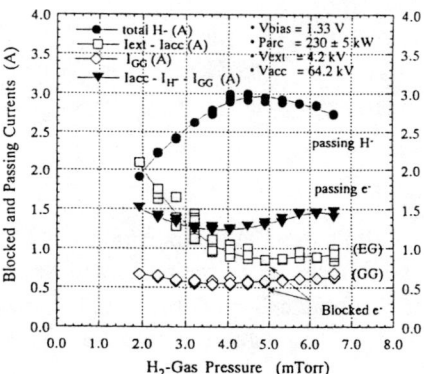

FIGURE 8. (a) Gas-pressure dependence of total H⁻ current, extraction and acceleration currents at an arc power of 230 kW. (b) Total H⁻ current, Blocked currents, Ieg and Igg, and passing current, Ip, as functions of filling gas pressure.

Figure 9a and 9b indicate relations of extracted currents to extraction current and acceleration current, respectively. The figures are converted from data in Fig. 7a and 7b with an addition of thermal load currents, Ieg and Igg. Current ratios are calculated by applying least square fitting and the values are shown in percentage on the right hand side of each figures. The ratios of H⁻ current to extraction current and to acceleration current are about 44 % and 56 %, respectively.

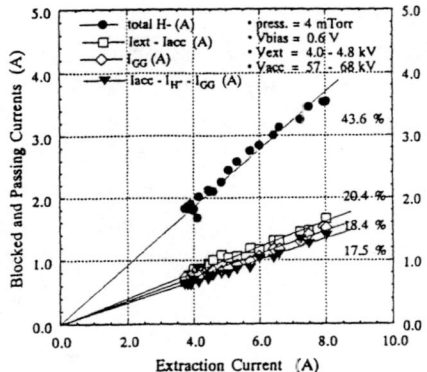 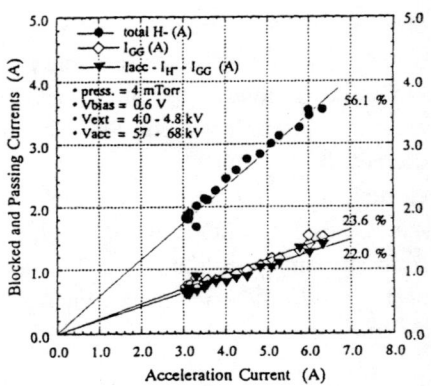

FIGURE 9. (a) Extracted currents plotted with respect to extraction current. (b) Acceleration-current dependence of total H⁻ current, thermal load current on grounded grid Igg and passing current through the grid Ip.

The current ratios in Fig. 9a and 9b are indicated schematically in Fig. 10. Values in round brackets are the ratios with respect to acceleration current. A ratio on Ieg is close to that obtained in an ion source with electron trapping structure in the extraction grid (10), while the ratios on Igg and Ip are as about twice large as the previous ratios

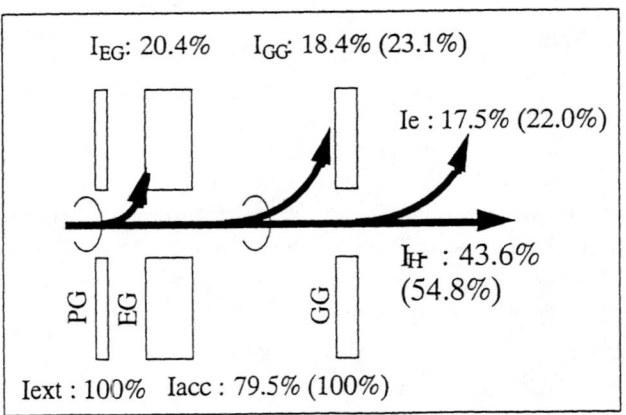

FIGURE 10. (a) Schematic figure indicating beam-current flow. Values in bracket show flow ratios based on acceleration current.

using the electron trap (10). It was reported that leakage of electron from extraction gap to acceleration region after scattering on inner surface of the extraction aperture (11). To reduce the electron leakage, it seems necessary to increase the strength of electron-deflection magnet in the extraction gap and to equip structures for electron trapping.

SUMMARY

A large scaled negative ion source has been designed and constructed as a prototype source for a neutral injector installed to the large helical device(LHD). The characteristics on its arc discharge were surveyed. Strong displacement of arc plasma was observed in the longitudinal direction by means of Langmiur probe and by monitoring the arc currents flowing to individual filaments. Direction of the displacement coincides to that of grad B drift. This plasma non uniformity was reduce by controlling variable resistor inserted between filament power supply and arc power supply. When filaments were attached from back-plate port, strong plasma nonuniformity appeared always. On the other hand, the nonuniformity decreased by attaching filaments from side plates.

Using the 1/5 segment of beam extraction area, which consists of 22(H) x 16(V) apertures of 9 mm in diameters, an H⁻ beam with the current of about 4 A and with the energy of 73 kV was obtained from 1/5 beam-extraction area at 250 kW of an arc power for 0.3 sec. The arc efficiency of the current density is 9.8 $(mA/cm^2)/(kW/cm^3)$, and is better than the efficiency obtained by an ion source with a rod filter. Beam components and beam current flow was also measured by monitoring electric current and thermal loads onto extraction and acceleration grids. It became clear that electron trapping structure is necessary to reduce the heat load onto the grounded grid.

ACKNOWLEDGMENTS

Authors would like to acknowledge Dr. Y. Okumura, Dr. T. Inoue and Dr. A. Hanada for their helpful comments and fruitful discussions in the IAEA technical meeting on neutral beam injector.

REFERENCES

1. Okumura, T., Fujiwara, Y., Inoue, T., Miyamoto, K., Miyamoto, N.,Nagase, A., Ohara, Y., and Watanabe, K., *Rev. Sci. Instrum.*, **67**, 1092 (1996).
2. Ando, A., Tsumori, K., Oka, Y., Kaneko, O., Takeiri, Y., Asano, E., Kawamoto, T., Akiyamo, R., and Kuroda, T., *Phys. Plasmas,* **1** 2813 (1994)
3. Takeiri, Y., Kaneko, O., Oka, Y., Tsumori, K., Asano, E., Akiyama, R., Kawamoto, T., Kuroda, T., and Ando, A., *Rev. Sci. Instrum.*, **67**, 1021 (1996).
4. Simonin, A., Bucalossi, J., Degranges, C., Fumeli, M., Jacquot, C., Massmann, J., Pamera, J., Riz, D., and Trainham, R., *Rev. Sci. Instrum.*, **67**, 1102 (1996).
5. Kaneko, O., Takeiri, Y., Tsumori, K., Oka, Y., Osakabe, M., Akiyama, R., Kawamoto,T., Asano, E., and Kuroda, T., *Proceedings of the 16th IAEA Fusion Engineering Conference,* Montreal, Canada, 1996 (unpublished), IAEA-CN-64/GP-9
6. Iiyoshi, A., Fujiwara, M., Motojima, O, Ohbiki, N., and Ymazaki, K., *Fusion technology* , **17**, 169 (1990)
7. Takeiri, Y., Ando, A., Kaneko, O., Oka, Y., Tsumori, K., Takanashi, T., Akiyama, R., Asano, E., Kawamoto,T., Kuroda, T., Tanaka, M., and Kawakami, H., *Rev. Sci. Instrum.*, **66**, 2541 (1995).
8. Takeiri, Y., Osakabe, M., Oka, Y., Tsumori, K., Kaneko, O., Takanashi, T., Asano, E., Kawamoto,T., Akiyama, R., and Kuroda, T., *Rev. Sci. Instrum.*, **68**, 2012 (1997).
9. Okumura, Y., Fujiwara, Y., Honda, A, Inoue, T., Kuriyama, M., Miyamoto, K., Miyamoto, N., Nagase, A., Ohara, Y., Usui, K., and Watanabe, K., *Rev. Sci. Instrum.*, **67**, 1018 (1996).
10. Takeiri, Y., Oka, Y., Osakabe, M., Tsumori, K., Kaneko, O., Takanashi, T., Asano, E., Kawamoto,T., Akiyama, R., and Kuroda, T., *Rev. Sci. Instrum.*, **68**, 2003 (1997).
11. Simonin, A., *Proceedings of the 7th International Symposium on Production and Neutralization of Negative Ions and Beams*, Brookhaven, 378 (1996).

Long Pulse Operation of the Kamaboko Negative Ion Source On the MANTIS Test Bed

R. Trainham, C. Jacquot, and D. Riz

DRFC/STID, CEA-Cadarache, EURATOM Association
13108 Saint Paul lez Durance, France

K. Miyamoto, Y. Fujiwara, and Y. Okumura

Japan Atomic Energy Research Institute, NAKA-MACHI, NAKA-GUN
Ibaraki-ken, 311-01 Japan

Abstract.
Advanced Tokamak concepts and steady state plasma scenarios require external plasma heating and current drive for extended time periods. This poses several problems for the neutral beam injection systems that are currently in use. The power loading of the ion source and accelerator are especially problematic. The Kamaboko negative ion source, a small scale model of the ITER arc source, is being prepared for extended operation of deuterium beams for up to 1000 seconds. The operating conditions of the plasma grid prove to be important for reducing electron power loading of the accelerator. Operation of deuterium beams for extended periods also poses radiation safety risks which must be addressed.

I INTRODUCTION

A significant obstacle for long pulse operation of negative ion sources used in thermonuclear fusion is the problem of electrons entering the beam accelerator. Accelerated electrons waste large powers, and stopping them produces X-rays which can damage sensitive materials such as HV insulators. Typically, special grids are installed at the beginning of the acceleration chain to trap the electrons, but the added electrode bulk and the complication of removing the energy dumped by the electrons compromises the ion beam optics. Furthermore, the various designs of electron suppression grids have not been able to tolerate the thermal loading during long pulse operation. It is, therefore, desirable to minimize the electron flux extracted from the ion source.

To this end, the Kamaboko source, developed by the Japan Atomic Energy Research Institute (JAERI), and currently residing at the CEA-Cadarache for deuterium testing on the MANTIS test bed, is being prepared for a campaign of 1000 second beam extraction experiments. Beam extraction of one minute has already been achieved at the low energy of 12 keV and ion current density of 4 mA/cm^2. Because of thermal loading of the plasma grid, longer operation times at higher current densities require active cooling of this grid, while, nevertheless, maintaining it at a temperature of approximately 300° C. The replacement of the plasma grid is under way and will the subject of a future paper. The focus of the present paper is on the importance of the conditions imposed upon the plasma grid for reducing the extracted electron current and, concomitantly, enhancing the negative ion flux. The key parameters are the barrier magnetic filter, the grid temperature, its electrical bias, and its cesium coverage. We shall also discuss radiation safety considerations for long pulse beams of deuterium ions and some of the modifications made to our operation software in order to conduct the experiment from a remote site.

To summarize the past results of the Kamaboko source, it has demonstrated that the ITER concept ion source is feasible. It has provided negative ion current densities of 30 mA/cm^2 of H$^-$ at a source pressure of 0.2 Pa, and 20 mA/cm^2 of D$^-$ at 0.35 Pa when operated with cesium seeding.

II EXPERIMENTAL SETUP

The source has been described previously [1], but, briefly, it is a 30 liter cylindrical chamber of machined oxygen free copper. Inside are 12 filaments mounted on water cooled bayonette type mounts. Efficient plasma confinement is achieved by 16 magnetic line cusps generated by SmCo magnets arranged in longitudinally machined channels on the outside of the chamber. Source cooling is achieved by countersunk waterlines brazed to the outside of the source next to the magnet channels. Typically, 70 kW of arc power are injected into the source to create a plasma with a density of 10^{13} cm^{-3} and an electron temperature of about 5 eV.

The extraction area is a rectangular plane measuring 250x340 mm cut into the side of the cylinder. Magnets along the sides of the opening provide a magnetic filter of up to 1800 Gauss-cm. For magnetic field homogeneity only the central 140x128 mm region is used for ion extraction. The molybdenum plasma grid has 45 holes, 14 mm each in diameter, arranged in a 7x7 matrix, excluding the 4 corner holes which are used for alignment of the three stage acceleration system. The plasma grid is electrically and thermally isolated. Extracted beams of H$^-$ or D$^-$ are accelerated up to energies of 30 keV.

III RESULTS

A Magnetic Filter

A magnetic filter of 800 Gauss-cm, originally optimized for hydrogen operation, proved to be inadequate for deuterium operation, initially allowing more than 13 amperes of electrons into the accelerator. Such currents are not acceptable for long pulse deuterium operation, and even in short pulse operation of a few seconds we observed local melting of the electron suppression grid. Fig. 1 shows the effect of several magnetic filters upon the negative ion current and the electron current for deuterium operation. One sees that while the ion current is only slightly affected, the electron current diminishes by more than an order of magnitude when going from a filter strength of 800 Gauss-cm to 1800 Gauss-cm.

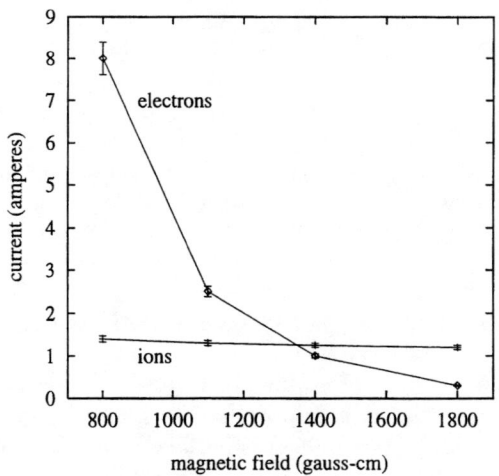

FIGURE 1. Effect of the magnetic filter on the extracted ion and electron currents. The extraction ion density is approximately 20 mA/cm^2.

B Grid Polarization

Fig. 2 shows the effect of electrically biasing the plasma grid. One sees that a judicious choice of bias voltage can dramatically reduce the electron current while having minimal impact on the negative ion current. The optimal bias voltage (here, approximately +5V) is close to the value of the plasma's floating potential as measured by a Langmuir probe. This suggests that it is perhaps not absolutely essential to connect the plasma grid to a bias power supply since the grid will float to this potential if left unconnected. This point is still being investigated. One point is clear, however, it is disadvantageous to ground the plasma grid.

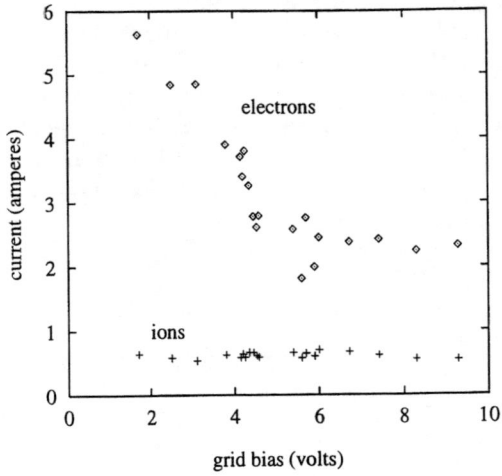

FIGURE 2. Effect of the plasma grid bias on extracted ion and electron currents. The extraction ion density here is approximately 10 mA/cm^2.

C Cesium Seeding and Grid Temperature

The enhancement effect of cesium in negative ion sources is well known, but the physical basis of it is still debated. There is considerable evidence, however, that the dominant process is surface production of negative ions on the plasma grid itself. We have recently reported [2] that the negative ion current has no direct correlation to the concentration of neutral cesium in the plasma. Furthermore, simulations of ion transport [3] indicate that volume processes more than about 2 cm from the plasma grid are completely irrelevant to ion extraction.

We find that for cesium seeded operation the negative ion current has a large dependence upon the temperature of the plasma grid. This is shown in fig. 3. In the absence of cesium, this behavior is greatly attenuated. This is further evidence in support of surface production, and can be understood by considering the surface coverage of cesium on the plasma grid. Optimal surface production occurs at the minimum of the composite surface's work function. For a clean tungsten surface this occurs at approximately 0.7 monolayers of Cs coverage, where the work function drops to 1.7 eV (compared to 4.6 eV for clean tungsten). Results of Taylor and Langmuir [4] show that this surface coverage is maintained by evaporation-condensation of cesium at the surface for temperatures ranging from 200 to 500°C. Lower temperatures result in larger surface coverage, with the work function of the composite rising to that of pure cesium, while at higher temperatures cesium evaporates from the surface resulting in a composite work function approaching that of pure tungsten. Although our plasma grid is made of molybdenum, evaporation of the tungsten filaments under steady source operation results in coating the plasma grid with tungsten, so the above considerations should remain valid.

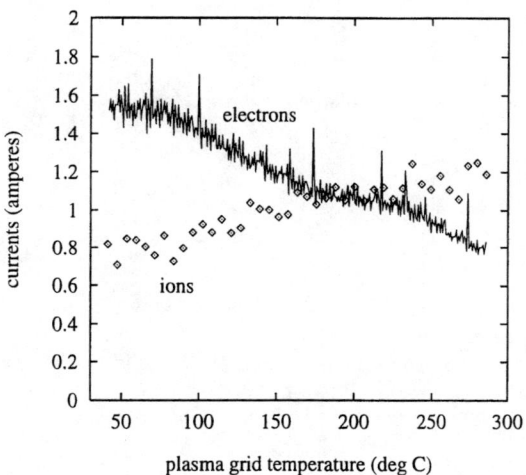

FIGURE 3. Evolution of the extracted ion and electron currents vs. the plasma grid's temperature during a pulse. The final extraction ion density is 16 mA/cm^2.

D Grid Thickness

In an attempt to further reduce the electron current a 12 mm thick plasma grid has been tested. One might expect that in addition to smaller electron currents the reduced gas flow into the accelerator would result in less stripping and thus larger ion currents on the beam target. Additionally, the extra surface area in the extraction holes may indeed further enhance the negative ion production.

The results shown in Fig. 4 do not support the above suppositions. For equivalent experimental parameters, the extracted electron current is reduced by 30% yet the ion current is reduced by 50%. A possible explanation for the degraded performance is that negative ion production from volume sources which incorporate magnetic filters at the plasma grid behave like fixed cathode sources. A plot of a Child-Langmuir style curve (I vs. $V^{3/2}$) shows that the extracted ion beam is space charge limited from a meniscus cathode withdrawn into the grid. For the 12mm thick grid the data suggest that the cathode is withdrawn about 8.5 mm into the grid.

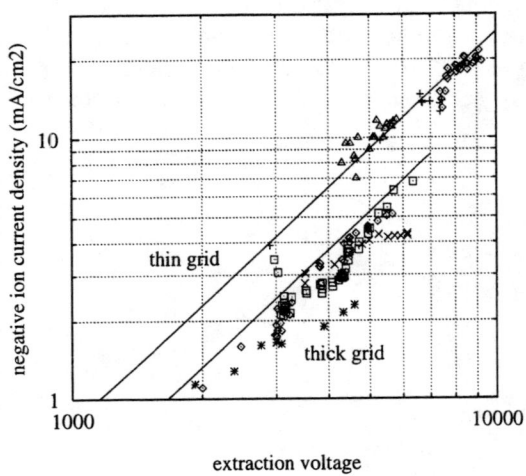

FIGURE 4. Comparison of the 12 mm thick plasma grid to the 2 mm grid. The thick grid substantially degrades the source performance.

IV RADIATION RISKS AND REMOTE CONTROL

Long pulse operation of 1.4 ampere deuterium beams at 25 keV can be problematic because of neutron generation from the deuterium-deuterium reaction on the beam target. We have calibrated the neutron production on MANTIS using a Cf^{252} source and have found that a 1 ampere beam at 25 keV generates approximately 10^8 neutrons per second. The control room for MANTIS is located 3 meters away from the beam target, and, unfortunately, there is no radiation shielding to protect the experimenters. For a 1000 second pulse the experimenter is expected to take a dose 0.1 milli-Sieverts. This dose rate would limit a single experimenter to no more than 10 pulses per year. Obviously it would be impractical and dangerous to operate in such an environment. We have thus chosen to run the experiment after hours, with the building evacuated, from a computer located in another building. We will now briefly discuss the control software needed to do this.

The MANTIS data acquisition system consists of four networked PC-class computers. Three of them are equipped with data acquisition cards and act as measurement stations. One computer also uses digital I/O to drive a flagboard for controlling the various power supplies. The measurement stations run without keyboards and monitors, but are accessible by ethernet and RS232 connections. A master computer coordinates them and also centralizes the data storage. The operating system is Linux [5], a clone of the UNIX operating system. The computers are networked by fiberoptic connections since each computer is vulnerable to high voltage breakdowns.

The control software is modular, being a combination of C programs and shell scripts. Inter-computer commands and data transfer are executed by remote procedure calls. Data acquisition and monitor procedures run in parallel as background processes. This is a powerful capability since a variety of surveillance programs can execute simultaneously to monitor critical parameters (e.g., plasma grid temperatures, extraction current, acceleration voltage, etc.) without interrupting the data acquisition process. These processes can trigger a shutdown procedure if something goes wrong. The modular software design allows the addition of new monitor tasks without interrupting tasks already in execution, and, most importantly, without rewriting existing code.

An important requirement of the software is the ability to completely control and monitor the experiment from an undesignated remote computer. This requires a user interface that is platform independent, preferably with graphics capability. Although X-window emulators are available for a variety of hardware platforms, we do not wish to make any software installations on the remote computer. We also want the freedom to operate from any (or even several) of a number of remote computers. This requirement has lead us to choose a browser program for the World Wide Web. The programs Netscape, Mosaic, and Internet Explorer seem to be universally installed applications with platform independent graphics capabilities. Thus the master computer on MANTIS has been configured as a http server using the Apache software [6]. MANTIS has a home page on the World Wide Web.

The experimental configuration is performed by standard HTML forms, and execution of the MANTIS programs is by means of the Common Gateway Interface (CGI) [7]. Real time monitoring of the experiment is performed by network sockets using the User Datagram Protocol (UDP) [8], and graphical display of the experimental parameters is performed by Java Applets. The software has been successfully tested from UNIX workstations, Windows PCs, Linux boxes, and MacIntoshes.

V   CONCLUSION

Investigations with the Kamaboko source have illuminated several key parameters to be optimized for long pulse (1000 s) negative ion source operation. We find that conditions of the plasma grid are perhaps the most important of all. Dramatic effects have been observed for the temperature, the electrical bias, and the magnetic filter imposed upon the plasma grid. It is also observed that optimal configuration for hydrogen operation does not assure optimal deuterium operation. Finally, the generation of neutrons from 1000 second Deuterium beams requires that the experimenters take special precautions to minimize the risks of radiation exposure. On the MANTIS test bed we have opted to pilot the experiment remotely by means of network capable software.

VI ACKNOWLEDGEMENTS

We would like to thank Jerôme Paméla, Michele Fumelli, Ron Hemsworth, Jerôme Bucalossi, and Gianfranco Tonon for many useful discussions, and Sophie Dutheil, Patrick VanCoilie, and Marc Rizotti for their willing and competent technical support.

REFERENCES

References

1. N. Miyamoto, H. Oguri, Y. Okumura, T. Inoue, Y. Fujiwara, K. Miyamoto, A. Nagase, Y. Ohara, and K. Watanabe, Experimental Results on ITER-NBI Concept Source, *7th International Symposium on the Production and Neutralization of Negative Ions and Beams; 6th European Workshop on the Production and Application of Light Negative Ions*, Brookhaven National Laboratory, Oct. 1995.
2. C. Jacquot, D. Riz, R. Trainham, K. Miyamoto, Y. Okumura, and M.B. Hopkins, Proc. of the 19th Symp. on Fus. Techn., Lisbon 1996.
3. David Riz, *ICIS 97*, also *Etude Expérimentale de sources à fort courant d'ions négatifs D^-/H^-. Analyse fondée sur la simulation du transport des ions dans le plasma de la source*, Ph.D. thesis Université Pierre et Marie Curie, Université Paris V (1996).
4. John Bradshaw Taylor and Irving Langmuir, Phys. Rev. **44**, 423 (1933).
5. Extensive documentation of Linux is available at: http://sunsite.unc.edu/mdw/linux.html
6. The Apache software can be downloaded from http://www.apache.org.
7. The CGI FAQ is available at http://htmlhelp.com/faq/cgifaq.html.
8. A nice tutorial on socket programming is available at: http://www.ecst.csuchico.edu/~beej/guide/net.

Status and Objectives of the DENISE Ion Source Project at DCU

McNeely, P., D. Boilson, N. Curran, M.B. Hopkins, D. Vender

Plasma Research Laboratory, School of Physical Sciences, Dublin City University, Glasnevin, Dublin 9, Ireland

Abstract: In support of the international collaborative effort into fusion research, work has begun to re-commission the Deuterium Negative Ion Source Experiment (DENISE) at Dublin City University (DCU). Modifications to the original system are nearing completion. The major goal of these modifications is easier access to both the ion source and the extraction region for diagnostic systems. A new source chamber has been installed. It can be used with both standard Langmuir probes, and a new laser photo-detachment probe currently under construction. The source chamber allows operation in any one of three modes: filament driven, CW RF driven, or Pulsed RF driven (up to 3 kW of 13.5 MHz RF at 250 Hz). A PC computer control system based on LabView™ is nearing completion and is intended for both system control and data logging. An investigation into the production and extraction of H^- ions has begun. Langmuir probe data has been collected at pressures from 5-200 mTorr, for CW RF powers between 25 and 600 W, and for 3 radial probe positions. This paper will present the current status of DENISE and the future plans for this versatile source.

1. INTRODUCTION

As a part of the Irish contribution to the ITER project, Dublin City University (DCU) is re-commissioning the DENISE (Deuterium Negative Ion Source Experiment) Ion Source (1) acquired from FOM Laboratory in the Netherlands. Additionally, a second plasma chamber simimilier to the one used on DENISE has been assembled on a vacuum test stand (2) to help support the work with DENISE. The inner diameter and attachment flanges of both the new DENISE plasma chamber and that of the second chamber are identical. This allows the chambers to use the same end plates. There is only a slight difference in the lengths of the two sources, thus any results obtained on one can be verified on the other. Theoretical efforts are now beginning to develop both PIC and Fluid Model codes to simulate the plasma. This paper will give the current status of the project and what the near future holds for it. Currently, DENISE is nearing a status where it will be possible to begin running the source again in filament driven mode. All major elements necessary for the commissioning of the source are in place. The plasma chamber on the vacuum test stand has been providing good data operating on the new high power RF source since early August, 1997. All of the results presented in this paper come from runs on that chamber.

2. EXPERIMENTAL SET UP

The main elements of the DENISE test bench are the ion source body, a large vacuum chamber and the attached Faraday Cup. The ion source is externally modified from the original source. Currently the original extraction electrode system is in use, and no plans for significant modification are contemplated.

A new source body has been built by Caburn-MDC. It is a water-cooled double walled cylindrical vacuum chamber [ϕ 20.0 cm x 31.2 cm] with 3 probe ports located on top of the chamber. Four additional ports (2 on either side) are provided at the midline for other diagnostics. Caburn also built a new endplate, for running filament discharges, it has water cooling channels and feed-throughs for 4 filaments. A second endplate exists which is designed to hold a RF antenna vacuum feed-through. The antenna consists of helically wound water-cooled copper tubing. The antenna is driven by a RF power supply. A 3.5 kW RF source with automatic matching unit has been installed. The RF source is capable of operating in both CW and pulsed mode. At present all tests have been run in CW mode and due to concerns about arcing in the antenna have been limited to powers of 600 W or less.

The large vacuum chamber supports the double lens extraction system. The chamber is pumped by a turbo-pump down to the cross-over vacuum at which point a cryo-pump is switched on. The pressure is monitored at 4 places. Thermo-couple gauges read the turbo fore-line vacuum and the chamber rough vacuum. A convectron gauge monitors the pressure in the source. A BA hot filament ion gauge reads the high vacuum in the vacuum chamber.

The extracted beam current is monitored by the faraday cup system. To facilitate easy instrumentation, the source body is at ground potential. This means that the Faraday cup must be at the extraction bias potential, thus the vacuum fitting for the cup must be isolated from grounded vacuum chamber. A Faraday cage surrounds the whole assembly. Reading the current on the Faraday cup requires that a personal computer with an ADC card be run in the faraday cage and its power supplied via a 30kV isolation transformer. The Faraday cup consists of 6 water-cooled copper bars. Each bar is insulated from the others by 1-mm thick ceramic insulator allowing the current in each bar to be read independently giving a rough indication of beam profile. A recirculating chiller provides the necessary water-cooling. This chiller is capable of dissipating 3.5 kW of heat and has a flow rate of 10 l/min. To prevent shorting de-ionized water is used throughout the system.

A new computer control system is in the final stages of testing and installation to control DENISE. The control system uses the LabView^{TM} program from National Instruments. The computer is equipped with four cards from National Instruments. It has a PCI-GPIB interface card for computer control of a HP 54600B digital scope. For general operation of the control system an AT-DIO96 digital input/output card, an AT-AO6 analogue output card, and an AT-MIO-16E10 analogue input card are required. The control program interfaces with an existing hardware controller (3) to monitor and remotely control the vacuum system. The RF power supply and the convectron and ion gauges are controlled by serial links to the control program. Direct links from the PC control the gas flow, the HV power supplies, and the filament supplies. The computer monitors the status on all safety interlocks to both protect the equipment and ensure the safety of personnel. The current from each Faraday cup bar is sent to an ADC in the computer that is in the faraday cage. After conversion the values for each bar are transmitted via a serial fibre optic link to the main data acquisition computer.

Data on the plasma is collected via a Langmuir probe. The probe is mounted on a linear vacuum feed-through allowing the radial position of the probe to be varied. The new vacuum chamber is designed so that the probe can be mounted in one of 3 positions: near the antenna, near the extraction system, or in the middle of the chamber. The probe (4) used was developed for commercial use by Scientific Systems. The Smart Soft analysis program is used to analyse the IV curves from the probe. It also allows for quick acquisition of the Electron Energy Distribution Function (EEDF).

3. RESULTS

The plasma chamber on the vacuum test stand running with the RF power supply has been in operation since early August, 1997. The initial results from the experimental runs are presented in this section. A systematic investigation on the effect of RF power, probe position, and gas pressure has resulted in the collection of sizeable body of information for Hydrogen. All of these runs were carried out without the use of a cusp magnetic field. In a preliminary test run to check the effect of the cusp field a factor of 5 improvement in electron density (N_e) was observed. The probe was not at the centre of the cusp where the brightest plasma was observed. For a viable ion source it was immediately apparent that the plasma would have to be in the inductive rather than capacitively coupled mode as the electron density in the capacitively couple regime was observed to be $10^8 - 10^9$ cm^{-3}. This is too low for efficient operation as an ion source. Immediately upon ignition of the inductive discharge the electron density jumps to 10^{10} cm^{-3}.

Shown in Fig. 1 are the probe IV traces for Hydrogen between the pressures 5 and 100 mTorr. Each trace was taken a probe position of 1.5 cm from the quartz tube shielding the antenna and at a RF power of 550 W. The transition from capacitive mode to inductive mode is clearly visible in the increase of the measured current in the probe. The sources eventually used on ITER will be operated at low pressure thus we are interested in investigating the properties of the source in this regime. There was a factor of 6 increase in the probe current from data collected at 300 W RF. Assuming the increase remains linear, for the source to operate in the inductive mode at a pressure in the 5-10 mTorr range the RF power will have to exceed 1 kW.

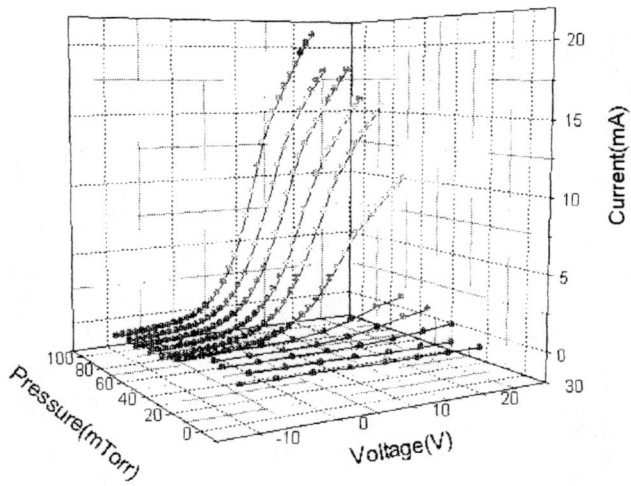

Figure 1. The Langmuir probe current collected for a Hydrogen discharge at 550 W of forward RF power between 5 and 100 mTorr. The probe was located at a radial position of 1.5 cm from the antenna.

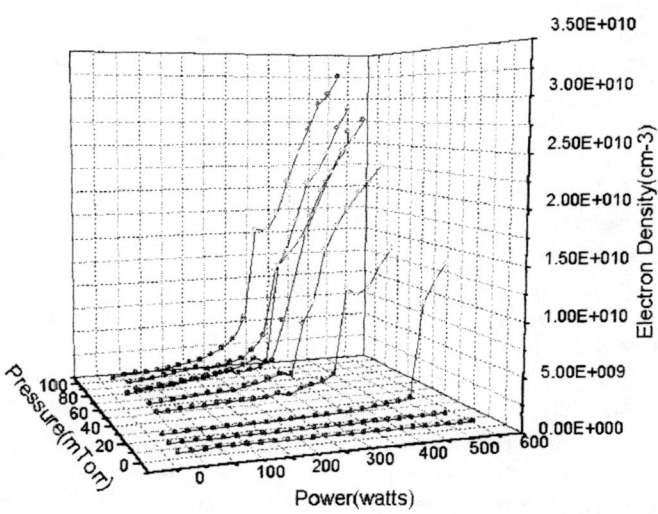

Figure 2. The value of electron density (N_e) determined by Langmuir probe measurements in a hydrogen plasma for differing source pressures and RF powers. The probe was located 1.5 cm from the antenna.

In Fig. 2 the effect of RF power and source pressure on N_e is demonstrated for the same probe position as Fig. 1. The transition from capacitive to inductive mode is clearly indicated. Once the transition has been made however, the density is fairly constant rising by only a factor of 3 compared to the factor of 10 increase inherent in the transition. Finally, in Fig. 3 a typical EEDF is shown for an inductive plasma at 47 mTorr and 450 W RF power at a distance of 4.5 cm from the quartz tube. The plasma parameters of this scan gave the plasma potential as 13.8 V and the electron temperature as 2.4 eV. The electron density was determined to be 1.8×10^{10} cm^{-3} and the ion density to be 1.2×10^{10} cm^{-3} the errors in all values were a few percent. The analysis program uses a procedure known to underestimate the ion density. This data was collected during a radial scan of the plasma.

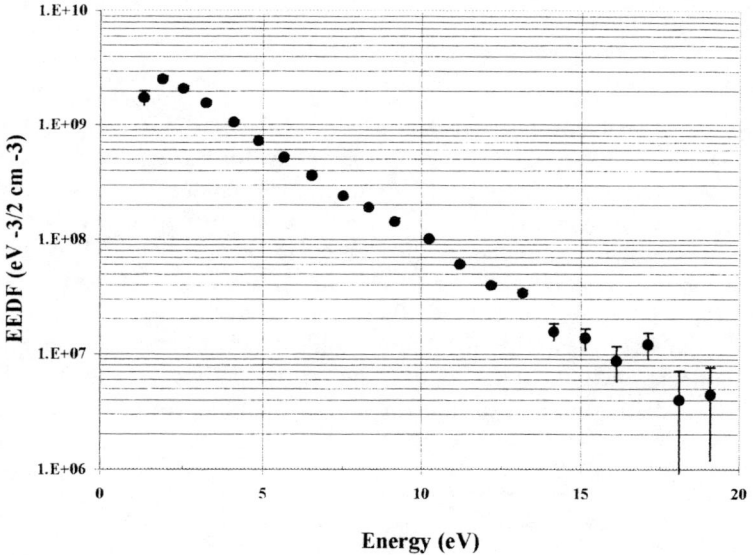

Figure 3. A typical EEDF obtained in a 47 mTorr Hydrogen discharge at 450 W RF power at a radial distance of 4.5 cm from the antenna.

4. FUTURE PLANS

The work both on and using DENISE is still in its initial stages. The first operational studies will focus on the filament driven mode with the new chamber. Their goal will be the characterisation of the plasma parameters. At the same time, work will continue on the vacuum test stand source in the RF driven mode. On the

RF source the first goal is to characterise the CW RF driven plasma and to use this information to verify fluid and PIC based computational models. The next stage will be to compare the plasma parameters between the pulsed and CW RF plasmas, the timing of this is dependant on modifications to both the antenna and the source to handle the full power output of the RF source. Work on the production of H$^-$ via pulsed RF will begin on the vacuum test stand at the same time as extraction studies are begun on DENISE. The studies on DENISE will focus on the relationship between plasma parameters and extracted ion beam current. By this time the laser photo-detachment probe (2) under design will be available and installed on DENISE. This probe will be used initially to study negative ion production with a filament driven source. Based on the experience gained with this type of running the probe will be used then with RF driven plasmas.

5. CONCLUSION

A great deal has been accomplished in the past months. This work has resulted in the DENISE source now being ready for experimental investigation. We have achieved good results with the vacuum test stand source running a CW RF driven Hydrogen discharge. We have been able to map the transition to the inductive mode of operation for a wide range of pressure and RF power. In the inductive mode we determined the EEDFs and the electron density. The source was not operated using magnetic confinement and at the low end of the available RF power, so it is clear that the future holds much promise. The information already collected will be now used to aid in modelling. It is hoped that the theoretical modelling will be sufficiently accurate that it will guide future stages of the investigation.

REFERENCES

1. Eenshuistra, P.J., M. Gochitashvilli, R. Becker, A. W. Kleyn, and H.J. Hopman, *J. Appl. Phys.* **67** (1) 85-96 (1990).
2. Mellon, K.N., B.P. Coonan, M.B. Hopkins, *J. Phys. D: Appl. Phys.* **27** 2480-2486 (1994).
3. Boilson, D., M.Sc. Thesis, Dublin City University Library, unpublished (1996).
4. Hopkins, M.B., *Journal of Research of the National Institute of Standards and Technology*, **100** (4), 415 (1995).

A LARGE-AREA RF SOURCE FOR NEGATIVE HYDROGEN IONS

P. Frank, J. H. Feist, W. Kraus, E. Speth, B. Heinemann, F. Probst

Max-Planck-Institut für Plasmaphysik, EURATOM-Association, 85748 Garching, Germany

R. Trainham, C. Jacquot

DRFC/STID, CEA-Cadarache, EURATOM-Association, 13108 Saint Paul lez Durance, France

Abstract. In a collaboration with CEA Cadarache, IPP is presently developing an rf source, in which the production of negative ions (H^- / D^-) is being investigated. It utilizes PINI-size rf sources with an external antenna and for the first step a small size extraction system with 48 cm^2 net extraction area. First results from BATMAN (Bavarian Test Machine for Negative Ions) show (without Cs) a linear dependence of the negative ion yield with rf power, without any sign of saturation. At elevated pressure (1.6 Pa) a current density of 4.5 mA/cm^2 H^- (without Cs) has been found so far. At medium pressure (0.6 Pa) the current density is lower by approx. a factor of 5, but preliminary results with Cesium injection show a relative increase by almost the same factor in this pressure range. Langmuir probe measurements indicate an electron temperature $T_e > 2$ eV close to the plasma grid with a moderate magnetic filter (700 Gcm). Attempts to improve the performance by using different magnetic configurations and different wall materials are under way.

1. INTRODUCTION

For high energy neutral beam injection systems based on negative ions, rf sources represent an attractive alternative to arc discharge sources, because the limited filament life-time requires a frequent filament replacement (for ITER: by remote handling). Furthermore the rf source relieves certain problems in the transmission line and might offer cost savings. Therefore a collaboration on high-frequency ion source development has been started in 1996 between CEA Cadarache and IPP Garching.

Testbed operation started in autumn 1996 with the novel compact rf source (1), called type III. This compact design with an external antenna and a Faraday screen integrated in the copper source body has advantages for tests with different magnetic filter systems and modifications of the rf antenna. However, for long term operation with Cs there is a high risk of damaging the vacuum sealing of the integrated Faraday screen. Therefore the rf source type II with a quartz vessel and a Faraday screen inlay (2) is foreseen for operation with Cs. Additional effects without the copper screen inlay can be studied, expecting a higher H_2 dissociation degree with quartz walls.

2. EXPERIMENTAL SET-UP

The two rf sources used for the first step are shown in Fig.1 (32x61 cm^2 cross section, 17 cm long, 30 l volume). The inner dimensions, the backplate and the front part of both types are identical, only the side walls have different designs. For the source type II a quartz wall and a removable Faraday screen inlay (copper) are used. The type III source has a vacuum-tight, self supporting Faraday shield made of electrodeposited copper as mentioned before. An rf power of up to 100 kW at a frequency of 1 MHz can be coupled to an external coil with six turns. The extraction system has three grids with 4x12 holes of 11 mm diameter, arranged in an area of a 6x20 cm^2 and is mounted inside a modified PINI structure. In order to reduce the electron current, a magnetic filter field is generated by two rows of Co-Sm permanent magnets placed close to the edge of the extraction area. The strength of the magnetic field parallel to the plasma grid, integrated in the perpendicular direction to the plasma grid is 700 Gcm (B_{max} = 110 G in the center between the magnet rows). For the purpose of biasing the plasma grid (PG) w.r.t. the source, the front cover plate and the source body are connected to each other and isolated against the PG. A Cesium oven or a Langmuir-probe can be mounted at the center of the backplate.

For beam diagnostics a moveable calorimter with 9 thermo-couples on a water-cooled flat copper plate is installed in 1.5 m distance. The electrons in the beam are deflected by a pair of Helmholtz coils.

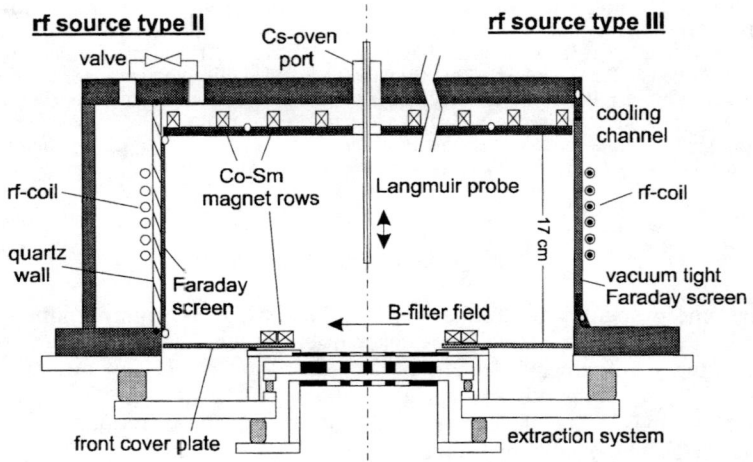

FIGURE 1. Schematic drawing of the rf sources and the extraction system.

3. EXPERIMENTAL RESULTS

Although both sources can run at 0.2 Pa, the operation range without Cs was restricted to pressures > 0.5 Pa, because otherwise the ion current is below the detection limit. In all cases hydrogen only has been used so far. The studies started with the source type III. Typical plasma parameters range from T_e = 3 eV, n_e = 10^{11}cm^{-3} (close to the plasma grid) to T_e = 8 eV, n_e = 4x10^{11} cm^{-3} (in the center of the source) at 30 kW and 0.6 Pa. By increasing the pressure the electron temperature decreases, but the electron density is only slightly influenced near the plasma grid (Fig.2).

Optimum beam profiles are achieved at a ratio of 2.5 between the acceleration and the extraction voltage, which corresponds roughly to well-known optimum ratio of electric fields (3). Biasing the plasma grid with a positive voltage of 18 V reduces the

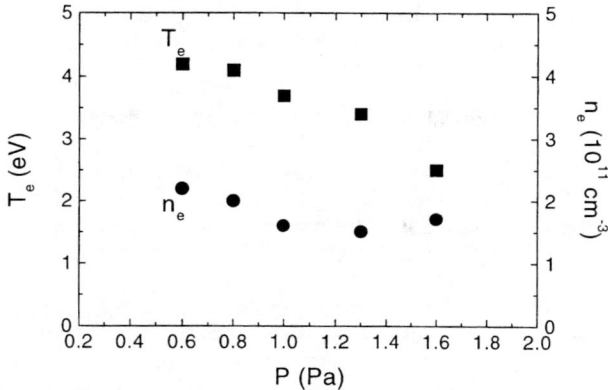

FIGURE 2. Electron temperature and density versus filling pressure (rf source type III, 40mm in front of the PG, $P_{rf} = 30$ kW)

electron current by a factor two, without changing the ion current significantly at 80 kW and 1.6 Pa (4). The floating potential is about +16 V. Under optimum conditions the electron current is a factor 8 higher than the negative ion current.

The rf power dependence of the negative ion current is shown in Fig.3. For the source type III (without Cs) the ion current shows a linear dependence of the rf power without any saturation and at the pressure of 1.6 Pa the H$^-$ current density reaches a maximum value of 4.5 mA/cm^2.

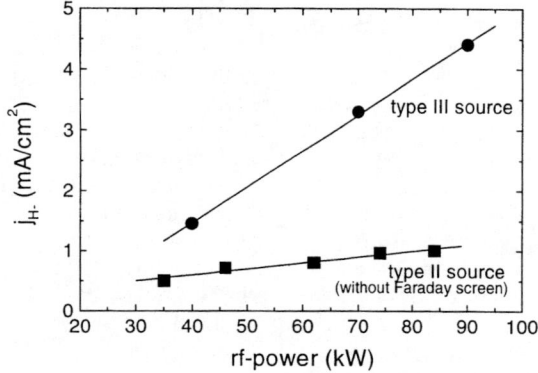

FIGURE 3. Extracted H$^-$ current density as a function of the rf power (without Cs, $P = 1.6$ Pa)

By decreasing the pressure to 0.6 Pa the current density is reduced by a factor of 5. But preliminary experiments with Cesium seeding into the type III source show a compensation of the decrease in this medium pressure range. However, due to the reasons explained above, those experiments with Cs seeding were not continued in this source. Instead the type II source was mounted and first experiments were conducted without the Faraday screen in order to study the effect of the quartz walls on the negative ion production. Without Cs and without the Faraday screen this source shows a strongly reduced ion current (Fig.3) compared to type III.

3. DISCUSSION

So far the rf source in its two versions has mainly been operated as a volume source, i.e. without Cs. The corresponding negative ion yield is still moderate, which is plausible, since the electron temperature in front of the plasma grid is too high for dissociative attachment becoming effective. Furthermore the destruction of negative ions by electron collisions is enhanced at those elevated temperatures. The reason for the elevated T_e is mainly the well-known property of inductively coupled rf sources having higher electron temperatures in the absence of a genuine magnetic filter compared to arc discharge sources. This effect is even more pronounced without the Faraday shield, as known from the positive ion rf source development, where higher ion flux densities and higher rf power efficiency are achieved without the Faraday shield (2). To what extent T_e is enhanced, has to be checked by probe measurements. The reduced recombination coefficient of the quartz wall would result in a reduced population of molecular species, thereby - considering volume effects - yielding a decrease of the negative ion density. Consequently, both the absence of a Faraday shield and the enhanced dissociation degree lead to a reduced negative ion yield in a pure volume source (compared to operation with a metallic Faraday shield), as observed. The situation may change in the case of Cs injection, where surface effects are thought to become dominant (5). Since a higher population of atomic species is associated with the reduced recombination on the quartz wall, this could possibly enhance the negative ion yield. Hence a Faraday shield coated by quartz or some suitable insulating material would be required.

5. CONCLUSION

First results of the large area rf source, operated as a volume source, show a linear increase of the H⁻ current density with rf power; no saturation has been observed. Compared to arc discharge sources a higher electron temperature in front of the plasma grid is found; consequently the negative ion yield has reached moderate 4.5 mA/cm^2 at a power density of 3 W/cm^3 and an elevated pressure (1.6 Pa). Further attempts of optimising the source performance including operation with Cs seeding are under way. In order to extend those experiments to the low pressure range < 0.6 Pa, the rf power efficiency, the plasma confinement and the magnetic filter field have to be improved. Furthermore the hardware of BATMAN will be upgraded by a modification of the rf generator for extending the power range up to about 150 kW, installing an extraction system optimised for electron suppression and a larger calorimeter for beam diagnostics.

REFERENCES

1. Heinemann, B., Feist, J. H., Kraus, W., Probst, F., Speth, E., Busch, M., Szcepaniak, W., in Proc. of the 19th Symp. on Fus. Techn., Lisbon 1996
2. Kraus, W., Feist, J. H., Speth, E., in Proc. of the 19th Symp. on Fus. Techn., Lisbon 1996
3. Kuroda, T., Kaneko, O., Takeiri, Y., *et al*, *Rev. Sci. Instr.* **67** (3), 1114-1119, (1996)
4. Kraus, W., Speth, E., Feist, J. H., Frank, P., Heinemann, B., Riedl, R., Trainham, R., Jacquot, C., 7th International Conference on Ion Sources, Taormina 1997, *to be published*
5. Pamela, J., *Plasma Phys. Control. Fusion* **32** A325-A336 (1995)

Efficient Production of H⁻ in High-Density Helicon Plasmas

D. Hayashi[a], K. Sasaki[a], K. Kadota[a],
[a] *Department of Electronics, Nagoya University,
Furo-cho, Chikusa-ku, Nagoya, 464-01, JAPAN*
Y. Oka[b], K. Tsumori[b] and O. Kaneko[b]
[b] *National Institute for Fusion Science,
Oroshi-cho, Toki-shi, Gifu, 509-01, JAPAN*

Abstract. This paper describes the machanism of efficient H⁻ production in a high-density hydrogen plasma produced by helicon wave discharge. An H⁻ source using helicon wave discharge has been developed. Spatial and temporal variations of H⁻ density in the helicon plasma were measured by laser photodetachment technique. A steep increase in H⁻ density was observed in the afterglow. The H⁻ ions were efficiently produced for the first 20μs after the termination of the discharge and mainly lost by mutual neutralization with positive ions. In the outer region of the plasma column, 20% of the negatively charged particles were H⁻ ions. In the region of T_e=1.5-2 eV, the effective attachment rate coefficient defined as the average value for hydrogen molecules of all energy levels was evaluated to be $0.5\text{-}1.3 \times 10^{-11}$ cm³/s, of which the value is 3-order higher than that for attachment to the $v = 0$ level of the ground state molecules.

INTRODUCTION

Recently, hydrogen plasmas produced by inductively-coupled discharges, wave-excitation discharges and ECR discharges draw attentions in developments of H⁻ sources for NBI heating systems of thermonuclear fusion reactors and accerelator experiments. These discharges have many advantages over dc filament discharges used in present H⁻ sources; (1) they have no erosive electrodes which have problems in long-time operations, (2) high-density ($> 10^{12}$ cm⁻³) and low-pressure (<10 mTorr) plasmas can be easily attained by relatively low applied power and (3) extraction bias can be controlled independent of the discharge system. In the present scheme for H⁻ sources, cesium feeding is inevitable to attain a required H⁻ current. For the practical use, however, it is very hard to control the H⁻ production by Cs feeding, because the control of amounts of Cs vapor introduced into the sources is difficult. Hence, it is of significant importance to develop an efficient H⁻ production system with pure hydrogen discharge.

In this context, we have developed an H⁻ source using helicon wave discharge, which is inductively coupled discharge in the frequency (ω) range of $\omega_{ci} \ll \omega \ll \omega_{ce}$ under a magnetic field. Spatial and temporal variations of H⁻ density were measured by probe-assisted laser photodetachment technique. Reaction kinetics of the charged particles in the afterglow was discussed. The effective rate coefficient defined as dissociative attachment to all kinds of hydrogen molecules in the region where the efficient H⁻ production occurs was estimated and compared with the value derived from other experiments.

EXPERIMENTAL SETUP

Figure 1 shows a schematic geometry of the plasma device and the diagnostic system. Plasmas were produced in a vacuum chamber consisting of a quartz discharge tube and a stainless-steel chamber. A uniform magnetic field of 250 G was applied in the longitudinal direction. The discharge was sustained by applying an rf (13.56 MHz) power of 0.8 kW to a helical antenna (m=1) wound around the discharge tube[1]. The discharge conditions were as follows: H_2 gas pressure was 10 mTorr, the gas flow rate was 10 ccm and the active discharge duration was 10 ms with the repetition frequency of 5 Hz.

The positive ion density (n_+) and the electron temperature (T_e) were measured by a Langmuir probe. The ratio of the negative ion density (n_-) to the electron density (n_e) was measured by probe-assisted laser photodetachment technique, which has been already detailed in ref. [2]. The absolute values of n_- and n_e were determined from n_+ and n_-/n_e under the assumption of the local charge neutrality. A pulsed dye laser pumped by an excimer laser [Lambda Physik] (wavelength=500 nm, pulse width=17 ns) was used to detach electrons from negative ions in the plasma. The detached electrons were collected by the Langmuir probe biased at +60 V against the chamber of the earth potential. It is well known that the dominant negative ion species in such high-density and low-pressure plasmas are H⁻. Hence, in this paper, the photodetachment efficiency[3] was determined by using the photodetachment cross section for H⁻. The space-resolved measurements of n_- (and plasma parameters) were carried

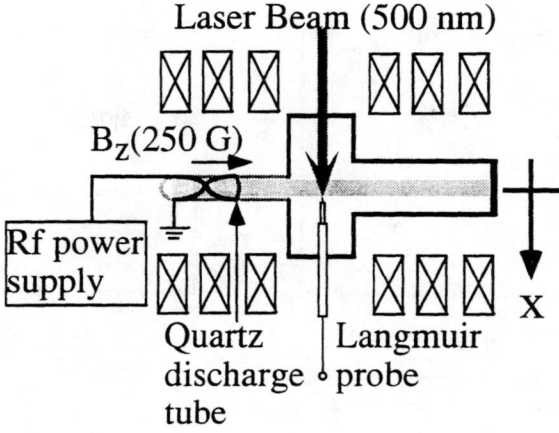

Figure 1: Experimental setup.

out by moving the Langmuir probe along the laser path. The spatial resolution was 5 mm. Both of the discharge system and diagnostic system were controlled by a pulse generator. Time-resolved measurements were done with the resolution of 1 μs.

RESULTS

Temporal Variations

Temporal variations of n_-, T_e, n_+ and n_e at the center of the plasma column are shown in Fig. 2. In the active discharge phase, H$^-$ ions were not detected under the detection limit of 1×10^{10} cm^{-3}. After the termination of the rf power, n_- steeply increased to 2×10^{10} cm^{-3} for the first 15 μs. After 20 μs, n_- changed into decrease. The decay structure of n_- has two different decay times, which were denoted as τ_- and τ_{d-} in Fig. 2 (a). As can be seen Fig. 2 (b), T_e decreased rapidly after the termination. For the first 10 μs, T_e decreased to 1 eV. In accordance with the decrease in T_e, n_- steeply increased. After 10 μs, T_e seemed to be nearly constant value of 0.8–1.0 eV. However, it is expected that T_e slowly decreases and becomes less than 0.5 eV in the later afterglow, since the measurable lowest value of T_e was 0.5 eV in the Langmuir probe measurements.

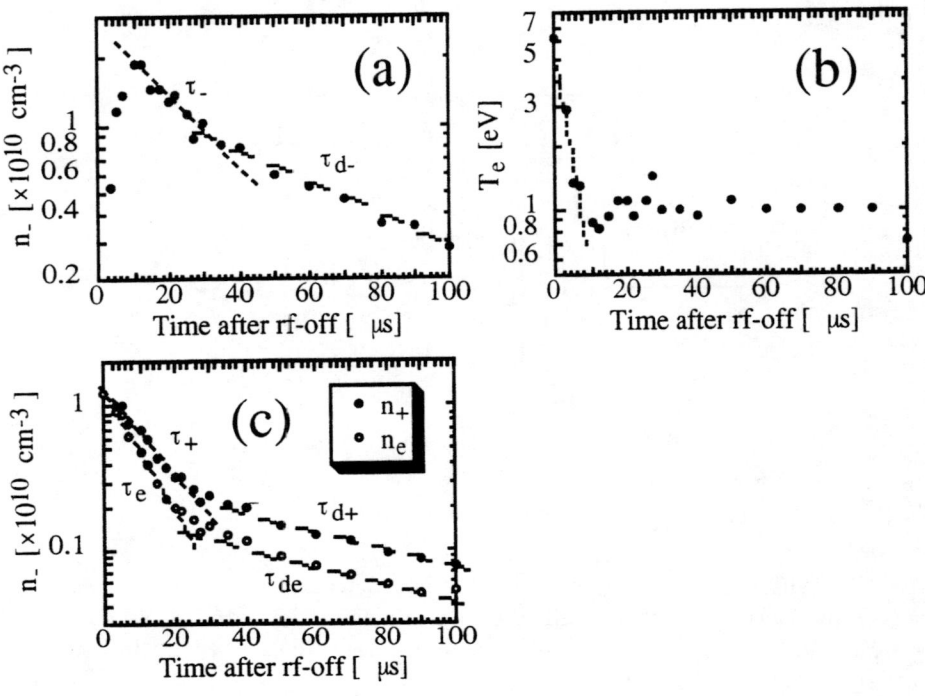

Figure 2: Temporal variations of n_-, T_e, n_+ and n_e.

Both the decays of n_+ and n_e also have the fast decay time of the order of 10 μs and the slow one of several tens μs. As shown in Fig. 2 (c), we call the fast decay times for the positive ions and the electrons τ_+ and τ_e, respectively (τ_{d+} and τ_{de} for the slow ones).

Spatial Variations

Figure 3 shows spatial variations of n_-, n_e, T_e and n_-/n_e in the active discharge phase. The H$^-$ ions were not detected in x=0–2 cm. The n_- had a peak value of 4×10^{10} cm^{-3} at x=3 cm. Both the spatial variations of n_e and T_e had peaks at the center of the plasma column. They decreased in x direction. The low-T_e plasma exsisted at x=3–5 cm, where H$^-$ ions were efficiently produced.

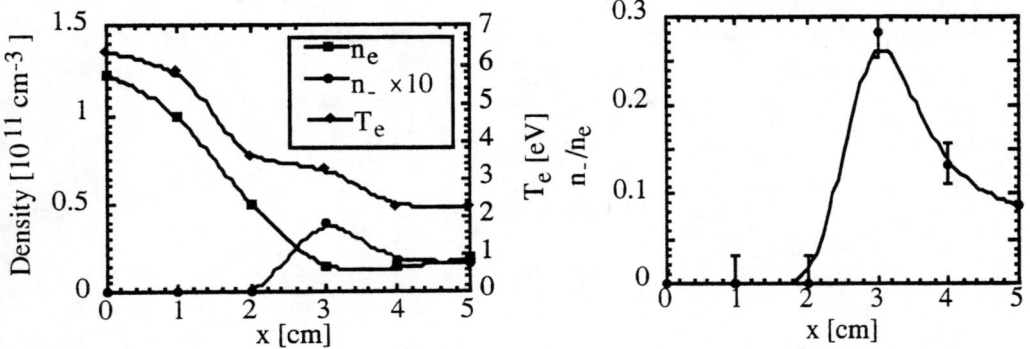

Figure 3: Spatial variations of n_-, n_e, T_e and n_-/n_e.

The n_-/n_e were 0.1–0.3 at x=3–5 cm, where 9–23% of the negatively charged particles were H$^-$. An intermediate electronegative plasma was formed in the outer region of the plasma column.

DISCUSSIONS

Reaction Kinetics in the Afterglow

As can be seen Fig. 2, the decay structures of n_-, n_+ and n_e exhibit two characteristic decay times. Here, based on these results, we discuss the reaction kinetics of the charged particles in the afterglow. We divide the afterglow into three periods; (I) attachment period ($0 < t < 20\mu s$), (II) recombination period ($20 < t < 50\mu s$) and (III) diffusion period ($50 < t < 100\mu s$).

Dominant processes related to the H$^-$, positive ions (H$^+$, H$_2^+$) and electrons in the afterglow are as follows,

$$\begin{aligned}
\text{(AT)} \quad & \text{H}_2^* + e^- \to \text{H}^- + \text{H}, \\
\text{(ID)} \quad & \text{H}^- + e^- \to \text{H} + 2e^-, \\
\text{(MN)} \quad & \text{H}^- + \text{H}^+, \text{H}_2^+ \to \text{H} + \text{H}^*, \text{H}_2^*, \\
\text{(DR)} \quad & \text{H}_2^+ + e^- \to \text{H} + \text{H}.
\end{aligned}$$

where reactions (AT), (ID), (MN) and (DR) stand for dissociative attachment to excited hydrogen molecules H$_2^*$, electron impact detachment of H$^-$, mutual neutralzation between H⁻ and the positive ions, and dissociative recombination between H_2^+ and the electrons, respectively. The rate equations for the H⁻, positive ions and electrons can be given as;

$$\frac{dn_-}{dt} = k_{at}^* n_{H_2^*} n_e - (k_{mn}n_+ + k_{id}n_e)n_- - \frac{n_-}{\tau_d}, \qquad (1)$$

$$\frac{dn_+}{dt} = -(k_{mn}n_- + k_{dr}n_e)n_+ - \frac{n_+}{\tau_d}, \qquad (2)$$

$$\frac{dn_e}{dt} = k_{id}n_- n_e - (k_{at}^* n_{H_2^*} + k_{dr}n_+)n_e - \frac{n_e}{\tau_d}, \qquad (3)$$

where k_{at}^*, k_{id}, k_{mn} and k_{dr} are the rate coefficients for reactions (AT), (ID), (MN) and (DR), and τ_d is the ambipolar diffusion time.

First, we discuss the decay processes of H⁻ in the recombination period (II). The H⁻ loss by reaction (ID) becomes insignificant in recombination period (II) and diffusion period (III) when T_e was less or nearly equal to the electron affinity of H⁻. Since τ_- is much shorter than the usual ambipolar diffusion time ($\sim 80\mu s$), the reaction (MN) dominates the decay structure of H⁻. The production yield by reaction (AT) becomes small because of the fast decrease in n_e. Therefore, the decay time for H⁻ is almost determined by the collision time for reaction (MN). The equation (1) is approximately given as

$$\frac{dn_-}{dt} = -k_{mn}n_+ n_- \approx \frac{n_-}{\tau_-}, \qquad (4)$$

where τ_- was 27.9 μs in this experiment and n_+ is roughly assumed to be constant value of $0.4\text{-}1.0\times 10^{11}$ cm⁻³. Then, k_{mn} is estimated to be $3.5\text{-}8.8\times 10^{-7}$ cm⁻³, which is consistent with the value for reaction (MN) with H_2^+ in the literature[4].

In the attachment period (I), the production yield of H⁻ increases by reaction (AT) with the decrease in T_e. The produced H⁻ ions are lost by the competing reactions of (ID) and (MN) for the early period and then the loss by the latter becomes dominant. The main loss processes of the positive ions are reactions (MN) and (DR). The decay time (τ_+) is much shorter than the diffusion time. No production of the positive ions by ionization occurs in the afterglow. Hence, these loss processes determine τ_+. Therefore, eq. (2) is given as

$$\frac{dn_+}{dt} = -(k_{mn}n_- + k_{dr}n_e)n_+ \approx \frac{n_+}{\tau_+}. \qquad (5)$$

Here, we assumed that the density of H_2^+ is approximately equal to n_+ and n_- is constant value of 10^{10} cm⁻³. The loss frequency of the positive ions ν_+ (=

$1/\tau_+$) is 6.3×10^4 s^{-1}. The first term of the parenthesis can be evaluated from the preceding result. The loss frequency by reaction (MN), $k_{mn}n_-$, becomes 3.5–8.8×10^3 s^{-1}, of which the value is less than ν_+ by 1-order. This means that the second term, that is, the loss by reaction (DR) predominates. The evaluated value for k_{dr} is 6.1–7.3×10^{-7} cm^3/s, which is consistent with the value for reaction (DR) with H_2^+ known in the literature[4]. The main loss processes of the electrons are reaction (AT) and (DR). The production by reaction (ID) is insignificant since n_- is at most 20% of n_e. The equation (3) is written as

$$\frac{dn_e}{dt} = -(k_{at}^* n_{H_2^*} + k_{dr}n_+)n_e \approx \frac{n_e}{\tau_e}. \tag{6}$$

We define the loss frequency of the electrons ν_e as $1/\tau_e$. By using the preceding results, the second term $k_{dr}n_+$ in the parenthesis is about 6.1–7.3×10^4 s^{-1} which is of the same order as ν_e (9.9×10^4 s^{-1}). Therefore, the frequency $k_{at}^* n_{H_2^*}$ for reaction (AT) is at most of the same order of them.

In the diffusion period (III), the decay times of the H$^-$, positive ions and electrons are about 70–80 μs. The ambipolar diffusion time for $n_-/n_e \sim 0.8$ is approximately 80 μs[5]. Hence, the diffusion loss of the charged particles becomes dominant in comparison with the reaction processes.

For summary, in the early afterglow, the charged particles are effectively lost by the recombination processes (AT), (MN) and (DR). The scheme for the loss processes in the afterglow is shown below.

$$H_2^+ + e^- \rightarrow (loss\, by\, DR),$$
$$e^- + H_2^* \rightarrow H^- + H \quad (conversion\, by\, AT),$$
$$H^- + H^+, H_2^+ \rightarrow (loss\, by\, MN).$$

In the active discharge phase, the plasma is constituted by the positive ions and electrons. In the afterglow, the electrons are mainly lost by reaction (DR). Some part of the electrons are attached to H_2^* to be converted into H^-. The produced H^- ions mainly recombine with the positive ions. Hence, all the charged particles decrease fast in the early afterglow. According to the decrease of the reactant particles, the diffusion loss becomes dominant and then decay times of the charged particles become longer.

The Effective Attachment Rate Coefficient

In the outer regions of the active discharge phase (cf. Fig. 3) where T_e is 1–2 eV, the main production process of H$^-$ is reaction (AT) and the dominant loss processes are reactions (MN) and (ID). In the steady state, from eq. (3) neglecting the diffusion term, n_-/n_e is given as

$$\frac{n_-}{n_e} = \frac{k_{at}^* n_{H_2^*}}{k_{mn} n_+ + k_{id} n_e} = \frac{\nu_{at}}{\nu_{loss}}, \qquad (7)$$

where ν_{at} and ν_{loss} are the electron attachment frequency and the loss frequency, respectively. The ν_{at} is given as the following equations (8)–(10) by considering all the attachment processes to many vibrational levels of the electronic ground state molecules.

$$\nu_{at} = \sum_{v=0}^{\infty} <\sigma_{at}(v)v_e> n(H_2(v)), \qquad (8)$$

$$= \sum_{v=0}^{\infty} <\sigma_{at}(v)v_e> p(v)n(H_2), \qquad (9)$$

$$= k_{at}^{eff} n(H_2), \qquad (10)$$

where $\sigma_{at}(v)$, $n(H_2(v))$ and $p(v)$ are the cross section for dissociative attachment to $H_2(v)$, the density and relative population of $H_2(v)$, respectively. The $n(H_2)$ is the total molecule density. The k_{at}^{eff} is defined as the effective rate coefficient for dissociative attachment to all kinds of hydrogen molecules. By substituting eq. (10) into eq. (6), k_{at}^{eff} is given as

$$k_{at}^{eff} = \frac{n_-}{n_e} \frac{\nu_{loss}}{n(H_2)}. \qquad (11)$$

In the outer region shown in Fig. 3, $n_-/n_e \sim 0.15$–0.3, $n(H_2)=3.6\times 10^{14}$ cm^{-3}, $n_+=2\times 10^{10}$ cm^{-3}, $n_e=1.2\times 10^{10}$ cm^{-3} and $T_e=1.5$–2 eV. The k_{id} at $T_e=1.5$–2 eV is 5–9×10^{-8} cm^3/s [4]. The estimated value for k_{at}^{eff} is 0.5–1.3×10^{-11} cm^3/s at $T_e=1.5$–2 eV. It is much higher than the attachment rate coefficient for $H_2(v=0)$ (1×10^{-14} cm^3/s at $T_e=2$ eV), and similar to the value for $H_2(v=3)$ [6]. It is concluded that higher excited states are responsible to the efficient production in the outer region. As an example, we estimated k_{at}^{eff} in the extraction region of a hybrid multicusp ion source reproted in ref. [7], where $n_-/n_e \sim 0.09$, $n(H_2)=3.6\times 10^{13}$ cm^{-3}, $n_+=3.5\times 10^{10}$ cm^{-3}, $n_e=3.4\times 10^{10}$ cm^{-3} and $T_e=1.3$ eV. The evaluated value for k_{at}^{eff} was about 5.3×10^{-11} cm^{-3} at $T_e=1.3$ eV, of which the value is comparable to that of our experiment.

CONCLUDING REMARKS

The H⁻ source using helicon wave discharge has been developed. The spatial and temporal variations of H⁻ density in the hydrogen plasma were measured by laser photodetachment technique. From the temporal measurements at the center of the plasma column, no H⁻ ions were detected in the active discharge phase. The steep increase in n_- was observed just after the termination of rf power. In the afterglow phase, H⁻ ions became approximately 20 % of the charged particles. The reaction kinetics concerning with H⁻ in the afterglow was investigated. During the first 20 µs, the electrons are greatly lost by dissociative recombination with the positive ions and efficient conversion of the electrons into H⁻ ions also occurs. The produced H⁻ are mainly lost by mutual neutralization with the positive ions. For the next 30 µs, production rate of H⁻ becomes small and the decay time of the electrons changes to longer one. After 50 µs, the ambipolar diffusion loss of the charged particles becomes dominant.

The H⁻ ions were efficiently produced in the outer region of the plasma column in the active discharge phase, where low-T_e condition was satisfied. We estimated the effective rate coefficient for electron attachment in the outer region. The coefficient is much higher by 3-order than that for the attachment to $H_2(v=0)$. The efficient production of H⁻ occurs in the outer region of the helicon hydrogen plasma. Therefore, helicon wave discharge can be a strong candidate for H⁻ sources in the next generation.

REFERENCES

1. Shoji, T., Sakawa, Y., Nakazawa, S., Kadota, K., and Sato, T., Plasma Sources Sci. Technol. **2**, 5 (1993).
2. Hayashi, D. and Kadota, K., submitted to J. Appl. Phys. (1997), Hayashi, D., Ishikawa, T., Sasaki, K., and Kadota, K., "Negative ion diagnostics by laser photodetachment in high-density oxygen plasmas" in *Proceedings of the 7th Int. Symp. on Laser-Aided Plasma Diagnostics, Fukuoka*, 1995, pp. 331–334.
3. Bacal, M., Hamilton, G. W., Braneteau, A. M., Doucet, H. I., Rev. Sci. Instrum., **50**, 6 (1979).
4. Massey, H. M. W., *Negative ions*, Cambridge Press (1976).
5. Thompson, J. B., Proc. Phys. Soc. **73**, 818 (1959).
6. Wadehra, J. M., J. Appl. Phys. Lett., **35**, 917 (1979).
7. Bacal, M., El Balghiti, F., Elizarov, L. I., and Tontegode, A. J., "Effect of

cesium seeding in volume negative ion sources" in *Proceedings of the 19th SOFT, Lisbon* 1996.

Development of a High Current H⁻ Source for ESS

A. Maaser, P. Beller, H. Klein, K. Volk, and M. Weber

Institut für Angewandte Physik, Universität Frankfurt
Robert-Mayer-Str. 2-4, 60054 Frankfurt a. M., Germany

Abstract. For the European Spallation Source (ESS), a volume source based on the HIEFS (**hi**gh **ef**ficiency **s**ource) is being developed. The source will be optimized to produce high current densities in pulsed operation. A pulse generator delivering 1 to 1.5 ms pulses was installed. Furthermore, cesium was supplied to the plasma generator from an external oven. The cesium injection was optimized for a low e/H⁻ ratio and a high current. We obtained a current density of 70 mA/cm². This way, with an aperture radius of 4.25 mm, an H⁻ current of 40 mA was extracted at an extraction voltage of 22 kV. After a description of the source and the experimental setup, measurements of the beam current density and the e/H⁻ ratio will be presented in this paper.

INTRODUCTION

For the European Spallation Source (ESS), the Institut für Angewandte Physik (IAP) in Frankfurt is developing a volume type H⁻ ion source. This type of source has in principle proven its capability for H⁻ ion production [1]. Its low emittance and the achievable high current densities make it one of the candidates for the ESS-injector [2,3]. The ion source is required to produce a 70 mA H⁻ beam in 1.3 ms pulses at a repetition rate of 50 Hz. The source must operate, while meeting the performance requirements, with a constancy and reliability over an acceptable period of time.

Several source prototypes have been tested to meet these requirements [4]. The present paper presents the results of the experimental and theoretical investigations on the latest version of the H⁻ source.

EXPERIMENTAL SETUP

A schematic cross-sectional view of the source and the extractor is shown in Fig. 1. The plasma generator is made of a water-cooled cylindrical copper chamber whose dimensions are 6.0 cm in diameter by 10.5 cm in depth. In the middle of the chamber, up to four tungsten cathodes can be mounted. The front end of the chamber is closed by the plasma electrode, which is negatively biased with respect to the anode. On the back the gas inlet is installed. A solenoid is surrounding the chamber to confine the plasma in longitudinal direction. An electro-magnet is installed in the flange of the plasma electrode. It serves as a filter to provide a variable transverse magnetic field (B_y).

For cesium injection, an external oven is mounted on the flange of the plasma electrode. By means of two small pipes, the cesium vapor is guided to the outlet electrode. An adjustable gas burner acts as heater for the oven. The pipes are heated by the arc discharge. The whole system is temperature controlled. In order to extract the H⁻ ions, a single hole diode extraction system is used. Due to the extension of the filter magnet field into the extraction region, the electron beam is deflected outside of the beam axis. Behind the extraction system, a water-cooled copper cup is installed for electron dumping. As the deflection angle of the electron beam depends on the filter field strength and the extraction energy, a so-called steerer magnet is necessary to contrive the electron beam in the dumping tube. In addition, this concept guarantees that no electrons reach the faraday cup. All the measurements were performed with the source and extraction operated in pulsed mode.

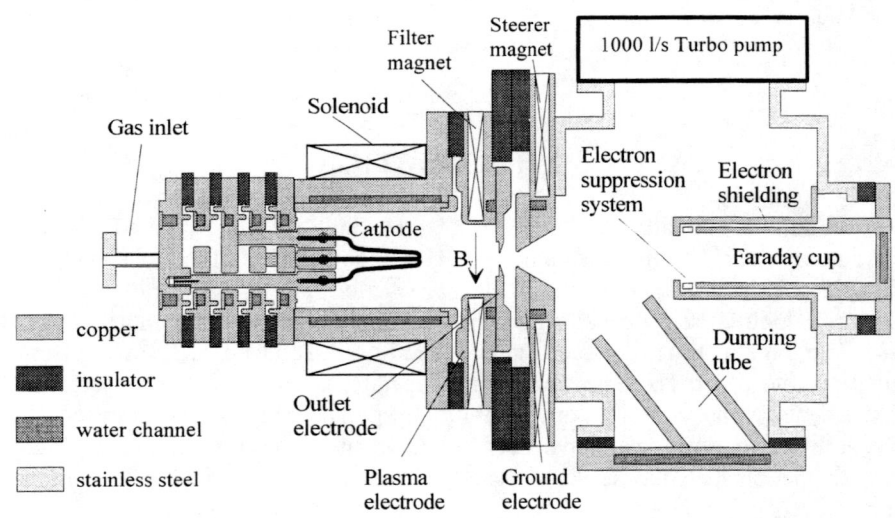

FIGURE 1. Schematic drawing of the experimental setup.

EXPERIMENTAL RESULTS

It is assumed that the dominant process for the volume production of negative ions is the dissociative attachment of electrons to hydrogen molecules in highly excited vibrational states (5). Highly excited molecules can be generated by collision of molecules with high energy electrons. However, the dissociative attachment of electrons to the highly excited molecules requires electrons with low energy. As the negative ion formation process needs electrons with different energies, it is recommended to use a transverse magnetic filter field, which divides the discharge chamber in two regions with different electron energies: first, in a region near the cathode, where high energy electrons produce highly excited vibrationally molecules; second, in a region near the outlet electrode, where the transverse magnetic filter field reduces the energy of the electrons. In this region, the negative ions are generated by the dissociative attachment of low energy electrons to the excited hydrogen molecules.

Figure 2 illustrates the H⁻ current and the e/H⁻ ratio as function of the flux density of the magnetic filter field. It can be seen that for flux densities up to 6 mT the H⁻ current increases while the e/H⁻ ratio decreases. Higher flux densities reduce the H⁻ current.

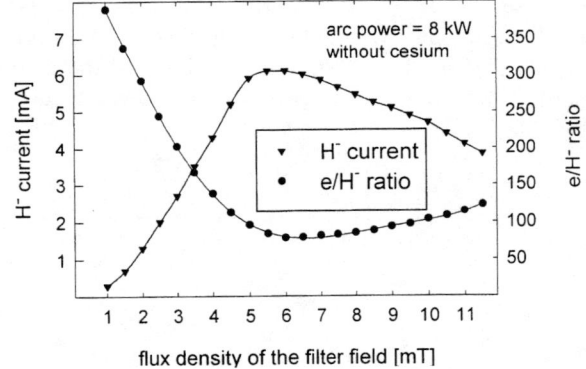

FIGURE 2. H⁻ current and e/H⁻ ratio vs flux density of the filter field.

The negative ion production efficiency can be increased by supplying cesium to the plasma chamber (6). In this process, positive hydrogen ions from the plasma are converted into negative ions on a cesium covered metal surface. The following paragraph describes our cesium injection system, the injection method, and the experimental results.

The cesiated surface should be near the outlet hole, because a shorter distance between the generation and the extraction region increases the possibility for H⁻ ions to reach the plasma meniscus. Therefore, two pipes lead from the cesium oven to the outlet electrode, which end 5 mm next to the extraction hole. After heating the pipes (300 °C) and the plasma chamber wall (80 °C), the cesium oven is heated up to about 150 °C. A cesium temperature of 150 °C corresponds to a vapor pressure of 1 Pa. The plasma electrode including the outlet electrode is kept cool (25 °C).

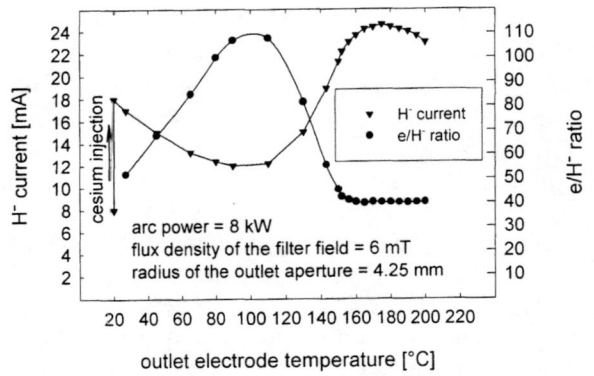

FIGURE 3. H⁻ current and e/H⁻ ratio vs plasma outlet electrode temperature.

Most of the cesium vapor has condensed on the outlet electrode, as it is the coldest part in the plasma chamber. In the next step, the cesium oven is cooled down and the outlet electrode is slowly heated up. The course of the H⁻ current and the e/H⁻ ratio as function of the outlet electrode temperature is presented in Fig. 3. For temperatures from 170 °C to 180 °C, the H⁻ current reaches its maximum. Compared to an operation without cesium, the H⁻ current is three times higher and the e/H⁻ ratio is three times lower. If all the details of the cesium injection are carefully carried out, these values are well reproducible.

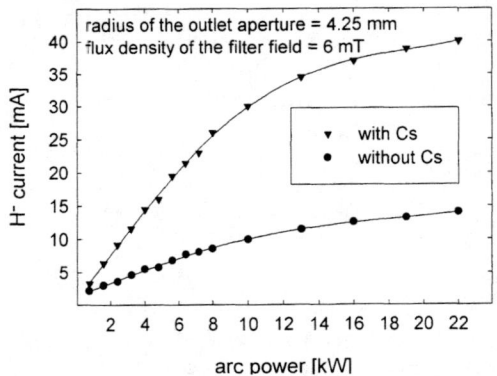

FIGURE 4. H⁻ current as function of the arc power.

In a further experiment, the influence of the arc power on the H⁻ current and the e/H⁻ ratio was investigated. Figure 4 shows the H⁻ current as a function of the arc power for cesiated and uncesiated source operation. The measurements were done using an outlet electrode with a radius of 4.25 mm. Without cesium, the ion source delivers about 14 mA H⁻ for an arc power of 22 kW. Operating the ion source with cesium, an

H⁻ current of 40 mA has been obtained for the same power.

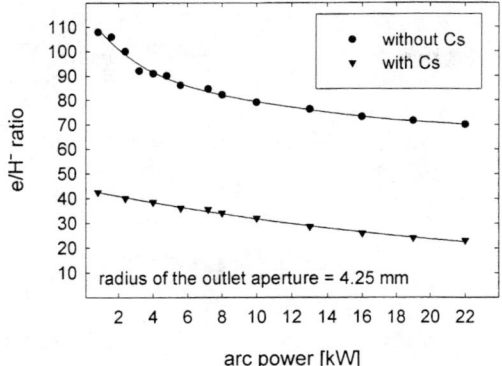

FIGURE 5. e/H⁻ ratio as function of the arc power.

According to the plot of the measured e/H⁻ ratio versus the arc power (Fig. 5), the e/H⁻ ratio decreases for an increasing arc power. Operating the source without cesium, the e/H⁻ ratio is reduced from 110 to 70. With cesium, a minimum e/H⁻ ratio of 23 was achieved.

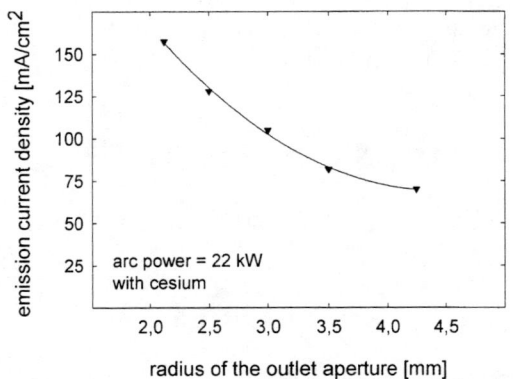

FIGURE 6. Emission current density vs radius of the outlet aperture.

In Fig. 6, the measured emission current densities for different outlet aperture radii are plotted. The measurements were made at an arc power of 22 kW and operation with cesium. As illustrated, the emission current density dramatically decreases as function of the outlet aperture radius. However, a calculation of the emission current shows a moderate increase. Hence, an essential current improvement by increasing the outlet aperture radius is not possible. A possible reason for this dependence is, that for larger outlet aperture radii the loss of neutral hydrogen molecules through the extraction hole of the plasma chamber is higher. For a constant gas flow, this leads to a reduction of the neutral gas pressure in the plasma chamber. Consequently, for lower neutral gas pressures the production rates of ions and electrons are reduced and both the plasma density and the emission current density decrease. In principle, it is possible to compensate this negative effect by increasing the gas pressure in the plasma chamber. However, a higher gas pressure in the plasma chamber results in a higher gas pressure in the extractor, too. A higher gas pressure in the extractor reduces the maximum extraction voltage that can be applied and hence the extractable H⁻ current. As a consequence of these results, for the generation of a specific high H⁻ current, the source should be operated with the maximum arc power in order to use the smallest possible outlet aperture radius.

CONCLUSION

In the present paper, the current status of the Frankfurt H⁻ volume source was described. The main emphasis of the investigations was put on the source operation with cesium. After a detailed description of the cesium injection, measurements of the H⁻ current and the e/H⁻ ratio were presented. Operating the ion source with cesium at an outlet electrode temperature of 180 °C, the H⁻ current is up to three times higher. At the same time, the e/H⁻ ratio is reduced by a factor three. With an arc power of 22 kW, an H⁻ current density of 70 mA/cm^2 has been achieved. Using an outlet aperture radius of 4.25 mm, an H⁻ current of 40 mA was extracted at an e/H⁻ ratio of 23.

Considering these results, the pulse generator will be enlarged in the next step for a higher arc power. Additionally, to reduce the gas flow through the extractor, a pulsed gas inlet system is being installed and tested.

ACKNOWLEDGMENT

The work was supported by FZ Jülich and BMFT under contract number 06 OF 841.

REFERENCES

1. K.N. Leung, G.J. DeVries, W.B. Kunkel, L.T. Perkins, D.S. Pickard, K. Saadatmand, A.B. Wengrow, and M.D. Williams, Proceedings of the Sixth European Particle Accelerator Conference, Sitges (Barcelona), 1996, p. 1513.
2. K. Volk, W. Barth, A. Lakatos, T. Ludwig, A. Maaser, H. Klein, K.N. Leung, Proceedings of the Fourth European Particle Accelerator Conference, London, 1994, p. 1438.
3. A. Maaser, K. Volk, M. Weber, and H. Klein, Rev. Sci. Instrum. **67** 1054 (1996).
4. K. Volk, H. Klein, A. Lakatos, A. Maaser, and M. Weber, Rev. Sci. Instrum. **67** (1995).
5. J.R. Hiskes, A.M. Karo, M. Bacal, A.M. Bruneteau, and W.G. Graham, J. Appl. Phys. **53** 3469 (1982).
6. J. Ishikawa, Nucl. Instrum. Methodes **B37/38** 38 (1989).

Studying of Negative Ions Source Based on Reflective Discharge in Regimes with Cesium Added and Without Cesium

I.A.Soloshenko, A.I.Shchedrin and A.V.Ryabtsev

*Institute of Physics of Ukrainian National Academy of Sciences
Prospect Nauki, 46, 252022 Kiev 22, Ukraine*

Abstract. The results of theoretical and experimental studies of physical processes in the source of hydrogen negative ions are presented. The source is based on reflective discharge with incandescent cathode and H⁻ ions extraction across magnetic field. Calculations of gas discharge plasma parameters for given current and energy of electrons emitted from the cathode are performed in theoretical part of the proceeding. Plasma parameters, including H⁻ ions concentration, are determined on a basis of Boltzman equation solution taking into consideration principal collision reactions in a volume and at a surface of the discharge chamber.

It is shown theoretically and experimentally, that:

1) effective operation of the source in non-cesium regime is realized with extraction of H⁻ ions not from the discharge column, but from drift plasma, where electron temperature does not exceed 1 eV;

2) emission current of H⁻ ions grows up with increase of specific power, introduced into the discharge, up to 500 W/cm^2; with subsequent increase of this specific power, current reaches its saturation as a result of increase of essentiality of H⁻ ions death during collisions with hydrogen atoms and plasma electrons;

3) introduction of cesium vapor increases gas economy of the source by one order of magnitude; the increase is not resulted with the growth of efficiency of H⁻ ions formation in the plasma, but is due to essential increase of flux of H⁻ ions, originated at cesium-doped surface of the source anode in a result of recharging of fast positive ions and atoms;

4) change of operation regime from stationary to pulsed-periodical one leads to the increase of plasma source efficiency due to the decrease of average concentration of atomic hydrogen which determines death of H⁻ ions at high specific power of the discharge.

Scheme of the ions source under study is shown in Fig. 1. Incandescent tungsten cathode 1 with diameter of 2 mm serves as electrons source. Before the anode chamber 3 the aperture 2 is placed which limits radial size of the plasma column 6. Reflection of electrons providing gas ionization is realized by anticathode 7 with potential equal to that of the cathode. Gas feed is performed through the holes in the anode 5. Magnetic field with strength of up to 2 kE is directed along the system axis. Extraction of ions is accomplished through the slit by field of electrode 4. Diameter of the discharge chamber comprises 5 mm, diameter of the aperture comprises 2.5 mm. Exactly with this configuration maximum ion yield is observed. Hydrogen pressure in the source chamber is varied in a range of $2 \cdot 10^{-2} - 2 \cdot 10^{-1}$ Torr; the discharge voltage - in a range of 100 - 200 V; the discharge current - in a range of 1 - 10 A; extracting voltage - in a range of 8 - 14 keV. For thorough comparison of experimental and calculated results measurements of the plasma parameters have been accomplished. Plasma density values, averaged along the plasma column radius, which have been obtained from measurements of ions flux on the anticathode, grow up proportionally to the discharge current and reach 10^{14} cm^{-3}. Plasma parameters outside the column could be measured by thin cylindrical probe introduced into the plasma through emission slit perpendicularly to magnetic field direction.

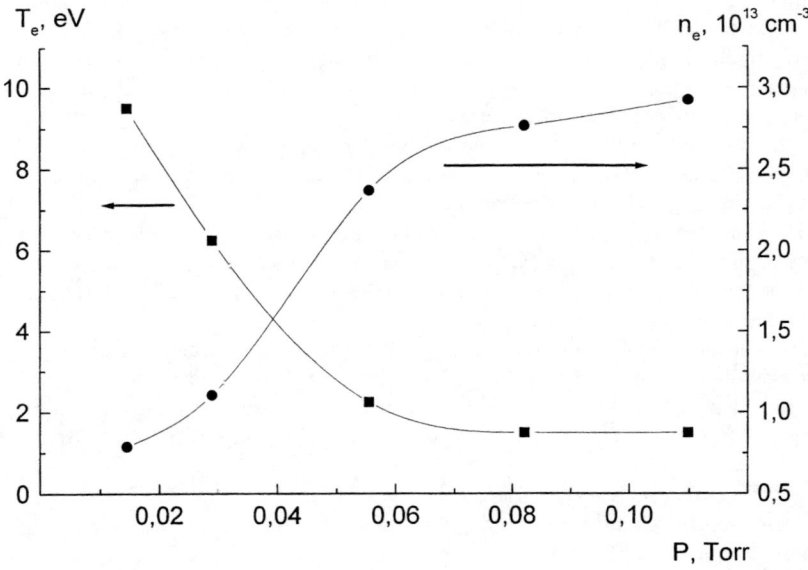

FIGURE 2. Dependencies of electrons temperature T_e and plasma concentration n_e in a space between anode and discharge column on pressure in the source chamber.

INTRODUCTION

Current proceeding presents a review of results of studies the source of hydrogen negative ions based on reflective discharge, which were obtained in the Institute of Physics of Ukrainian National Academy of Sciences. The review is based on thorough theoretical and experimental studies of the processes, which take place in the source, and their influence on magnitude of extracted flux of H⁻ ions in two regimes: with cesium added and without cesium. The investigations are carried out mostly in stationary regime, however, theoretical treatment embraces pulsed-periodical regime as well.

The source with parameters most close to those, required by thermo-nuclear setups, is currently two-chamber source, in which zone of plasma generation and zone of negative ions formation are separated by magnetic barrier [1]. In the source under consideration there is no special magnetic barrier, however, analogous separation of mentioned zones is implemented practically. For this reason, the results, obtained in our proceedings, can relate immediately to two-chamber sources.

DESCRIPTION OF THE SETUP AND THE DISCHARGE PARAMETERS

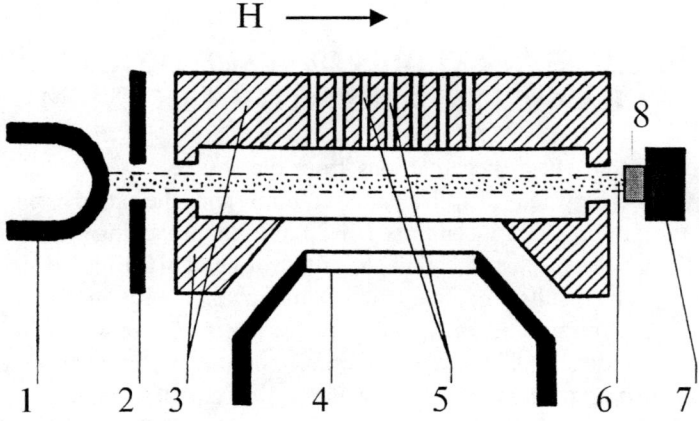

FIGURE 1. Scheme of the source. 1 - heated cathode, 2 - cathode aperture, 3 -anode chamber, 4 - extracting electrode, 5 - holes for the gas supply, 6 - discharge column, 7 - anticathode, 8 - dichromated cesium tablet.

Fig. 2 presents the dependencies of concentration (curve 1) and temperature (curve 2) on the gas pressure, measured by mentioned probe. One can see that the plasma concentration in periphery is less than that in column by more than one order of magnitude, and temperature of electrons at $p = 1\cdot10^{-1}$ Torr is about 1 eV. Thus already from these data one can see, that the discharge under study represents natural two-step system. In the discharge column fast electrons create dense plasma and produce excited molecules, whereas outside column the conditions being optimal for H⁻ ions formation are realized. Electrons cooling is realized at that by their diffusion across magnetic field.

The data about composition of ion component of the plasma are important as well. These data have been obtained by extracting positive ions through emission slit. Ion components were separated in transverse magnetic field and measured by narrow moveable collector. The results of component structure measurements for positive ions are brought together in Table 1.

TABLE 1. Positive Ions Composition Structure.

p, Torr	I_d, A	$I(H^+_1)$ mA-%	$I(H^+_2)$ mA-%	$I(H^+_3)$ mA-%	I_+, mA
$5\cdot10^{-2}$	8	3.75 - 89.3	0.40 - 9.5	0.05 - 1.2	4.20
$1.1\cdot10^{-1}$	8	3.44 - 81.9	0.56 - 13.3	0.20 - 4.8	4.20
$1.6\cdot10^{-1}$	8	3.10 - 73.8	0.80 - 19.0	0.30 - 7.2	4.20
$1.1\cdot10^{-1}$	6	2.70 - 77.0	0.43 - 12.3	0.32 - 9.2	3.5
$1.1\cdot10^{-1}$	3	1.60 - 65.0	0.46 - 19.0	0.40 - 16	2.46

EXPERIMENTAL STUDIES OF DISCHARGE EMISSION CHARACTERISTICS WITHOUT CESIUM

Emission current of negative ions from the source depended essentially on size of a gap between the discharge column and the anode. In particular configuration optimal gap comprised 1.25 mm (approaching of emission slit immediately to the discharge column resulted in about 10 times decrease of H⁻ ions current). The results presented below were obtained with this optimal gap.

Principal emission characteristics are presented in Figures 3 and 4. Fig. 3 shows the dependencies of H⁻ extracted current on the discharge current, obtained at various gas pressures. One can see from the figure, that with ion current grows up only at the increase of I_d up to 5 A; in subsequent the increase of power, introduced into the discharge, does not result in the current increase. With pressure increase the maximum current value increases as well.

Fig. 4 exhibits dependencies of H⁻ current on gas pressure obtained at various width of emission slit. One can see, that optimal pressure comprises $1\cdot10^{-1} - 2\cdot10^{-1}$ Torr, at which maximum current is reached. Maximum current

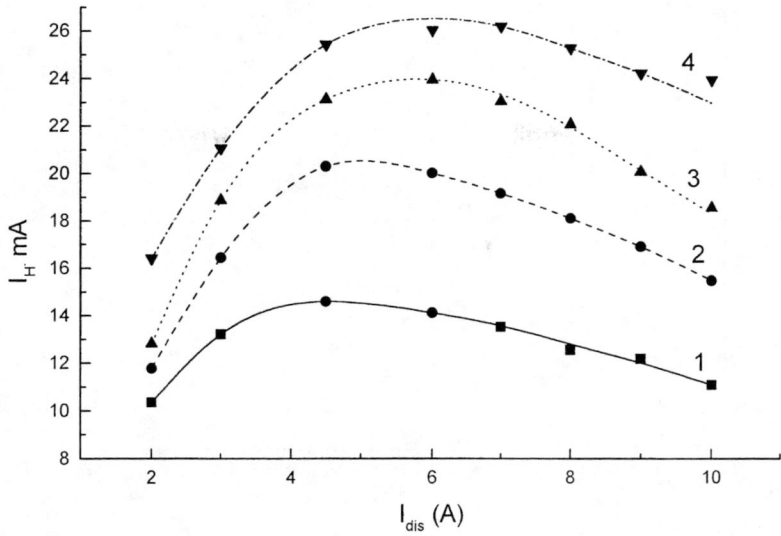

FIGURE 3. Dependencies of H⁻ ions current on the discharge current. Extracting voltage $U_0 = 14$ kV, emission slit 1x40 mm². Pressure in the source (p, Torr): 1 - $6.5 \cdot 10^{-2}$; 2 - $8.7 \cdot 10^{-2}$; 3 - $1.1 \cdot 10^{-1}$; 4 - $1.4 \cdot 10^{-1}$.

increases proportionally with slit width increase; it gives evidence to volume character of H⁻ ions formation. At optimal discharge parameters emission current density of $j^-_{max} \approx 80$ mA/cm² can be obtained. At that the ratio of ions current to electrons current comprises 0.2, and gas economy is 0.2 %.

The degree of current modulation and value of normalized emittance are also important characteristics of the source. Measurements have shown, that at low gas pressure rotation instability is developed in the column, leading to strong modulation of beam current (modulation coefficient reaches 20 %). Normalized emittance of the beam in this regime is also intolerably high. With pressure increase up to $p = 1 \cdot 10^{-1}$ Torr the instability is stabilized, and modulation coefficient becomes less than 1 %. Normalized emittance value reaches small values in this case ($E_{\parallel} = 3 \cdot 10^{-5}$ cm·rad in direction along magnetic field, and $E_{\perp} = 1 \cdot 10^{-5}$ cm·rad in transverse direction). These values of normalized emittance correspond to ions temperature of 0.8 eV.

In conclusion we should mention that limit value of H⁻ ions current is reached at specific power, introduced into the discharge, of ~ 500 W/cm³. Subsequent increase of introduced power does not lead to ions current increase.

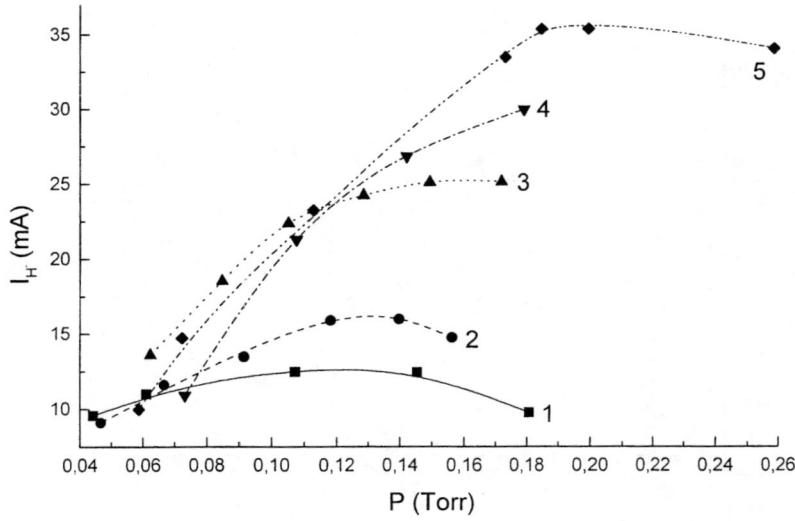

FIGURE 4. Dependencies of H⁻ ions current on gas pressure. $U_0 = 14$ kV, discharge current is optimal, emission slit length 40 mm. Width (Δ, mm): 1 - 0.6; 2 - 0.7; 3 - 1; 4 - 1.2; 5 - 1.5.

CESIUM VAPOR INFLUENCE ON NEGATIVE IONS YIELD

As it was shown in a set of proceedings (see e.g. [2,3]), cesium adding to the plasma source results in both essential increase of H⁻ ions current density, and in increase of gas economy of the source. Optimal usage of this phenomenon required establishing of its mechanism. For this reason we performed corresponding theoretical and experimental investigations. Let us concentrate at first on experimental results.

Introduction of cesium was accomplished by sputtering of dichromated cesium tablet 8 (see Fig. 1) placed on reflector 7. When the discharge was glowing with tablet installed, its voltage fell down to 50 V, and current of negative ions, extracted from the source, decreased. After several hours of operation the voltage returned back to value typical for pure hydrogen regime, and current of H⁻ ions raised up dramatically. Subsequent measurements were performed in this regime.

Fig. 5 exhibits the dependence of current of H⁻ ions, extracted from the source, on pressure with extracting voltage of 14 kV. One can see from the figure, that introduction of cesium results in significant changes of emission properties of the source. Principal changes are the following:

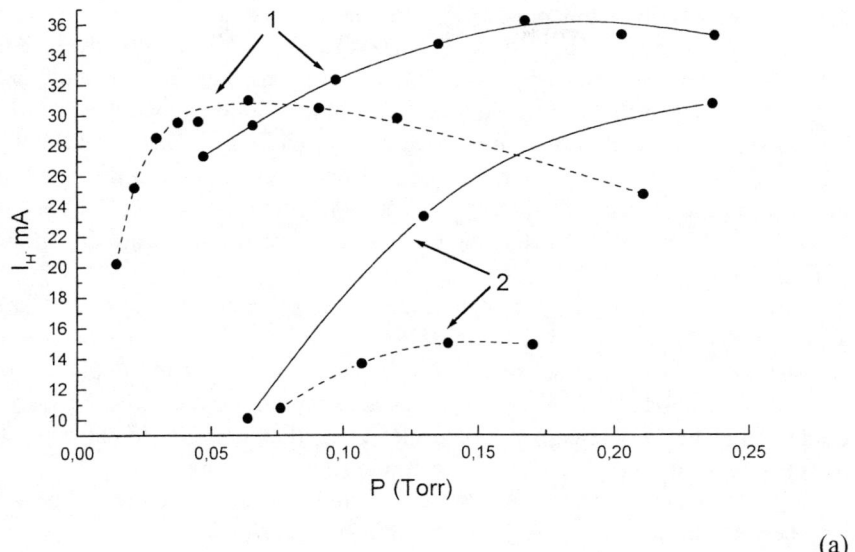

(a)

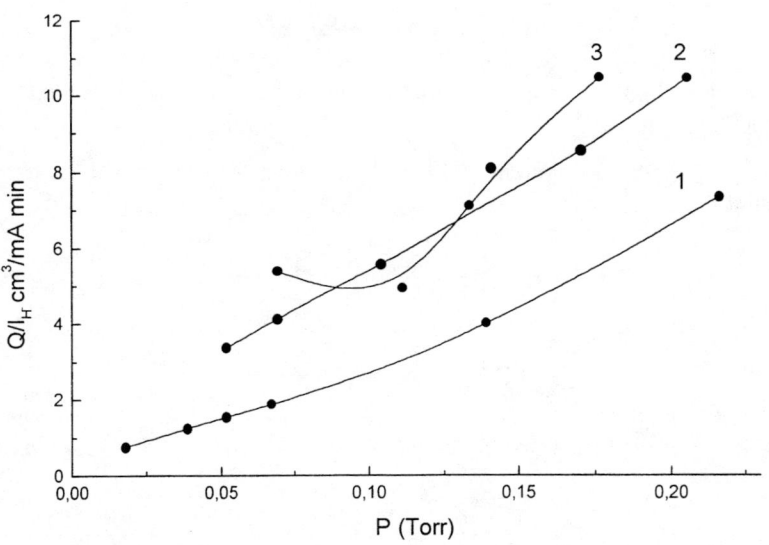

(b)

FIGURE 5. Emission characteristics of the source: a - dependence of extracted current of H⁻ ions on pressure in the source chamber at $U_0 = 14$ kV (1 - with Cs, 2 - without Cs, solid curve for emission slit 1.5x40 mm², dash curve - 0.7x40 mm²); b - dependence of gas expense per unit of H⁻ ions extracted current on pressure (1,2 - source with Cs for emission slit 0.7x40 and 1.5x40 mm², respectively; 3 - source without Cs, emission slit 0.7x40 mm²).

1) increase of maximum value of H⁻ ions current density; 2) increase of gas economy, especially at low pressure; 3) influence of cesium adding decreases with increase of slit width. Three those peculiarities are exhibited clearly in Fig. 5b showing the dependencies of gas flux to negative ions current ratio on hydrogen pressure. Attention should be paid to changes in mentioned dependencies: in pure hydrogen regime optimal pressure value exists providing the best gas economy of the source ($p \approx 10^{-1}$ Torr), whereas in regime with cesium added the economy is inversely proportional to H_2 pressure. Maximum achieved value of gas economy comprises ~ 2 %, which exceeds the value in pure hydrogen regime by one order of magnitude.

Principal issue consists in determining the mechanism of source efficiency growth with cesium vapor added. This effect could be due to either with efficiency increase of volume processes of H⁻ ions formation, or with additional creation of H⁻ ions at the anode surface, which possesses small withdrawal work value due to cesium film. In the last case it would mean that the source is transformed from plasma one to surface-plasma one, which was studied thoroughly in [4].

For clarifying this issue numeric simulation of hydrogen negative ion source with cesium and in pure hydrogen regime has been performed.

NUMERIC SIMULATION

For determining current density of H⁻ ions and calculating concentrations of both charged ($n_e, N_{H^-}, N_{H_2^+}, N_{H^+}, N_{Cs^+}$) and neutral ($N_{H_2}, N_H, N_{H_2(v)}, N_{Cs}$) mixture components the system of kinetic equations (1) - (3) is solved together with Boltzman equation [5]

$$\frac{\partial N_i}{\partial t} + \text{div}\, \Gamma_i = \sum_i K_i N_i + \sum_{i,j} K_{ij} N_i N_j, \quad (1)$$

$$\Gamma_i = \mu_i E N_i - D_i \frac{\partial N_i}{\partial r}, \quad (2)$$

(μ_i is taken with corresponding charge sign),

$$n_e + N_{H^-} \approx N_{H_2^+} + N_{H^+} + N_{Cs^+}, \quad (3)$$

where E is an electric field, μ_i and D_i are respectively mobilities and diffusion coefficients of charged components.

Spatial distribution of neutral components is assumed to be uniform, since in experimental conditions their free run paths are comparable or more than the chamber radius. Thus $\text{div}\,\Gamma_i = 0$ is valid for them.

Rates of non-elastic processes K_{ij}

$$K_{ij} = \sqrt{\frac{2q}{m}} \int_0^\infty \varepsilon Q_{ij}(\varepsilon) f_0(\varepsilon) d\varepsilon \tag{4}$$

are calculated taking into consideration electron energy distribution function $f_0(\varepsilon)$. In (4) ε is electron energy (eV), m is electron mass (g), $q = 1.602 \cdot 10^{-12}$ Erg/eV, $Q_{ij}(\varepsilon)$ is cross section of corresponding non-elastic process (cm^2).

Electron energy distribution function (EEDF) in discharge column is determined from Boltzman equation with assumption of weak dependencies of concentrations of charged components on a radius. The last is proved by probe measurements. As it was shown in previous proceedings [6-8], EEDF in discharge column region possesses flat plateau shaped dependence on input beam energy ranging from ~ 100 eV down to ~ 20 eV, and has practically Maxwellian shape at low energies ($T_e = 3 \div 5$ eV). High-energy part of EEDF is absent between cathode and anode due to strong magnetizing of electrons along a radius, and electron temperature is essentially smaller than in column region ($T_e \leq 1$ eV). Due to this EEDF is assumed to be practically Maxwellian f_{0m} with radially dependent temperature, which is determined from the equation of energy balance:

$$\frac{3}{2} V_e \frac{dT_e}{dr} = -\varepsilon_v \sqrt{\frac{2q}{m}} N_{H_2} \int_0^\infty \varepsilon \cdot f_{0m}(\varepsilon) Q_v(\varepsilon) d\varepsilon,$$

$$f_{0m} = \frac{2}{\sqrt{\pi}} T_e^{-\frac{3}{2}} e^{-\frac{\varepsilon}{T_e}}.$$

(5)

Here V_e is velocity of electron drift across magnetic field taking into account ambipolar field, $Q_v(\varepsilon)$ is excitation cross section of the first vibration level of H_2 (since vibration excitation is principal channel of electron energy losses between cathode and anode), ε_v is an energy of vibration quantum (eV).

Mobilities and diffusion coefficients are calculated taking magnetizing into account. At that it is assumed that in whole discharge chamber ions (H_2^+, H^+, H^-, Cs^+) posses Maxwellian distribution. Principal mechanism determining transfer coefficients at pressure $p \sim 0.1$ Torr and plasma concentration $n \sim 10^{13} - 10^{14}$ cm^{-3} for electrons consists in scattering on ions.

It is assumed, like in [7], that temperature of atomic hydrogen T_H and that of negative ions T_{H^-} are equal to $4 \cdot 10^3$ K.

During solving the equations (1) - (3) and Boltzman equation elementary processes listed in Table 2 are taken into consideration.

Rate coefficients K_i describe linear on concentration processes (4, 9, 18, 19, 20 in Table 2).

It should be noted that in calculations the mechanism of negative ions formation due to Rydberg states is not taken into account since, as it follows from numeric estimations, its contribution in conditions of real experiments does not exceed several percent.

TABLE 2. Elementary Processes

1	$H_2(v) + e \rightarrow H_2^+ + e + e$, $v = 0,...,14$
2	$H + e \rightarrow H^+ + e + e$
3	$H_2 + e \rightarrow H + H + e$
4	$H + H(wall) \rightarrow H_2$
5	$H_2^+ + e \rightarrow H + H$
6	$H_2 + e \leftrightarrow H_2(v) + e$, $v = 1,2,3$
7	$H_2 + e \rightarrow H_2^* (B^1\Sigma_u^+, c^1\Pi_u) + e \rightarrow H_2(v) + e + h\omega$, $v = 1,...,14$
8	$H_2(v) + H \rightarrow H_2(v') + H$
9	$H_2(v) + wall \rightarrow H_2(v')$, $v = 1,...,14$, $v' = 0,...,v$
10	$H_2(v) + e \rightarrow H^- + H$, $v = 1,...,14$
11	$H + e \rightarrow H^-$
12	$H_2^+ + e \rightarrow H^- + H^+$
13	$H^- + H_2^+ \rightarrow H_2 + H$
14	$H^- + H^+ \rightarrow H + H$
15	$H^- + H_2 \rightarrow H + H_2 + e$
16	$H^- + H \rightarrow H + H + e$
17	$H^- + e \rightarrow H + e + e$
18	$H^+ + e(wall) \rightarrow H$
19	$H_2^+ + e(wall) \rightarrow H_2$
20	$Cs + e \rightarrow Cs^+ + e + e$
21	$Cs^+ + Cs + M \rightarrow Cs_2^+ + M$
22	$Cs_2^+ + e \rightarrow Cs + Cs$
23	$Cs^+ + H^- \rightarrow Cs + H$
24	$Cs_2^+ + H^- \rightarrow Cs + Cs + H$
25	$Cs^+ + e(wall) \rightarrow Cs$
26	$H + e(wall) \rightarrow H^-$

RESULTS OF NUMERIC MODELING

As it was mentioned in previous section, principal goal of numeric simulations consists in determining the influence of Cs in a volume of the discharge chamber and Cs absorbed at the surface on emission properties of hydrogen negative ions source based on reflective discharge. Thus numeric simulation is performed for three basic situations: 1 - pure hydrogen discharge (considered processes are 1 - 19 in Table 2); 2 - cesium is present only in a volume of the discharge chamber (processes 1 - 25); 3 - cesium is present only at the surface (processes 1 - 19, 26).

Fig. 6 shows the dependencies of current density for hydrogen negative ions j_H^- at the anode on pressure for three mentioned cases. One can see from Fig. 6 (curve 2) that introducing Cs vapor in discharge chamber volume, in condition of artificial maintaining of the discharge voltage, results only in inessential increase of j_H^- at high pressure as compared to the case of pure hydrogen regime (curve 1).

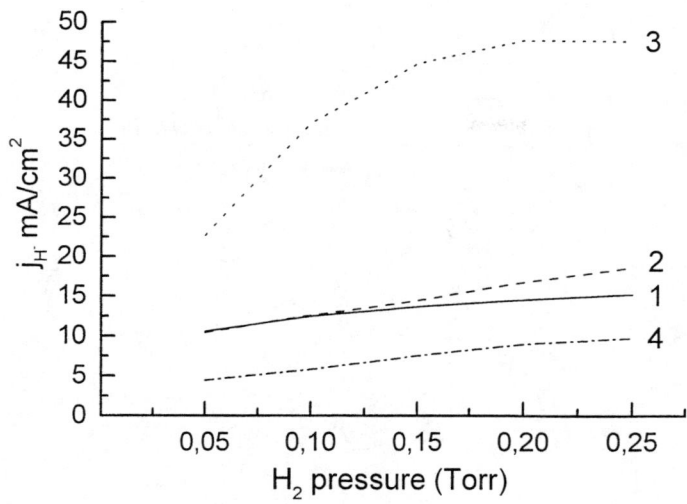

FIGURE 6. Dependence of H⁻ ions current density on pressure. 1 - pure hydrogen discharge; 2,4 - discharge in mixture H_2/Cs, $N_{Cs} = 1.6 \cdot 10^{13}$ cm^{-3}, processes 1-25 are considered; 3 - discharge in hydrogen with taking into account conversion H to H⁻ at cesium doped surface. Discharge voltage 120 V (1-3); 50 V (4).

Principal mechanism of hydrogen negative ions formation in this case, as well as in the case of pure hydrogen discharge, consists in dissociative sticking of electrons to vibrationally excited hydrogen molecules (process 10). Weak increase of negative ions current with Cs added is caused by the following reasons. Cesium performs both appropriate and inappropriate roles in volume processes. Due to low energies of electron state excitation and ionization of Cs the number of electrons with energy exceeding Cs excitation energy is decreased. Electron temperature T_e ($T_e = -1/(d \ln f_0/d\varepsilon)$ makes sense only for energy ≤ 20 eV) falls down at that (Fig. 7,f), which is in agreement with experimental data. From one hand, the decrease of amount of fast electrons and temperature of majority of electrons causes decrease of rate of vibrationally excited molecules formation and, consequently, H⁻ ions. From the other hand, decrease of T_e leads to lowering of rate of electrons unsticking (process 17; cross section of this process depends on temperature) and, as well, to lowering of H_2 dissociation rate and, respectively, to decrease of atomic hydrogen concentration (Fig. 8), which participates actively in taking off H_2 vibration excitation. The last two factors promote the increase of H⁻ ions concentration. Mentioned appropriate and inappropriate effects resulting from Cs introduction into the discharge practically compensate each other (Fig. 6), however, in condition of artificial maintaining discharge voltage at the same level,

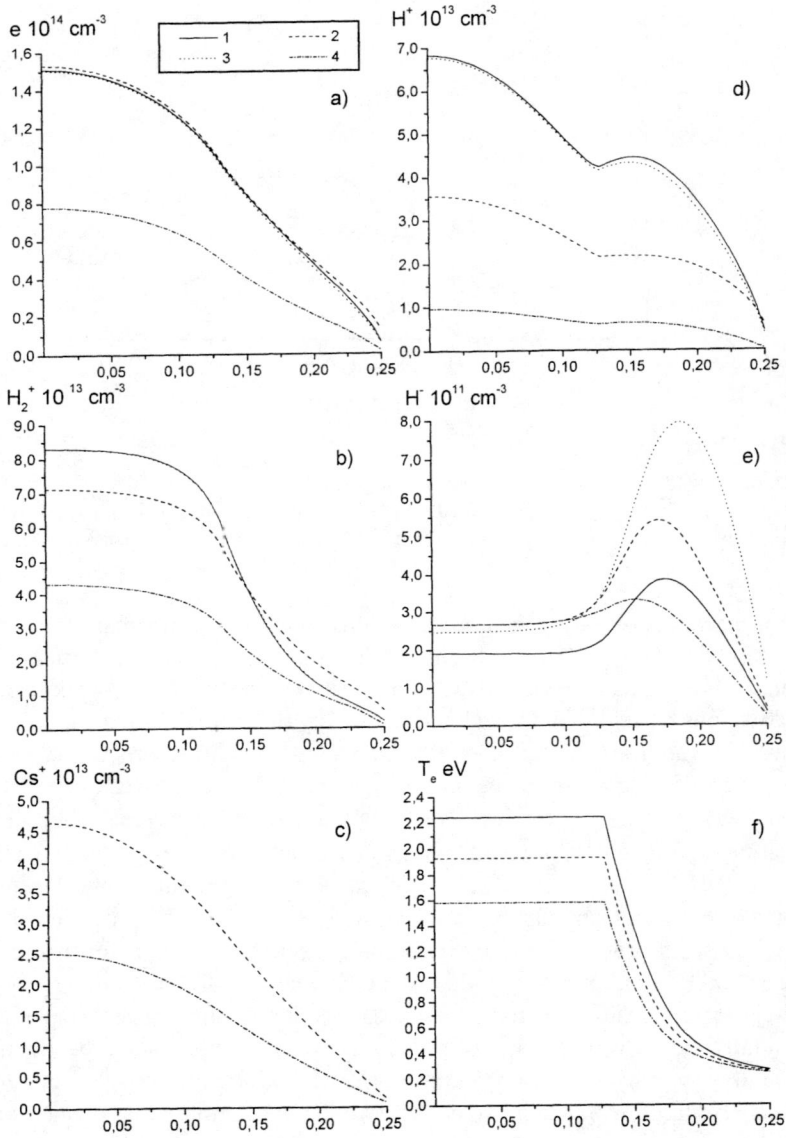

FIGURE 7. Distributions of electron and ion concentrations and electron temperature along radius of the discharge chamber. Labels 1 - 4 correspond to those in Fig. 6.

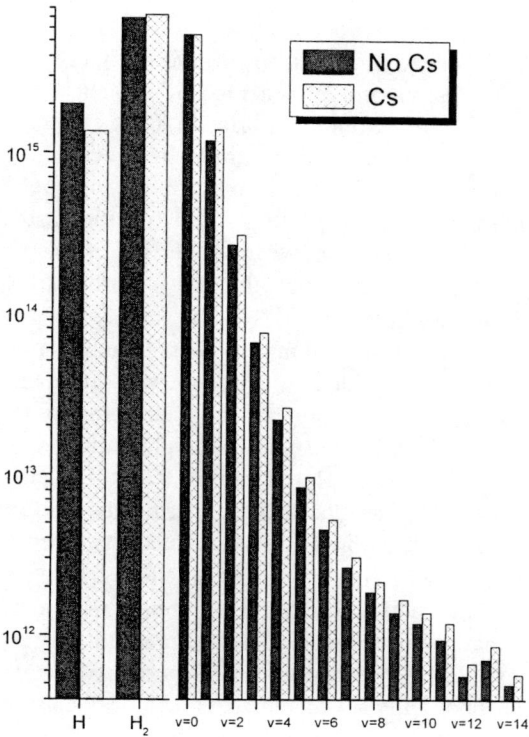

FIGURE 8. H, H_2 and $H_2(v)$ concentrations in the discharge column with and without Cs volume processes. Pressure p = 0.15 Torr, voltage 120 V.

appropriate effects slightly overcome inappropriate ones. Decrease of calculated value of the discharge voltage down to 50 V (which is practically observed in experiments with Cs introduced into a volume) leads to diminishing of H^- ions current. Thus, in accordance with experiment, the theory shows that Cs in a volume can not increase current density of H^- ions extracted from the source.

On the contrary to this, considering H to H^- conversion at the anode with adsorbed cesium leads to increase of H^- ions current (Fig. 6, curve 4) and volume concentration of H^- (Fig. 7,e) by a factor of 2 - 4 already at conversion coefficient $\gamma = 10^{-3}$. We shall not proceed with detailed comparison now, but mention that this increase is in agreement with increase of H^- current observed experimentally (Fig. 5). We shall also mention that in conditions of experiments no special care was taken for checking or optimization of Cs coating. However, as it was mentioned in [9], conversion coefficients for molybdenum and tungsten in conditions of ion sources are usually in a range of 10^{-3} to 10^{-1}, depending on hydrogen atoms temperature and degree of surface coating by cesium. One can see from the calculation, that already at rather low conversion coefficient essential increase of negative ions yield at source surface is observed.

Concentration of negative ions in pure hydrogen regime, as well as with cesium added, reaches its maximum in a region between discharge column and anode. This fact, which is of extreme importance for the experiment, is explained in the following way. N_H^- value is determined by processes of originating and death of H^- ions, at that the principal mechanism of the death consists in unsticking of electron due to collisions of H^- with fast plasma electrons (process 17). As it was mentioned above, cross section of this process depends on energy and possesses threshold behavior: the cross section equals zero at electron energy $\varepsilon \leq 1.25$ eV and reaches value of $\sim 2 \cdot 10^{-15}$ cm^2 at energy $\varepsilon \approx 10$ eV. Since electrons temperature outside the column decreases along a radius (Fig. 7,f), rate of mentioned process decreases dramatically, thus resulting in H^- concentration growth in spite of certain decrease of rate of their formation. If one artificially considers rate of process 17 as a constant, the maximum disappears. One can see from Fig. 7,e that for discharge column radius $R_c = 0.125$ cm the maximum is achieved at $R/R_c = 1.8 - 2.2$. Results of the experiments, performed with various diameters of anode chamber, are in a good agreement with results of calculations: optimum in regime without cesium is reached at $R/R_c \approx 2$.

In conclusion of the section devoted to stationary regime of source operation we shall list principal mismatches of calculations and experiments, and shall present the reasons of these mismatches. In pure hydrogen regime principal mismatches are the following: 1. Calculated values of density of H^- ions current on the anode are $2 \div 3$ times smaller than those observed in the experiment. As it was mentioned in [11], it is due to the following: at first, square of plasma emission surface near the slit is significantly larger than square of slit surface; at second, drift approximation used in the calculations at analysis of H^- ions motion is not sufficiently applicable to the experimental conditions. 2. In both regimes (with and without cesium) the calculation can give only saturation of H^- ions current density with pressure growth (see Fig. 6), and can not explain in any way decrease of j^- with pressure growth observed experimentally at high values of p (see Figs. 4,5). This mismatch is easily explained by the fact, that at $p \geq 0.1$ Torr the process of negative ions recharging at gas stream, outgoing through emission slit, becomes essential. Estimations show, that at high pressure this process can lead to death up to 30 % of total amount of H^- ions, outgoing through the slit, and this fact is completely sufficient for appearance of maximum at experimental dependence $j^-(p)$.

Thus, on a basis of experimental and calculated results it can be stated for sure, that cesium in a volume of the source under study can not lead to the increase of current of H^- ions. Observed growth of this current with cesium introduction is due to conversion of hydrogen atoms at discharge anode surface, covered by cesium. In other words, adding of cesium results in transformation of the source of H^- ions of volume kind to the source of surface-plasma kind.

SOURCE OF HYDROGEN NEGATIVE IONS IN PULSED-PERIODICAL REGIME WITHOUT CESIUM

Since temporal evolution of each process of originating and death of the particles after initiating the discharge possesses its specific features, the best parameters of volume source might expected to be achieved in pulsed-periodical operation regime of the source. Principal considerations allowing to expect higher efficiency in this regime consist in the following.

Due to slow process of H regeneration on chamber walls, the longest time is spent for relaxation of atomic hydrogen concentration N_H to its stationary value. Since atomic hydrogen plays important role in taking down vibration excitation of the molecules, concentration of H^- ions can exceed its stationary value at initial stage of the pulse. For checking this assumption numeric simulation of kinetic processes in pulsed-periodical operation regime of the discharge is performed. The calculations are accomplished in a wide range of discharge current, frequency and duration of the pulse with presumably rectangular shape.

Fig. 9 exhibits temporal evolution of concentrations of H^- (curves 1 - 3), H (multiplied by 10^{-4}, curve 4), $H_2(v=9)$ (curve 5) and electron temperature T_e (curve 6) for single pulse with duration of $2.5 \cdot 10^{-5}$ s. Two maxima are observed in curve $H^-(t)$. For elucidation of the mechanisms resulting in appearance of these

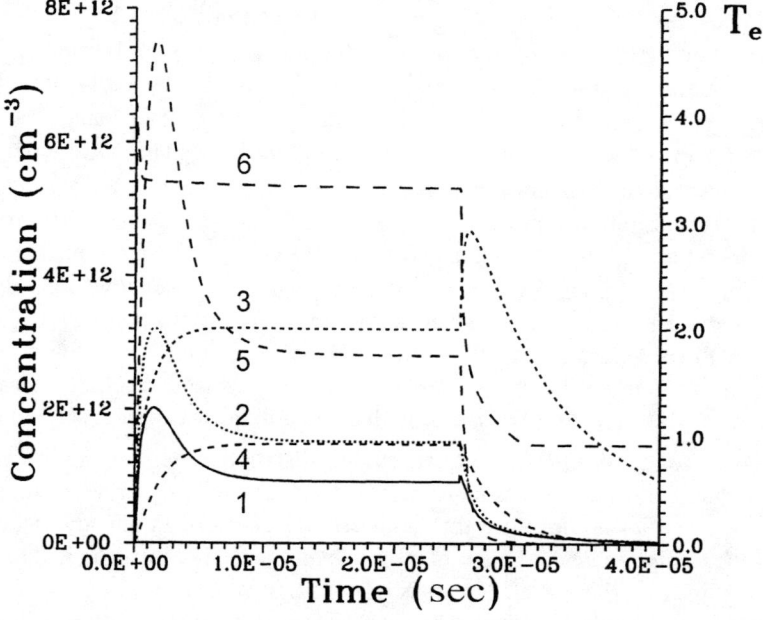

FIGURE 9. Temporal evolutions of concentrations of H^- (curves 1 - 3), H (multiplied by 10^{-4}, curve 4), $H_2(v=9)$ (curve 5) and electron temperature T_e (curve 6) in plasma of the source.

maxima two more calculations are performed; the first does not take into account negative ions losses due to collisions with electrons (process 17), the second considers the coefficient of molecular hydrogen regeneration on the chamber walls to be artificially increased by a factor of 10^6. Temporal dependencies of H^- ions for two those concentrations are depicted by curves 2 and 3, respectively. One can see from comparison of curves 1 and 2, that missing of consideration of H^- ions losses due to collisions with electrons leads to disappearing of maximum at rear edge of the pulse, and to the increase of N_{H^-} during the pulse. Thus, appearance of maximum at rear edge of the pulse is due to sharp cooling of the electrons (curve 6) with smoother changing of concentrations of electrons and excited molecules. Increase of negative ions density in after-discharge stage was observed in experimental proceeding [10]. As it follows from comparison of curves 1 and 3, increase of H_2 regeneration rate (that is, decrease of atomic hydrogen concentration) leads to disappearing of the second maximum and to the increase of H^- ions concentration during the pulse. It can thus be stated, that appearance of the maximum behind a front of the pulse is due to slower growth of hydrogen atoms concentration (curve 4), than that of plasma and excited molecules densities. One can see from Fig. 9, that at pulse discharge current $I_d = 25$ A maximum of H^- ions density is observed at delay $\tau_H = 2 \cdot 10^{-6}$ s after initiation of the discharge. This time by the order of magnitude is close or shorter than specific time of hydrogen atoms concentration coming to stationary level. Plasma decay time τ_d after switching off the pulse comprises $\approx 3 \cdot 10^{-6}$ s. Choice of duration Δt and period T of the pulses is determined by two these specific times. From one hand, relations $\Delta t \ll \tau_H$ and $\Delta t / T \ll 1$ must be obeyed to provide the amount of atomic hydrogen being small enough; from the other hand, $T \ll \tau_d$ is required for the plasma to have not enough time to decay. Thus, $\Delta t = 10^{-7}$ s and $T \approx 10^{-6}$ s are taken for the calculation.

Fig. 10 exhibits dynamics of variations of H^-, $H_2(v = 9)$, H concentrations and temperature T_e during single period in pulsed-periodical regime at quasi-equilibrium stage (time of coming to this stage is $\approx \tau_H$). Negative ions density decreases during current pulse, which is due to high electron temperature at this time (see Fig. 10) and intensive death of H^- in reaction 17. On the contrary, growth of H^- ions concentration after pulse of the current, despite decrease of N_{H_2} and n_e (n_e is electrons density), is due to cooling of electrons. It should be noted, that average electron temperature both in stationary regime, and in pulsed one, is practically the same at equal energy deposits. In stationary discharge H^- and $H_2(v)$ densities are lower at the same average current, as already mentioned above, due to higher concentration of atomic hydrogen. This situation is illustrated clearly in Fig. 11, where average values of densities of various components of the mixture and plasma density are shown in arbitrary units for pulsed-periodical and stationary regimes.

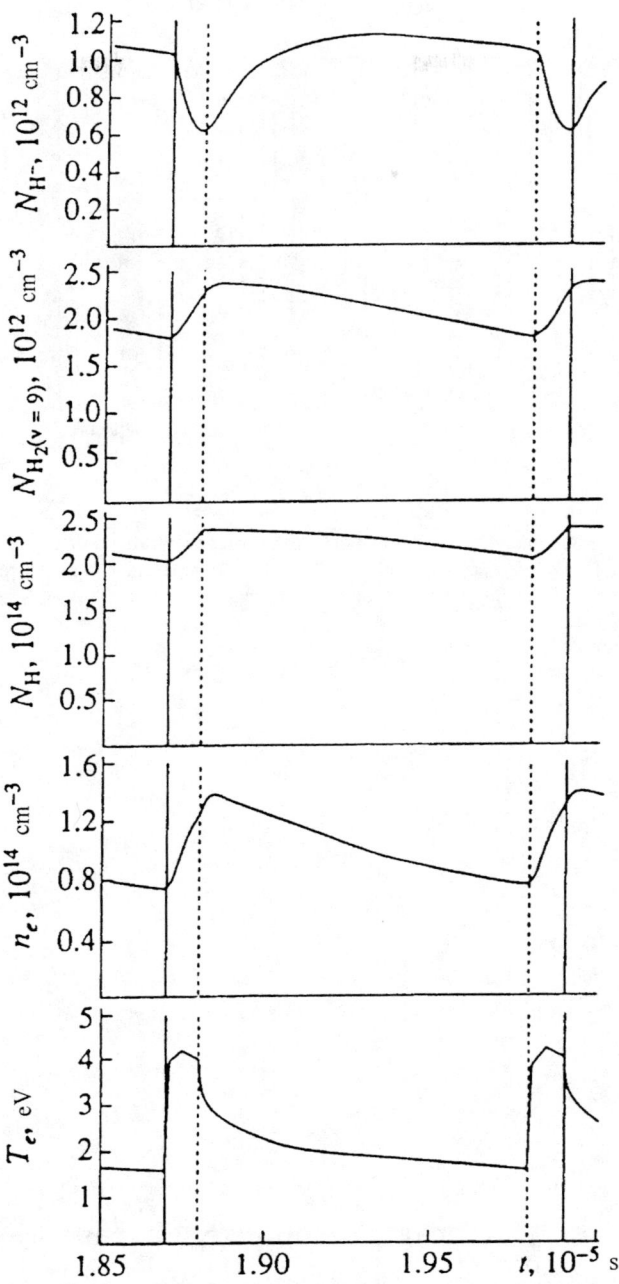

FIGURE 10. Dynamics of variations of H⁻, $H_2(v=9)$, H and T_e during single period.

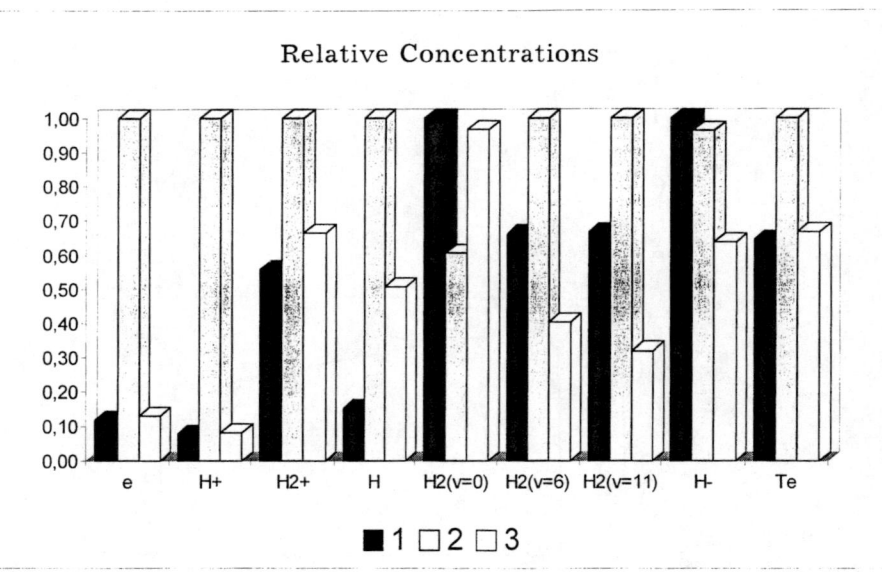

FIGURE 11. Comparative histogram of time averaged concentrations of the plasma components in the source at 120 V discharge voltage: 1 - pulsed regime, discharge current (pulse) 25 A, average current 2.5 A; 2 - stationary regime, discharge current 25 A; 3 - stationary regime, discharge current 2.5 A.

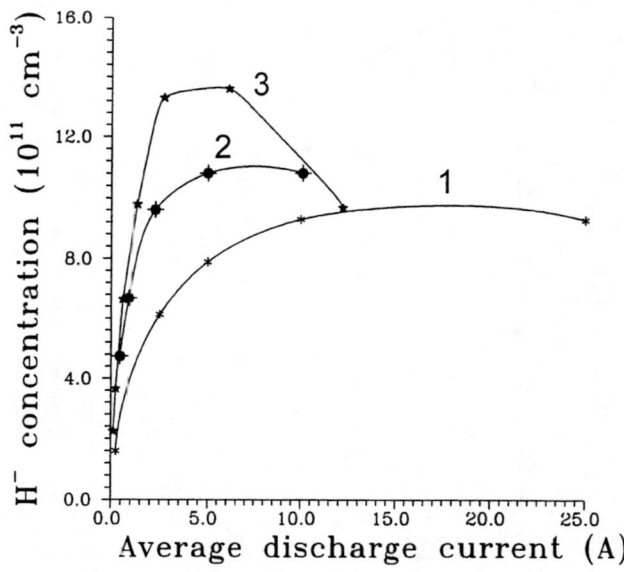

FIGURE 12. Dependence of H^- ions concentration in the source plasma on average discharge current at $U_d = 120$ V. 1 - stationary regime; 2 - discharge pulse duration 0.1 μs, pause duration 1 μs; 3 - discharge pulse duration 0.1 μs, pause duration 4 μs.

Concentration of H⁻ depends non-monotonously on average discharge current (Fig. 12); at big energy deposits the differences between pulsed-periodical and stationary regimes disappear, because H density in the discharge reaches its limit value $N_H \approx N_{H_2}$ during current pulse time. Analogous situation takes place at small energy deposits, however, for the reason of low H concentration and absence of process 8 contribution both in stationary, and in pulsed regimes.

CONCLUSION

On the basis of theoretical and experimental results, obtained in the proceeding, it is possible to conclude the following.

1. Adding of cesium to initial plasma source of H⁻ ions can result in essential increase of current density and gas economy of the source. This effect is due to the principal mechanism of H⁻ ions formation becoming the recharging of positive ions and fast atoms at a border of emission slit, that is, cesium adding transforms the plasma source into surface-plasma one with anode generation of the ions.

2. Transfer from stationary operation regime to pulsed-periodical one can lead to plasma source efficiency growth. It is due, first of all, to possibility of decreasing average concentration of hydrogen atoms, which determine death of H⁻ ions at large specific power in the discharge.

REFERENCES

1. Leung K.N., Ehlers K.W., Bacal M. *Rev.Sci.Instr.* **54**, 56 (1985).
2. Okumura Y., Hanada M., Inoue T. et al. // *Production and Neutralization of Negative Ions and Beams. Fifth Intern. Symp.* Brookhaven, N.Y. 1990, p. 169.
3. Walters S.K., Leung K.N., Hunkel W.B. *J.Appl.Phys.* **64**, 3424 (1988).
4. Bel'chenko Yu.I., Dimov G.I., DudnikovV.G. // *Preprint No 39-73.* Novosibirsk, Inst. of Nuclear Physics, Siberian Branch Acad. Sci. USSR, 1973.
5. Shkarovskii I., Johnston T., Baginskii M. // *Kinetics of the Plasma Particles.* Moscow, AtomPress, 1969.
6. Golovinsky P.M., Goretsky V.P., Mosijuk A.N. et al. // *Production and Neutralization of Negative Ions and Beams. Fifth Intern. Symp.* Brookhaven, N.Y. 1990, p. 341.
7. Goretsky V.P., Ryabtsev A.V., Soloshenko I.A. et al. *Zh.Tech.Fiz.* **63**, 46 (1993).
8. Golovinsky P.M., Goretsky V.P., Ryabtsev A.V.et al. *Zh.Tech.Fiz.* **61**, 46 (1991).
9. Seidl M., Cui H.L., Isenberg I.O. et al. // *Production and Neutralization of Negative Ions and Beams. Sixth Intern. Symp.* Upton, N.Y. 1992, p. 25.
10. Hopkins M.B., Bacal M., Graham G. *J.Appl.Phys.* **70**, 2009 (1991).
11. Goretsky V.P., Ryabtsev A.V., Soloshenko I.A., Tarasenko A.F. and Schedrin A.I. *Rev.Sci.Instr.* **67**, 67 (1996).

Negative ion beams from compact surface plasma sources and their merits for new applications

S. K. Guharay

Institute for Plasma Research
University of Maryland
College Park, Maryland 20742

Abstract. In recent times, special merits of negative ion beams have been recognized in the context of novel materials science applications, and there is a plenty of room for further growth of the field. Various kinds of negative ion sources are currently available. Compact surface-plasma type sources, namely, Penning, planotron and semi-planotron sources, are discussed with special emphasis on the merits of these sources for applications in the area of microlithography, ion microscopy, and surface modifications. The potential of these sources for high-current applications, such as spallation neutron source, is also discussed.

INTRODUCTION

Experiments on the production of intense negative ion beams started around the early seventies due to the demand in the magnetic fusion community for high energy neutral beams (>100 keV) and in the high energy accelerator community for intense, high-brightness H⁻ beams (beam current of ~ several tens of mA, and normalized beam brightness of ~ 10^{12} A/(m-rad)^2). Since then, many laboratories around the world have been engaged in this area of research, and the basic understanding and the technology have advanced to a matured level (1-8). The direction of research can be followed from the conference proceedings series of Production and Neutralization of Negative Ions and Beams. The boundary of negative ion beam applications has been steadily expanding. Recent pioneering experiments on implantation of negative ions in insulating and powdery materials by the Kyoto group demonstrated distinctive merits of negative ion beams in materials science applications (9). While irradiating a surface with ion beams, especially an insulated surface, secondary electrons are emitted and they contribute to charge-up of the surface. This often causes serious problems in many critical technologies, and complex charge-neutralization schemes are required to maintain the striking surface close to charge-balanced condition. In comparison with positive ions, the charge-up problem is much less serious for negative ions. For beams with a typical energy of ≥10 keV, the charge-up voltage is only ~ a few volts for negative ions, compared to ~ full beam voltage for positive ions. Furthermore, the low electron affinity of negative ions (~ 1 eV) can be very effective in applications on crystallinity control and development of new materials (10). What is needed to further push the technologies is the development of intense, bright sources, especially for heavy negative ions, namely, O⁻ and B⁻. The sputter-type sources, which are used currently (11-14), have rather large gap, ~ several cm, between the emitter surface and the extraction aperture, and this limits the beam intensity as well as the beam brightness. It will be worthwhile to develop compact surface plasma sources for the production of heavy negative ion beams and compare the performance with the existing sputter sources.

In this article, negative ion beams from surface plasma sources and their relevance to several new applications are discussed. High quality negative ion beams can be very attractive for ion projection lithography (IPL), and analytical instruments including Particle Induced X-ray Emission (PIXE) and Secondary Ion Mass Spectrometry (SIMS). The major ion source issues in these applications are the requirements for small virtual source (≤ 1 μm), high beam intensity ($\geq$ several tens of mA/sr) and low energy spread (≤ 2 eV). The special needs for IPL are described in a review article by Melngailis et al (15). It is projected to achieve sub-0.13 μm design rules with throughput of 75, 300 mm wafers per hour. A beam current of ~90 μA within ±3° will be required from the source. For 0.13 μm design rules, the critical dimensions (CD) budget limits the axial energy spread at the ion source to less than 2 eV. The demand for the above beam intensity and the energy spread is beyond the present state-of-the art of the ion sources. In the frontier of the analytical instruments, namely, PIXE and SIMS, there exists a need for improved ion sources to augment the resolution and throughput (16)--- lighter ions, namely, hydrogen or helium ions, are desired for PIXE, while oxygen ions are normally used in SIMS. High-quality ion beams, especially negative ions, will be useful for surface modifications, for example, local treatment of polymers and ceramics for improved material characteristics and for adhesion to other materials. Finally, the need for high current, high brightness H⁻ sources exists for advanced accelerator programs, namely, pulsed spallation neutron source projects (17).

SURFACE PLASMA SOURCES AND HIGH-BRIGHTNESS NEGATIVE ION BEAMS

The discovery of the catalytic role of cesium in lowering the work function of cathode surfaces and thus enhancing the H⁻ yield was the major triumph in the technical run of the surface plasma source activities. A major development in this area occurred at the Budker Institute of Nuclear Physics (BINP) in Novosibirsk in connection with the program of the Moscow meson factory (18-22). H⁻ beams with emission current density of $j > 2 A/cm^2$, beam current of 100mA and perpendicular ion temperature at the emission surface of $T_\perp < 1$ eV were demonstrated in long (> 300 hours) operations at a duty factor of ~ 2%. The high level of success with Penning sources at BINP triggered intense effort elsewhere (23-25); the activities at Oak Ridge had, however, started independently. Important results on scalings of H⁻ beam intensity with discharge dimensions were reported by Smith et al. for Penning SPS sources (26). The scaling laws for high-density plasmas (27) were applicable in the case of Penning SPS sources. These results provide good guidance for designing a Penning SPS source. At Maryland, H⁻ beam characteristics have been studied using both Penning and magnetron SPS sources; the Penning SPS source was developed in collaboration with BINP. The results have been discussed in detail earlier (28,29). Some notable results are: (a) Negligibly small beam noise was achieved in both magnetron and Penning SPS sources cases by appropriately setting the source parameters, mainly by adjusting the gas pressure pulse and the magnetic field. (b) H⁻ beams from the Penning source are brighter, by more than an order of magnitude, than the magnetron case -- the normalized beam brightness for almost the full beam in the Penning case is $\sim 2 \times 10^{12}$ A/(m-rad)^2.

In order to demonstrate the merits of negative ion beams from Penning SPS sources for IPL and microbeam applications, some illustrative experimental results on critical beam characteristics, namely, energy spread, beam intensity and focused beam parameters, are presented in the following sections.

Beam characteristics from a Penning SPS source for small current operation: merits for Focused Ion Beam applications

Virtual source size, energy spread and focused beam spot size

The following results were taken at the Maryland negative ion beam test stand. The Penning SPS source at Maryland runs at a pulse length of ~1ms and repetition rate of 6-10 Hz; the electrodes are cooled using forced air. In typical operations, the discharge voltage and discharge current are ~ 250 V and 40 A, respectively, and the H⁻ current through an emission aperture of 0.6 mm is ≥1 mA at 10 kV. Details about the source and its operating conditions have been reported earlier (28). Given the beam emittance ϵ, beam current I, and beam's geometrical parameters at certain locations, namely, beam size R and slope of the beam envelope R', the virtual source size can be estimated by solving the well-known envelope equation

$$R'' - K/R - \epsilon^2/R^3 = 0.$$

The beam waist is a measure of the virtual source size. Here, $K = I/I_0 \beta^3 \gamma^3$ is the generalized beam perveance, I_0 is the characteristic beam current ($\sim mc^2/30q$), $\beta = v/c$, and $\gamma = 1/(1-\beta^2)^{1/2}$. From measurements of the beam envelope and beam emittance, the radius of the virtual source is estimated to be ~ 2-5 μm. Since the beam current through the emission aperture of diameter = 0.6 mm is typically more than 1 mA, there is ample room for reducing the source aperture and extracting enough beam current for IPL and other microbeam applications. It will be important to examine the effect of reducing the emission aperture on the virtual source size and other beam characteristics. A critical technical issue will be to build the emission aperture as small as possible.

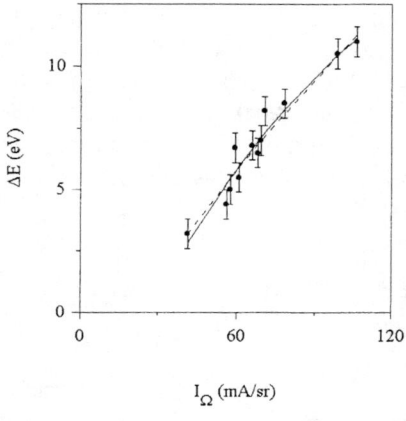

Figure 1. Energy spread of H⁻ beams versus angular intensity for $V_b = 5$ kV. Measured values (dot), polynomial fitting (solid line), and curve for $\Delta E \propto I_\Omega^{2/3}$ (dashed line).

In Fig.1 is shown the variation of energy spread of H⁻ beams with the angular beam intensities over a range of 40 to 100 mA/sr. The error bars represent uncertainties in polynomial fitting of the raw experimental data. It is noted that the dashed curve representing $\Delta E \propto I_\Omega^{2/3}$ is close to the solid line which is obtained by regression analysis of the energy spread data. The minimum energy spread ΔE (FWHM) of ≤ 2.5 eV in the SPS source has been obtained for angular beam intensity I_Ω of about 40 mA/sr. This value of ΔE is close to the resolution of the energy analyzer. A similar dependence, $\Delta E \propto I^{0.6}$, was reported earlier in the case of LMIS sources (30). The angular beam intensities in the above cases are about three orders of magnitude higher than that for LMISs, but the virtual source size here is much larger, ~ 2-5 μm versus ~ 50 nm for LMIS. Of noteworthy, the virtual source size of ~ 2-5 μm, beam intensity of ≥40 mA/sr and energy spread of ≤2.5 eV seem to be highly attractive in view of the demands for IPL. The parameter space of the above measurements needs to be extended toward lower angular intensities, and the limiting conditions for stable source operation must be determined. This will allow to make a good comparison of the beam from SPS sources with that of LMIS sources.

Preliminary experiments have been performed to focus the core H⁻ beam from the Penning SPS source and study the characteristics of microbeams. Using a single focusing lens, a target current density of about 5 mA/cm² over a spot size of ≤ 9 μm is obtained. If the beam from the Penning SPS source is further demagnified, at least by a factor of 10, using another lens, the target spot size will approach ≤1 μm and the current density will be ~0.5 A/cm². Figure 2 shows a comparison of the focused ion beam current density versus spectral brightness for beams from different sources. Since the spectral brightness, defined as B/ΔE, is commonly used

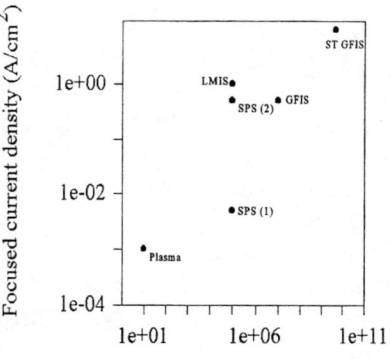

Figure 2. Focused beam current density versus spectral brightness for various sources. The labels represent: existing plasma sources (Plasma), liquid metal ion source (LMIS), gas field ionization source (GFIS), and supertip GFIS (ST GFIS). SPS Penning source results using a single focusing lens is shown by SPS(1), and expected results with a two lens system is marked as SPS(2).

in evaluating the beam qualities for FIB applications, this parameter has been used in the comparison. The focused beam results for sources other than the SPS source follow the article of Wilbertz, et al (31). It is noteworthy that the expected focused current density with the Penning SPS source can approach the LMIS results if the focusing system consists of two lenses. An experimental demonstration of such beam spots will generate strong interests for new applications of negative ion beams from SPS sources.

Penning SPS source for high current operation: merits for pulsed spallation neutron source applications

A comprehensive study of the performance of various sources has been made in view of choosing an appropriate ion source for the US pulsed spallation neutron source program, and the experience of seven facilities, namely, ISIS at Rutherford Appleton Laboratory, SINQ at Paul Scherrer Institute, ESS in Europe, ETA (BTA) in Japan, LANSCII in Los Alamos National Laboratory (LANL), IPNS upgrade in Argonne National Laboratory, PSNS proposal at Brookhaven and AUSTRON in Austria, was discussed (17). Penning and volume sources were recommended as potential candidates for the pulsed spallation neutron source program. The general requirements for pulsed spallation neutron source are: H^- beam current ≥ 100 mA, emittance (90%) ~ 0.1 mm-mrad, duty factor ~ 1% (6% for ~35 mA beam current operation), high reliability, and ~ 500 hours of operation. The BINP/LANL/ISIS experience on Penning SPS sources is very encouraging. Some preliminary studies with magnetron and Penning sources have been reported by Guharay and Reiser (32). The space-charge limited H^- beam current is noted to increase proportionately with the discharge current. It is estimated that the Penning SPS source at Maryland can deliver a beam current of ≥ 40 mA if the emission aperture is increased from 0.6 mm, as currently used, to ~2.4 mm. The Penning source has been running reliably over more than 500 hours. It would be important to compare the scalings of the beam characteristics and source parameters for Penning, magnetron and semiplanotron geometry. The Penning SPS source has certainly a strong merit to deliver intense, high-brightness beams. However, one of the key issues in this high duty factor and high current application is the long reliability of the source.

CONCLUSIONS

Negative ion beams from SPS sources have strong merits for many important applications. Further work on SPS sources should be continued and the following issues need to be investigated: (a) cw operation for low beam current (<1mA) and detailed studies of microbeam characteristics, (b) operation at a duty factor of ~6% for beam current of ~40 mA and studies of source reliability over long time, (c) studies of beam characteristics for heavier ions, namely, O^- and B^-, and (d) studies of limits on virtual source size and energy spread.

ACKNOWLEDGMENTS

The author acknowledges valuable discussions with J. Orloff, J. Melngailis, M. Reiser, V. Dudnikov, G. Derevyankin and J. Alessi. Thanks are due to E. Sokolovsky for his participation in Maryland experiments reported in this article. The research at Maryland is supported by the Department of Defense Microelectronics Research Laboratory Grant No. MDA904-94-H-2039.

REFERENCES

1. K. Prelec and Th. Sluyters, Rev. Sci. Instrum. **44**, 1451 (1973).
2. J. G. Alessi, AIP Conf. Proc. No. 253 (American Institute of Physics, N.Y., 1992), p. 193, and references therein.
3. C.W. Schmidt, Proc. Linear Accelerator Conf., 1990, p. 259, and references therein.
4. Yu. Belchenko, Rev. Sci. Instrum. **64**, 1385 (1993).
5. J.R. Hiskes and A.K. Karo, J. Vac. Sci Technol. A**2**, 670 (1984).
6. R. McAdams, A.J.T. Holmes, M.P.S. Nightingale, L.M. Lea, M.D. Hinton, A.F. Newman and T.S. Green, AIP Conf. Proc. No. 158 (American Institute of Physics, N.Y., 1987), p. 298.
7. K.N. Leung, O.A. Anderson, C.F. Chan, W.S. Cooper, G.J. DeVries, C.A. Hauck, W.B. Kunkel, J.W. Kwan, A.F. Lietzke, P. Purgalis and R.P. Wells, Rev. Sci. Instrum. **61**, 2378 (1990).
8. C.F.A. vanOs, K.N. Leung and W.B. Kunkel, Appl. Phys. Lett. **57**, 861 (1990).
9. J. Ishikawa, Rev. Sci. Instrum. **65**, 1290 (1994).
10. J. Ishikawa, Rev. Sci. Instrum. **67**, 1410 (1996); Surface and Coatings Technol. **65**, 64 (1994).
11. T. Takagi and J. Ishikawa, Rev. Sci. Instrum. **61**, 517 (1990).
12. N. Kishimoto (private communication).
13. G. D. Alton, Nucl. Instrum. & Meth. in Phys. Res. **B37/38**, 45 (1989).
14. A. Takagi and Y. Mori, Rev. Sci. Instrum. **65**, 1223 (1994).
15. J. Melngailis, A.A. Mondelli, I.L. Berry and R. Mohondro, J. Vac. Sci. Technol. B (submitted).
16. G. J. F. Legge, G. R. Moloney, R. A. Colman, and G. L. Allan, Rev. Sci. Instrum. **67**, 909 (1996).
17. J. Alonso, Rev. Sci. Instrum. **67**, 1308 (1996); and Proceedings of the Workshop on Ion Source Issues Relevant to a Pulsed Spallation Neutron Source, ed. L. Schroeder, K.-N. Leung and J. Alonso, Lawrence Berkeley Laboratory, CA, October 24-26, 1994.
18. G. I. Dimov, G. E. Derevyankin, V. G. Dudnikov, IEEE Trans. Nucl. Sci. **NS-24**, 1545 (1977).
19. V. G. Dudnikov, in Proc. 2nd Int'l Symp. on Production and Neutralization of Negative Ions and Beams, Brookhaven Nat'l Lab. Rept. BNL51304, 1980, p.387.
20. G. E. Derevyankin and V. G. Dudnikov, Instruments and Experimental Techniques, p. 523, 1987.
21. V. G. Dudnikov and G. E. Derevyankin, Proc.on Production and Neutralization of Negative Ions and Beams, AIP Conf. Proc., No.287, 239 (1994).
22. G.E. Derevyankin, Proc. Eur. Particle Accelerator Conf. 1992, p.1450.
23. H. V. Smith, N. M. Schnurr, D. H. Whitaker, and K. E. Kalash, IEEE Catalog No. 87CH2387-9, 301 (1987); H. V. Smith, P. Allison, C. Geisik, D. R. Schmitt, J. D. Schneider, and J. E. Stelzer, Rev. Sci. Instrum. **65**, 1176 (1994).
24. R. Sidlow and N. D. West, Proc. European Particle Accelerator Conf., p.1005, 1992.
25. W. K. Dagenhart, C. C. Tsai, W. L. Stirling, P. M. Ryan, D. E. Schechter, J. H. Whealton, and J. J. Donaghy, AIP Conf. Proc., no. 158, 366 (1987).
26. H.V. Smith, P. Allison and J.D. Sherman, AIP Conf. Proc. No. 287 (American Institute of Physics, N.Y., 1994) p. 271.
27. C.E. Muehe, J. Appl. Phys. **45**, 82 (1974).

28. S. K. Guharay, W. Wang, V. Dudnikov, M. Reiser, J. Orloff, and J. Melngailis, J. Vac. Sci. Technol. B **14**, 3907 (1996).
29. S. K. Guharay, E. Sokolovsky, M. Reiser, J. Orloff, and J. Melngailis, Microelectronic Engineering **35**, 435 (1997).
30. L. W. Swanson, G. A. Schwind, and A. E. Bell, J. Appl. Phys. **51**, 3453 (1980).
31. Ch. Wilbertz, Th. Miller and S. Kalbitzer, "Recent progress in gas field ion source technology", Proc. SPIE Conf. on Electron-beam, X-ray, and Ion Beam Technology: Submicrometer Lithographies, 1994.
32. S. K. Guharay and M. Reiser, Proc. Particle Accelerator Conf., Vancouver, May 12-16, 1997 (to be published).

Production and Control of ECR Plasmas for Negative Ion Sources

Osamu Fukumasa* and Naoki Tsuda

*Department of Electrical and Electronic Engineering,
Faculty of Engineering, Yamaguchi University, Ube 755, Japan*

Abstract. Production and control of electron cyclotron resonance (ECR) plasma for volume sources have been studied and preliminary results are presented. Effects of magnetic filter in ECR plasmas are discussed by comparing with ones in DC plasmas. Electron energy distribution in DC plasma is well controlled with the use of the magnetic filter. In ECR plasmas, however, electron cooling due to the magnetic filter in the extraction region is not so effective as ones in DC plasmas.

INTRODUCTION

In a design of a neutral beam injection (NBI) system for future large fusion experimental devices, such as ITER, a deuterium negative ion source with an energy of several hundreds keV is proposed to be used. In the present positive ion sources for the NBI system, the source plasma is generated by the arc discharge with a hot filament as a cathode, as well as in the negative ion sources under development. Lifetime of the ion source is limited to several hundred hours due to erosion and fatigue of the cathode-filaments and damage of the filaments by the anomalous arc discharge. Contamination of the plasma source by the evaporated filament materials could be also a problem in the cesium-seeded operation in the negative ion source. In the future fusion reactor the device materials are radio-activated due to irradiation of the neutrons yielded by the fusion reaction, and accessibility to the device is extremely limited. Thus, a long lifetime ion source is required for the future NBI system. A microwave-discharge ion source[1] and an rf-driven ion source[2,3] are promising for the long lifetime ion source, because these sources have no filaments.

In the present work, a new production method using permanent magnets is proposed[4] as one possibility for a large diameter high density uniform microwave plasma. The microwave power is launched into the circumference of a chamber by an annular slot antenna and the magnetic field of ring-typed permanent magnets. Advantages of using permanent magnets are that the magnetic field can be applied in the local region, where plasmas can be efficiently generated if the electron cyclotron resonance (ECR) condition is satisfied, and that an almost magnetic free condition can be achieved on the extraction grid.

In this paper, we report the structure of the microwave (ECR) -discharge negative ion source, the characteristics of the ECR plasma, the optimization of the plasma parameters in the extraction region, comparison of the ECR plasma with DC discharge plasma from a viewpoint of the magnetic filter effect and the results of the negative ion extraction.

CONCEPT OF AN ECR PLASMA SOURCE

A schematic diagram of the ECR hydrogen negative ion source is shown in Fig.1. The plasma source chamber (213 mm in diameter and 300 mm in length) made of stainless steel is a conventional multicusp volume source equipped with both a magnetic filter and a plasma grid. The microwave power (2.45 GHz, 100~600 W) is transferred to the annular slot antenna through a coaxial waveguide, and the microwave is launched from the antenna into the circumference of the cylindrical vacuum chamber (213 mm in diameter and 100 mm in length) through a fused silica plate window (30 mm in thickness). The width of the annular slot is 15 mm (the outer and the inner diameters are 200 and 170 mm, respectively). Two ring-typed samarium-cobalt permanent magnets are located just outside the chamber with facing north and south polarities to the radial direction at a separation of 4 mm, as shown in Fig.1. The magnetic field provided by this arrangement confines the plasma generated in the circumference of the chamber and diffuses it toward the central region. The permanent magnets used provided the resonance magnetic flux density 875 G for 2.45 GHz inside the chamber, i.e., the annular region 15 mm inside from the chamber wall.

Plasma parameters are measured by two Langmuir probes movable in axial and radial directions, respectively. The right end plate, i.e. the plasma grid, has a single hole (5 mm diameter) through which negative ions were extracted from the source. A Faraday cup with deflection magnet was used for measurement of the extracted H$^-$ current.

DC PLASMA OPERATION

At first, we show the characteristics of the source plasma produced by DC discharge. To this end, instead of th microwave launcher, the resonance magnet and the silica window, the filament flange is set in the source chamber shown in Fig.2.

H$^-$ ions are generated by the dissociative attachment[5,6] of slow plasma electrons ($T_e \sim$ 1eV) to highly vibrationally excited hydrogen molecules H$_2$(v"), where these H$_2$(v") are primarily produced by the collisional excitation of fast electrons with energies in excess of 20-30 eV. With the use of the magnetic filter, the electron energy distribution is optimized for the two-step process of H$^-$ formation[7].

Figure 3 shows the typical example of spatial distributions of the plasma parameters (electron density n_e and electron temperature T_e). According to the magnetic filter effect (M.F. is the position of the magnetic filter), plasma parameters in the downstream region (i.e. H$^-$ production region) is changed markedly compared with ones in the upstream region (i.e. source region or production region of highly

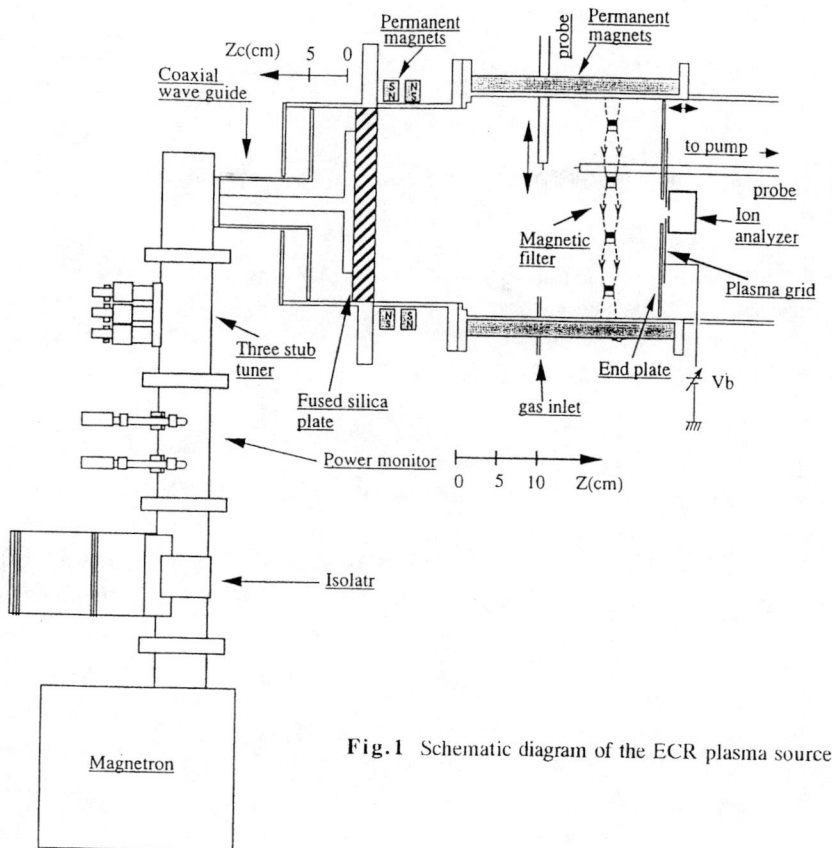

Fig.1 Schematic diagram of the ECR plasma source.

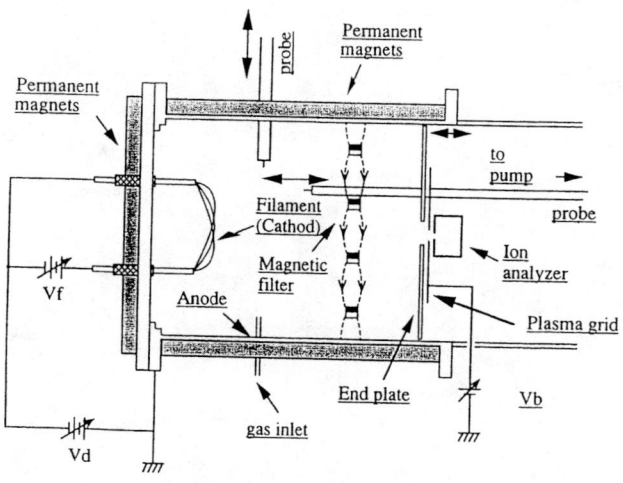

Fig.2 Schematic diagram of the DC plasma source.

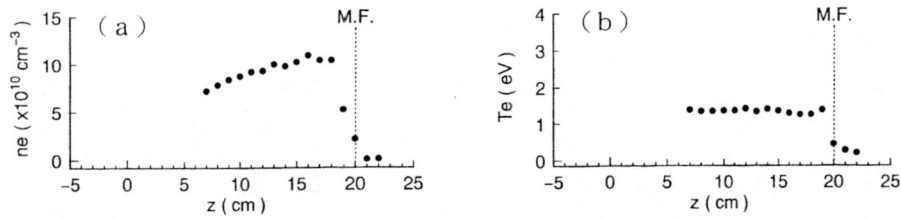

Fig.3 Axial distribution of plasma parameters in DC discharge plasmas : (a) electron density n_e, (b) electron temperature T_e. Experimental conditions are as follows : discharge voltage V_d = 80V, discharge current I_d = 3A and hydrogen gas pressure $p(H_2)$ = 2mTorr. M.F. is the position of the magnetic filter. Extraction grid is set at z = 22 cm.

vibrationally excited hydrogen molecules $H_2(v'')$).

With the use of magnetic filter, plasma parameters are well optimized for the two-step H⁻ volume production. Figure 4 shows the extracted H⁻ current as a function of discharge power, where discharge voltage V_d is kept constant at 80 V and discharge current I_d is varied. With increasing discharge power, extracted H⁻ current increases linearly.

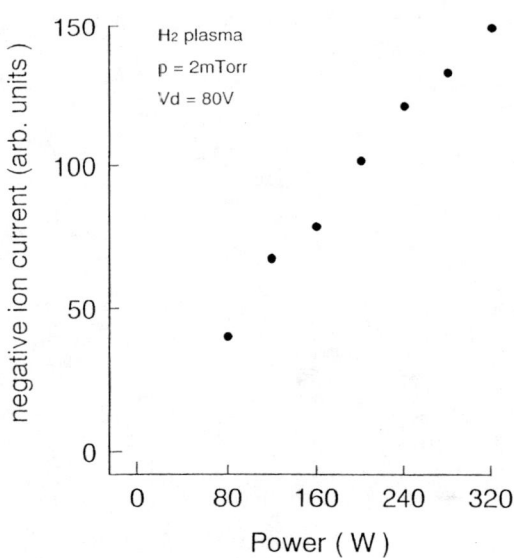

Fig.4 Extracted H⁻ current as a function of DC discharge power. Experimental conditions are as follows: V_d = 80V, $p(H_2)$ = 2mTorr, extraction voltage V_{ex} = 600V, and I_d is varied from 1 to 4A.

PRODUCTION AND CONTROL OF ECR PLASMAS

Experiments concerning ECR discharge is now begun. Preliminary results are reported here. At first, we show some characteristic features on production of ECR plasmas.

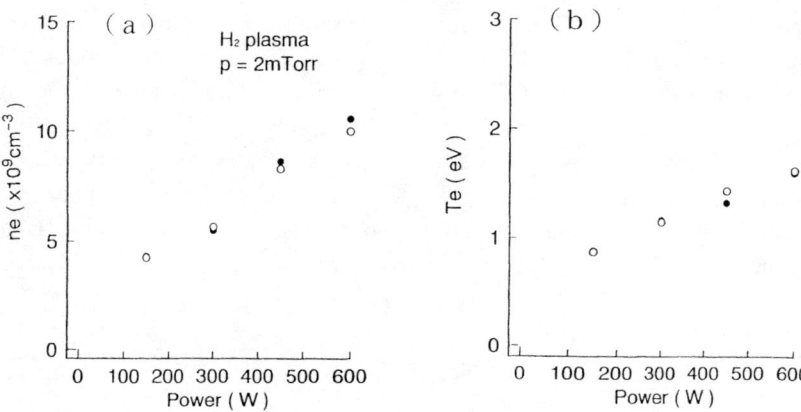

Fig.5 Plasma parameters in ECR plasmas as a function of microwave power P_μ : (a) n_e , (b) T_e. Experimental conditions are as follows : P_μ is estimated as the difference between incident power P_f and refrected power P_r, and $p(H_2) = 2$mTorr. Plasma parameters in the center are measured at the axial position $z = 11$ cm (i.e. in the source region).

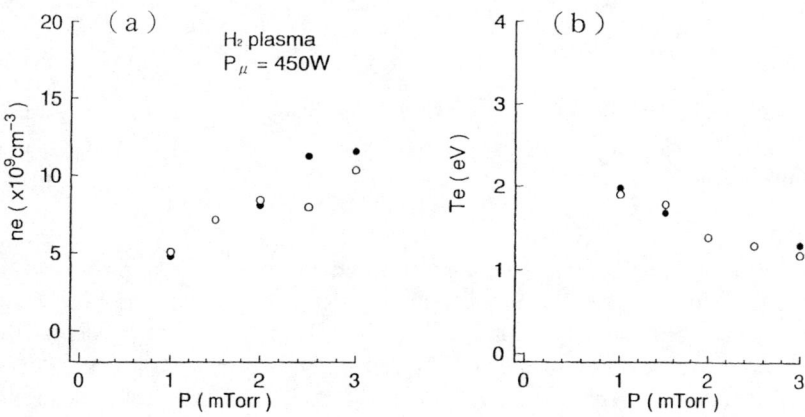

Fig.6 Plasma parameters in ECR plasmas as a function of hydrogen gas pressure : (a) n_e , (b) T_e. Experimental conditions are the same as ones in Fig.5 except that $P_\mu = 450$W.

Figure 5 shows plasma parameters (n_e and T_e) of ECR hydrogen plasmas as a function of microwave power $P\mu$. $P\mu$ is estimated by $(P_f - P_r)$, where P_f is the power of forward wave and P_r is the power of refrected wave. With increasing $P\mu$, both n_e and T_e increase. Figure 6 shows plasma parameters (n_e and T_e) as a function of hydrogen gas pressure p. With increasing p, n_e increases and T_e decreases. These features of plasma parameters shown in Figs.5 and 6, are nearly the same as ones in DC plasmas except that n_e in ECR plasmas is much lower than one in DC plasmas.

Next, we will discuss the axial distribution of plasma parameters (n_e and T_e), i.e. effect of magnetic filter. Figure 7 shows the typical example of axial distributions of n_e and T_e. For reference, n_e and T_e in DC plasmas are also plotted by dotted circles. The ring-typed permanent magnets for ECR are set at z = -5 cm. In this region, T_e is high and decreases markedly along the axial direction. In the source region (z = 0~20), T_e keeps nearly constant value. As is shown clearly, however, T_e does not decrease or change markedly across the magnetic filter as T_e does in DC plasmas. Namely, T_e in the extraction region is rather high and two-step process of H^- production is not optimized. Then the extracted H^- current also very small.

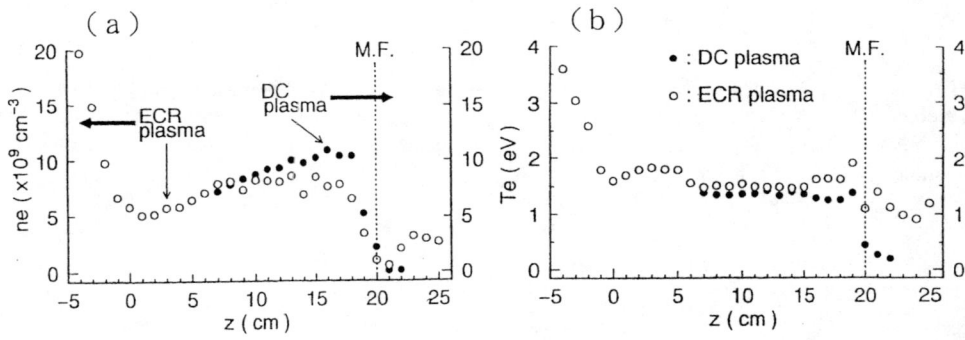

Fig.7 Axial distribution of plasma parameters in DC discharge plasma and ECR discharge plasma : (a) n_e, (b)T_e. Plasma parameters of DC plasma case is the same as ones in Fig.3. In ECR plasma case, $P\mu$ = 450W and $p(H_2)$ = 2mTorr.

The role of the magnetic filter is not well clarified. Previously, we proposed the following model[8,9] for the preferential reflection of high-energy electrons. Electrons entering the magnetic filter are trapped in filter field ***B***. Then, most electrons cross the magnetic filter due to ***E***×***B*** drift where ***E*** represents thermally excited low frequency electrostatic fluctuations. Therefore, ***E***×***B*** drift for electrons decreases with the increase of electron velocity (i.e., energy), since fluctuating electric fields are averaged over their finite Larmor radii. Namely, high-energy electrons diffuse at a lower rate than do low-energy electrons.

In the above model, fluctuating field E plays an important role and E in DC plasmas should be different from that in ECR plasmas. Then, it can be explained the difference of magnetic filter effect in controlling plasma parameters between DC plasmas and ECR plasmas. This point is now under study.

In summary, we have newly designed the ECR plasma source for NBI system. Characteristic features of the preliminary results of ECR plasmas are reported. Production of high density ECR plasma, optimization of plasma parameters for H^- volume production, clarifying the effect of magnetic filter and extraction of H^- ions are under study.

ACKNOWLEDGMENTS

The authors would like to thank Profs. T.Kuroda and O.Kaneko, the National Institute for Fusion Science (NIFS) of Japan, and Prof. H.Fujita, Saga University, for their discussion and encouragement.

This work was supported by a Grant-in-Aid for Scientific Research from the Japanese Ministry of Education, Science, Sports and Culture. This work was also carried out as a collaboration research of the NIFS.

REFERENCES

1) K. Hashimoto and S. Asano : Fusion Engineering and Design **26**, 495 (1995).
2) Y. Takeiri, T. Takanashi, O. Kaneko, Y. Oka, A. Ando, K. Tsumori and T. Kuroda : Fusion Engineering and Design **26**, 501 (1995).
3) T. Takanashi, Y. Takeiri, O. Kaneko, Y. Oka, K. Tsumori and T. Kuroda : Rev Sci. Instrum. **67**, 1024 (1996).
4) T. Ikushima, Y. Okuno and H. Fujita : Appl. Phys. Lett. **64**, 25 (1994).
5) O. Fukumasa : J. Phys. **D 22**, 1668 (1989).
6) O. Fukumasa : J. Appl. Phys. **71**, 3193 (1992).
7) J. R. Hiskes and A. M. Karo : J. Appl. Phys. **56**, 1927 (1984).
8) O. Fukumasa, H. Naitou and S. Sakiyama : Trans. IEE Jpn. **111A**, 1057 (1991).
9) O. Fukumasa, H. Naitou and S. Sakiyama : J. Appl. Phys. **74**, 848 (1993).

The DESY RF Driven H⁻ Volume Source

J. Peters

Deutsches Elektronen-Synchrotron DESY,
Notkestraße 85, 22607 Hamburg, Germany

Abstract. Since 1994 an RF driven H⁻ volume source has been operated in the DESY source test stand. In 1995 the source was operated successfully as injector for LINAC III and DESY III. The source is planned as a Cs free low emittance replacement of the magnetron. H⁻ sources for HERA have to run for long uninterrupted periods (300 days) with a low duty factor (0.01%) and a high reliability.
New Al_2O_3 ceramic antenna constructions are presented in order to improve the antenna lifetime. Recent measurements have demonstrated H⁻ currents of up to 80mA in our uncesiated volume source. This was possible due to a negative bias of the collar, a tantalum coating on the collar inside and an optimization of collar and extraction geometry.

INTRODUCTION

The original RF driven volume source was manufactured by AccSys (1) using plans from LBL (2). Since 1994 it was completely rebuilt at DESY. Fig.1 shows the redesigned bucket. Techniques developed at DESY made it possible to reach an H⁻ current of 80 mA without using Cs. For details of the source see (8). The results of the LINAC III –DESY III test are published (7).

A. Antenna design

Sputtering of antenna coating to the source bucket limits the lifetime of the antenna. Insulating coating building up on the bucket reduces the production of H_2^* on the walls. Al_2O_3 ceramic has a very low sputtering rate seven times lower than glass. Three different antenna designs with Al_2O_3 insulation are at present under test at DESY. A rectangular coiled antenna covered with pipes and bends. A circular coiled antenna which is covered by two shells connected to two pipes and an antenna situated at the wall of the source completely covered by L shaped Al_2O_3 ceramic (see Fig.1).

B. Collar shape and bias

A collar was first introduced into the source in 1989 for electron-suppression (3). At DESY the diameter of the collar was varied in the range between 5 mm and 250mm. The collar length was changed between 0 and 30mm. Beam current, beam size and beam divergence were measured. An insulated collar of 13mm diameter and 18mm length showed the best results. By using a negative bias for the plasma electrode an increase in H⁻ current of up to 5 % was reported (4). It was explained as a cleaning of the plasma electrode by positive ion bombardment. This effect was extensively studied at DESY using different collar sizes.

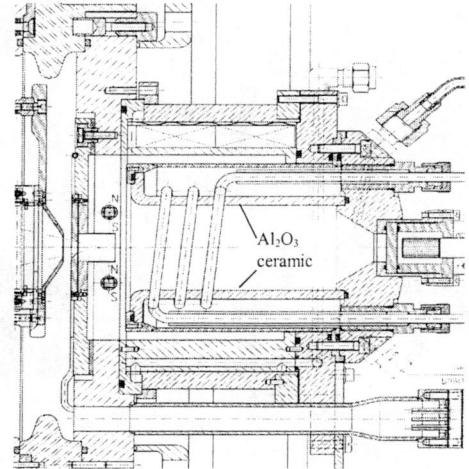

FIGURE 1. Volume source with Al_2O_3 antenna cover.

It turned out that the H⁻ current rises reproducibly up to 30 %. Fig.2 shows the power dependent effect for 35 kV extraction voltage and a plasma aperture of 6mm.

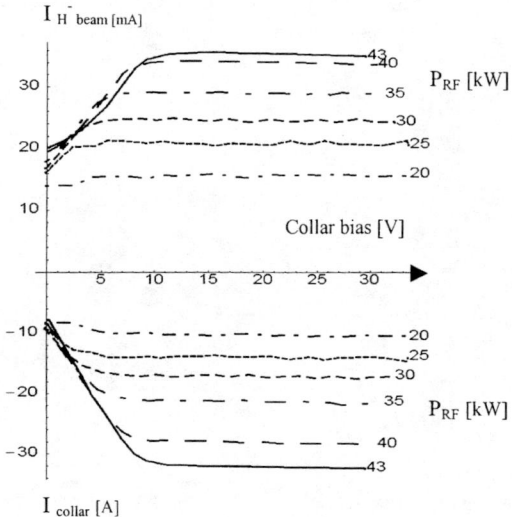

FIGURE 2. H⁻ beam and collar current as function of collar bias.

The upper curves show the dependence of the H⁻ current on the collar bias, the lower curves that of the bias current.

The reproducibility of the volume source is about 20% after putting it up to air. In order to discover where the increase in H⁻ current comes from, the collar and plasma plate were split and individually biased. With this technique it was possible to find out that the increase of H⁻current depends on the change of potential distribution and wall current on the inside collar surface. There is also a focussing effect due to the collar bias (see Fig.3). The σ values are given in arbitrary units.

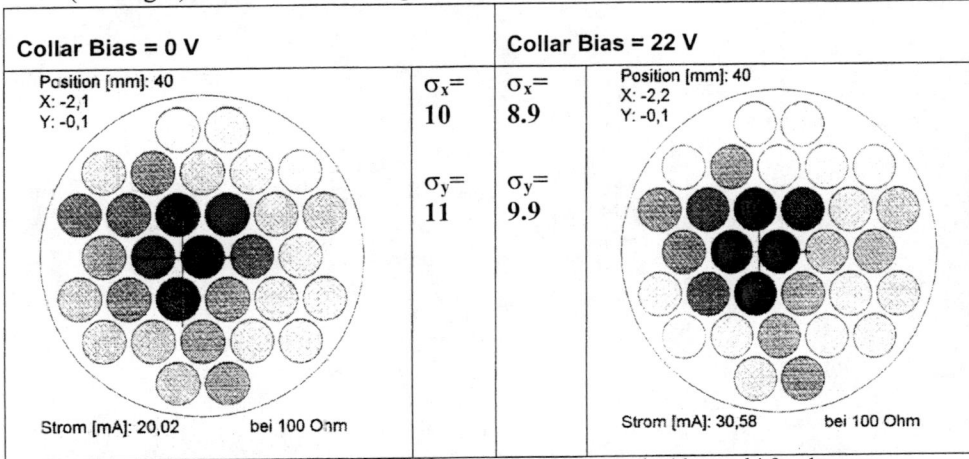

FIGURE 3. H⁻beam without and with a collar bias of 22V measured with a multi faraday cup.

C. Tantalum collar

In 1988 it was proved (5) that hydrogen atoms desorbed in tantalum can produce vibrational excited hydrogen atoms.

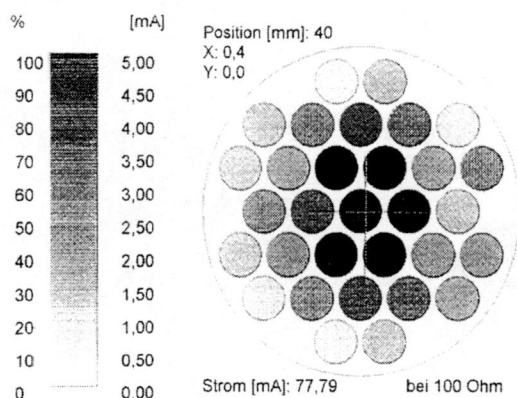

FIGURE 4. H⁻beam of the tantalized and biased source measured with a multi faraday cup.

H+H+ tatalum surface --> $H_2^*(v)$.
H⁻ ions are then produced in the volume reaction
$H_2^*(v)+e(1eV)->H^-+H$.

A microwave source experiment (6) demonstrated the production of 4 mA H⁻ ions due to sheets of tantalum. At DESY a sheet of tantalum was inserted into the steel collar of the source. The H⁻ current jumped at the source start-up from 20mA to 40mA. It then rose slowly with time an additional 20 %. It is obvious that vibrational excited hydrogen molecules were formed in the collar and produced H⁻ ions together with the low energy electrons present in this area. After the tantalum cylinder was removed the source started with about 30 mA and then rose to the old maximum current of 48 mA. The explanation is that a small layer of tantalum covered the whole source, which was subsequently distributed onto the collar surface. After optimizing collar size and bias an H⁻ current of 80 mA was reached (see Fig.4.).

CONCLUSION

A collar biased tantalum treated volume source has delivered 80mA, the highest H⁻ current measured to date in an uncesiated source. After improving the antenna lifetime it will be a strong candidate for accelerators.

ACKNOWLEDGMENTS

The author is grateful for the contribution of the following colleagues at DESY: N.Holtkamp, I.Hansen, H.Sahling and R.Subke. I wish to thank the technical groups at DESY for their support, the students T.Butschkat (FHH) and O.Krücken (FHW), T.Warner (FHW) for their programming and J.Maidment of DESY for helpful suggestions to wording of the report.

REFERENCES

1. AccSys,Pleasonton, California 94566
2. K.N.Leung, D.A Bachman, and D.S.McDonald, AIP Conf. Proceedings No. 287, pp 368-372.
3. K.N.Leung ,Rev.Sci.Instrum. 61 (3), March 1990.
4. private communication with K.Saadatmand and K.N.Leung.
5. R.I.Hall, I.Cadenz, M.Landau, F.Pichou and C.Schermann, Phys. Rev. Lett. 60, 337(1988).
6. D.Spence, K.R.Lykke and G.E.McMichael, Proceedings of the XVIII International Linear Accelerator Conference (August 1996).
7. J.Peters, Proceedings of the XVIII International Linear Accelerator Conference (August 1996).
8. J.Peters, DESY-HERA 98-01, under preparation.

NEGATIVE IONS ACCELERATION

Multi-stage, multi-aperture electrostatic accelerator for H⁻ ions

Kazuhiro Watanabe, Yukio Fujiwara, Masaya Hanada,
Takashi Inoue, Kenji Miyamoto, Naoki Miyamoto,
Yoshihiro Ohara, and Yoshikazu Okumura

Japan Atomic Energy Research Institute
801-1 Mukoyama, Naka-machi, Naka-gun, Ibaraki-ken 311-01 Japan

Abstract. A multi-stage, multi-aperture electrostatic accelerator for hydrogen negative ions has being developed to demonstrate 1 MeV, ampere-class H- beam acceleration. The accelerator consists of an extractor and a five stage accelerator supported and insulated by fiber reinforced plastic (FRP) cylinders. Forty nine apertures were drilled on each electrode. Diameter of the aperture is 14 mm in the extractor and 16 mm in the accelerator. Whole dimensions of the accelerator are 2 m in diameter and 2.5 m in height. The accelerator was conditioned up to a voltage of 920 kV without beam. A highest beam energy of 868 keV with an acceleration current of 19 mA for a duration of 1 s was obtained from nine apertures. An optimum beam perveance agreed well to the beam trajectory simulation.

INTRODUCTION

High energy and high power neutral beams are required for ITER (International Thermonuclear Experimental Reactor) to heat and sustain steady state fusion plasma. A powerful neutral beam injection system of 1 MeV, 50 MW from three modules is designed in the Engineering Design Activity [1]. To realize such system, demonstration of ampere class negative in beam acceleration up to the energy of 1 MeV is an essential milestone. For the high energy beam acceleration experiment, a prototype ion source/accelerator (MeV accelerator) has being tested in the MeV Test Facility (MTF)[2].

In the previous experiment, a high energy beam of 805 keV, 150 mA for 1 sec. was obtained and it was confirmed that there is no voltage degradation due to beam acceleration [3,4]. However, sustained voltage of the accelerator column did not exceed 820 kV even after a long time conditioning. Further, degradation of voltage holding was observed after many breakdowns. Finally the accelerator could hold the voltage of only 500 kV. Breakdown damages appeared at the inner surface of the fiber reinforce plastic insulators (FRP) which support the accelerator electrodes and vacuum. The reasons of many breakdowns and damage of the accelerator column were considered to be due to "glow-like discharges" during the high voltage applying and energy input at breakdown from a 1 MV, 1 A Cockcroft-Walton generator that has a big capacitor of 0.14 uF.

The accelerator was modified to increase gas pumping conductance to suppress pressure increase by out gassing during the conditioning to avoid the glow-like discharges. To detect the glow-like discharge and cut of the power supply, an interlock system of the intermediate grid current was installed to a 800 kV line. Further, energy suppression resistors of 400 kilo-ohm to 1 MV line, 300 kilo-ohm to the 800 and 600 kV lines, 200 kilo-ohm to the 400 and 200 kV lines were connected respectively. After these modifications, the conditioning of the accelerator was performed. Beam optics of the accelerator was investigated by using intermediate grid current monitor system and a video camera to observe beamlets in the beam line chamber.

In the present paper, configuration of the accelerator is described. The voltage holding characteristics of the accelerator and results of the beam optics study are also reported.

ACCELERATOR

Figure 1 shows a cross-sectional view of the accelerator. Dimensions of a semi-cylindrical multi-cusp plasma generator called KAMABOKO source is mounted on the accelerator. Produced negative ions are extracted by an extractor that consists of three grids called a plasma grid (PG), an extraction grid (EXG) and an electron suppression grid (ESG). ESG is electrically connected to EXG. The electrons are separated from the negative ions by the magnetic field generated by the permanent magnets embedded in EXG. The negative ions are scarcely deflected by the magnetic field and injected to the accelerator.

The accelerator consists of five grids forming five stage electrostatic acceleration gaps. Acceleration gap lengths are 104, 94, 87, 78, and 72 mm from the extractor side. The same voltage of DC 200 kV is supplied to each gap, and the electric field in these gaps become stronger in downstream of the accelerator. The H$^-$ ions are accelerated and converged by electric field lens at each stage.

These grids are supported by cylindrical insulators of fiber reinforced plastic (FRP) whose dimensions are 1.8 m in inner diameter and 33 cm in height. Electric field tends to concentrate at triple junction points of insulator, vacuum and metal. Stress shields are mounted inside of the accelerator to relax the electric field concentration. Shield rings are also mounted on the grid support to shield the insulator surface from charged particles and photons from the beam to prevent creepage breakdown.

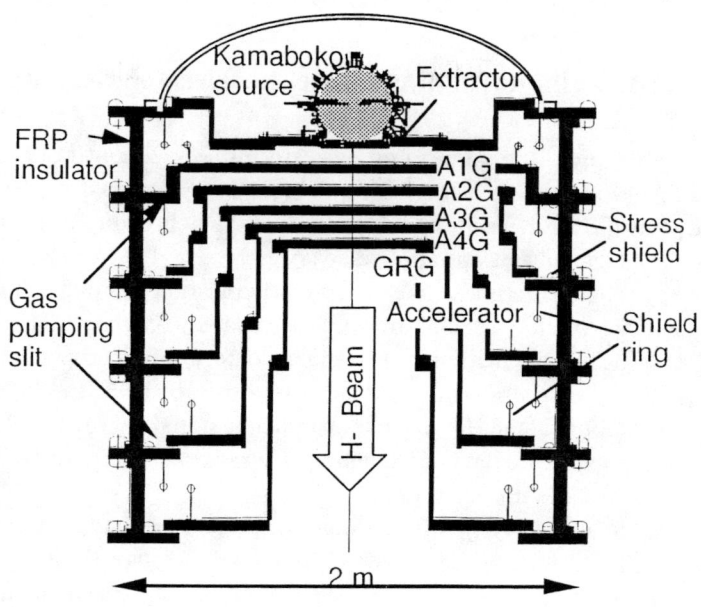

FIGURE 1. A cross-sectional view of the MeV accelerator that has five-stage electrostatic multi-aperture accelerator.

A typical result of the beam trajectory simulation is shown in Fig. 2. The H⁻ current density of around 24 mA/cm^2 at the plasma surface is an optimum current density that gives a small beam divergence. The extraction voltage of 7 kV and the acceleration voltage of 1000 kV were assumed.

*MeV FRP33cm VACC=1000KV VEXT=7.KV 24.MA/CM2 H- GAP (5.8/2/104/94/87/78/72)

```
Current Density    = 2.4000E+01 (mA/cm2)
Total Current      = 3.7166E-02 (A)
Perveance          = 3.6779E-11 (A/V**1.5)
Minimum Potential  = 0.0000E+00 (V) AT Z= 5.7912E-01 (m)
Divergence (RMS)   = 2.4213E-01 (Deg)
Electron Temperature = 0.0000E+00 (eV)
Ion Temperature    = 0.0000E+00 (eV)
```

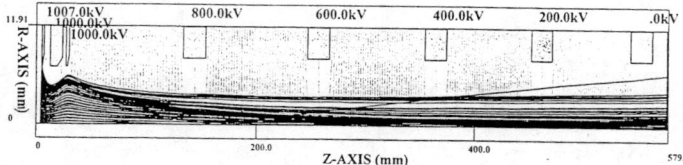

FIGURE 2. A typical result of beam trajectory simulation.

EXPERIMENTAL RESULTS AND DISCUSSION

I. Voltage Holding of the Accelerator Column

Conditioning of the accelerator was performed without beam acceleration. Applied high voltage was increased until breakdown occurred. Once breakdown occurred, the high voltage power supply was switched off. After one or two minutes, the high voltage was applied again.

The previous experiment, we observed degradation of voltage holding of the accelerator. One of the reasons of such degradation on the voltage holding is considered to be due to energy input to the insulator from the capacitor of the high voltage generator of the Cockcroft Walton power supply at the breakdowns. A capacitor of 0.14 uF in the HV generator has a large stored energy of 70 kJ at 1 MV. To dissipate the stored energy, high impedance resistors were connected between the power supply and the accelerator.

Figure 3 shows the result of conditioning operation with two cases of resistors. Withstand voltage of about 700 kV was obtained with using 40 kilo-ohm resistor. However, many breakdowns occurred around 600 kV and it was hard to increase higher voltage. After this operation, we changed the resistor from 40 kilo-ohm to 400 kilo-ohm. The withstand voltage could be increased to 800 kV by only two hour operation. Breakdown number was only 12 shots until 800 kV.

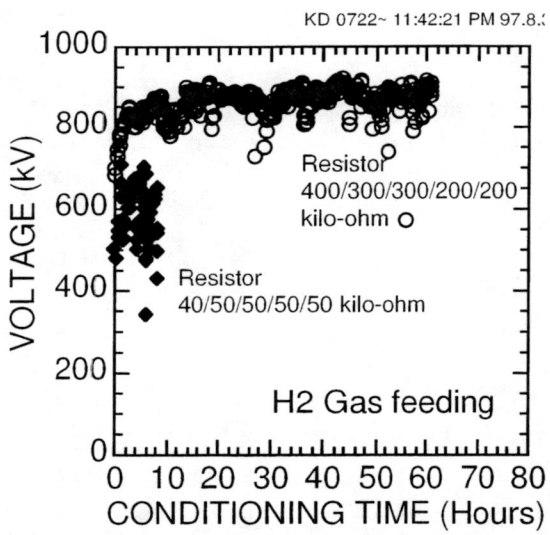

FIGURE 3. Conditioning of the prototype MeV accelerator.

During the conditioning, a large amount of out gassing phenomena were observed synchronized with a small current pulse discharge or long pulse discharge from several tens millisecond to several second,(i.e. glow-like discharge). Analysis of the gas species showed that a main component of the exhausted gas was hydrocarbon. This means that there are many micro-discharges and glow-like discharges originated at the triple junction. By continuing the conditioning, the highest voltage of 920 kV was sustained. Introduction of hydrogen gas from the plasma generator was effective to prevent breakdowns in the accelerator during the conditioning higher than 800 kV.

Dark current of the accelerator was measured with increasing applied voltage. The dark current as a function of the voltage is shown in Fig. 4. The current increased linearly to the applied voltage and was about 3 mA at 800 kV. The glow-like discharges and pulse discharges could be suppressed well by feeding hydrogen gas from the plasma generator. However the dark current did not change at all. Therefore, a greater part of the current is considered to be flown in the water choke system which supply the pure water to high voltage part.

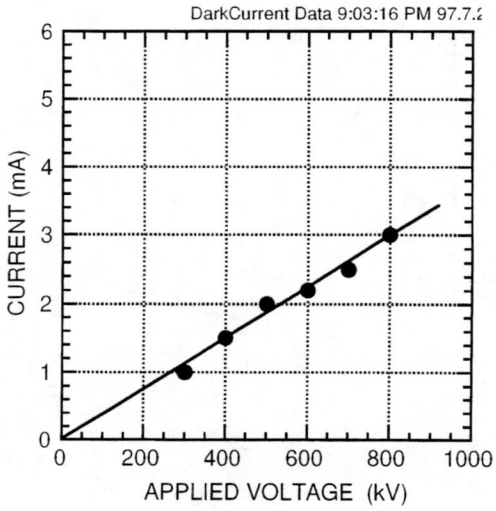

FIGURE 4. Dark current of the accelerator.

II. Beam Optic Study of the Five Stage Accelerator

Beam optics were experimentally investigated at a low energy of 400 keV. Since the ion source was operated without Cs injection, the current density could not be increased to satisfy a perveance required at a higher energy. Acceleration drain current of Iacc and, intermediate grid currents of IA2G, IA3G and IA4G as a function of the extraction voltage is shown in Fig. 5. A positive polarity of the intermediate grid currents shows that the negative particles flow into the grid, and negative polarity means the electron emission from the grid. By increasing the extraction voltage from 2.2 kV to 4 kV, the beam optics is changed and a maximum H⁻ beam current was obtained at the extraction voltage of about 2.8 kV. Other intermediate grid currents showed the minimum at the same point. A clearly separated beamlets were observed by using a video camera at the position of 2.2 m downstream from the grounded grid at this condition.

Taking the calorimetric error into account, and considering the negligible small intermediate grid currents at the optimum condition, it can be assumed that a greater part of the accelerated drain current is the negative ion current. A stripping loss of the H⁻ ions in the accelerator was estimated to be about 30 % for the 400 keV beam. The accelerated drain current of 55 mA is converted to an

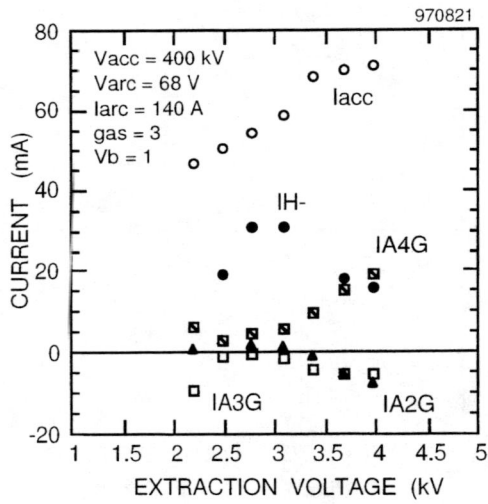

FIGURE 5. Accelerator currents as a function of the extraction voltage.

extracted current of 78.6 mA. This corresponds to a current density of 5.7 mA/cm^2 at plasma surface. From the beam trajectory simulation, an optimum current density of 6.1 mA/cm^2 can be evaluated for the 400 keV operation. This value agrees well to the current density at the plasma surface evaluated the experimental data. An optimum beam perveance of 2.25×10^{-11} A/cm^2/V$^{3/2}$ is estimated. It corresponds to a current density of 23 mA/cm^2 at 1 MV operation. A good beam optics will be obtained at high energy region with keeping the optimum perveance by injecting Cs vapor to the source.

Figure 6 shows the accelerator currents as a function of the acceleration voltage. The acceleration voltage of 400 kV was confirmed to be the optimum for the same condition of the arc power and the extraction voltage showed in Fig. 5. The beam optics seems not to be changed for the acceleration voltage variation of +/-10% from 400 kV. This characteristics satisfies the ripple design value of +/-5% for the ITER NBI acceleration power supply[5].

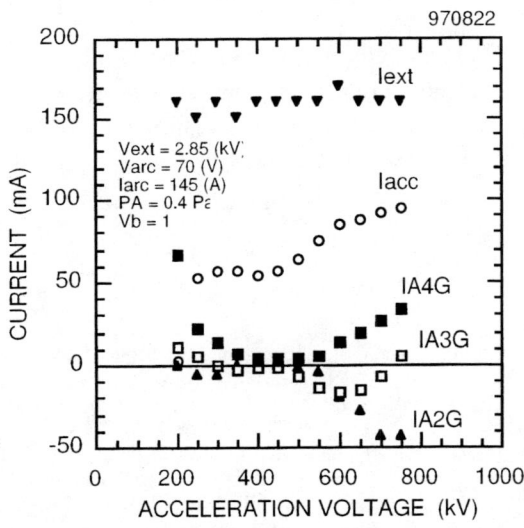

FIGURE 6. Accelerator currents as a function of the acceleration voltage.

III. High Energy Beam Acceleration

After the conditioning of the accelerator without beam, a high energy beam acceleration was performed. The highest energy beam of 868 keV with an accelerated drain current of 19 mA was obtained for 1 s. The accelerator was operated at very low perveance, because the plasma generator was operated without cesium. It is planed to optimize the perveance by increasing the H⁻ current density by seeding cesium to the plasma generator.

CONCLUSION

The MeV accelerator was conditioned up to 920 kV. The high energy beam of 868 keV with an accelerated drain current of 19 mA was obtained for 1 s. The beam optics study showed that the optimum perveance of the accelerator is 2.25×10^{-11} $A/cm^2/V^{3/2}$ and agreed well to the beam trajectory simulation.

ACKNOWLEDGMENT

The authors would like to express their gratitude to other members of the NBI Heating Laboratory for valuable discussion. They are also grateful to Dr. M. Ohta and Dr. H. Kishimoto for their continuous encouragement and support. They are grateful to Dr. R. Hemsworth, Dr. P.L.Mondino of JCT-Naka for their cooperation in ITER R&D.

REFERENCES

[1] Hemsworth, R.S., Feist, J.H., Hanada, M., Heinemann, B., Inoue, T., Kussel, E., Krylov, A., Lotte, P., Miyamoto, K., Murdoch, D., Nagase, A., Ohara, Y., Okumura, Y., Pamela, J., Panasenkov, A., Shibata, K., Tanii, M., and Watson, M., "*Neutral beams for ITER,*" Rev. Sci. Instrum. 67(3), 1120 (1996).
[2] Inoue, T., Hanada, M., Maeno, S., Miyamoto, K., Ohara, Y., Okumura, Y., and Watanabe, K., "*Design study of prototype accelerator and mev test facility for demonstration of 1 MeV, 1A negative ion beam production*", Japan Atomic Energy Research Institute Report JAERI-Tech 94-007(1994).
[3] Hanada, M., Inoue, T., Mizuno, M., Ohara, Y., Okumura, Y., Suzuki, Y., Tanaka, H., Tanaka, M., and Watanabe, K., "*Negative ion production in a large semicylindrical ion source*", Rev. Sci. Instrum. 64(4), 2699(1992).
[4] Inoue, T., Fujiwara, Y., Miyamoto, K., Miyamoto, N., Nagase, A., Ohara, Y., Okumura, Y., Watanabe, K., and Yokoyama, K., "*High energy acceleration of H- ion beam at MeV test facility*", 7th Int. Symp. On the Production and Neutralization of Negative Ions and Beams and 6th European Workshop on the Production and Application of Light Negative Ions, BNL,(1995).
[5] Watanabe, K., Higa, O., Kawashima, S., Ohara, Y., Okumura, Y., Ono, Y., Tanaka, M., and Yasutomi, S., "*Design of ITER NBI power supply system*", Japan Atomic Energy Research Institute Report JAERI-Tech 97-034(1997).

Results of the 1 MV SINGAP Experiment

C. Desgranges, J. Bucalossi, M. Fumelli, R. S. Hemsworth,
P. Massmann, J. Paméla, and A. Simonin

Association EURATOM - CEA
DRFC / STID, CE Cadarache,
13108 St Paul-lez-Durance, France

Abstract. Injection of intense neutral beams based on the neutralisation of negative deuterium ions up to energies of 1 MeV is an important option for plasma heating and non-inductive current drive in future thermonuclear fusion machines.
The objective of the SINGAP experiment is to demonstrate the acceleration of 100 mA of D⁻ to 1 MeV in an electrostatic accelerator concept which is substantially simpler than the multi-aperture, multi-grid accelerator system foreseen for the ITER neutral beam injectors.
Voltage holding without beam has been demonstrated at above 1 MV after only 35 min of integrated voltage on-time. H⁻ beams have been produced up to 860 kV, 40 mA in 1 s pulses and without caesium seeding of the ion source. With caesium admixture, D⁻ beams of 105 mA have been accelerated to 630 keV for 1 s. These results were obtained at energies lower than 700 keV because of the bushing deterioration. The measurements of the beam profiles are in good agreement with the results obtained from 3D trajectory calculations.

1 INTRODUCTION

Magnetic thermonuclear fusion machines have benefited extensively from neutral beam injection for plasma heating, and the physics of the generation of non inductive current by neutral beam injection, which could be important for the next generation of machines, is well established. In the majority of the present day machines the high energy neutrals are produced by neutralisation of an accelerated positive ion beam in a gas cell (the "neutraliser"). The high density large sized fusion plasmas of the next generation of machines will require neutral beams of even higher energy to reach fusion conditions, e.g. ITER requires 50 MW of 1 MeV D^0 beams [1]. The very low neutralisation efficiency of positive ions at such high energies means that these beams can reasonably only be produced from an accelerated beam of D⁻ ions [2, 3].

The SINGAP configuration typically consists of a positive electrostatic lens at the exit of the (multi-aperture) pre-accelerator followed by a negative lens formed by the single aperture post-accelerator electrode. The SINGAP principle has been proposed [4] as an alternative to the <u>M</u>ulti-<u>A</u>perture, <u>M</u>ulti-<u>G</u>rid (MAMUG) [5] beam acceleration system of the ITER injector reference design. The fact that no intermediate post-accelerator potentials have to be transmitted and maintained could result in a substantial cost reduction due to the simpler high voltage power supplies and transmission lines as well a welcome reduction in size of the high cost 1 MV ceramic insulators.

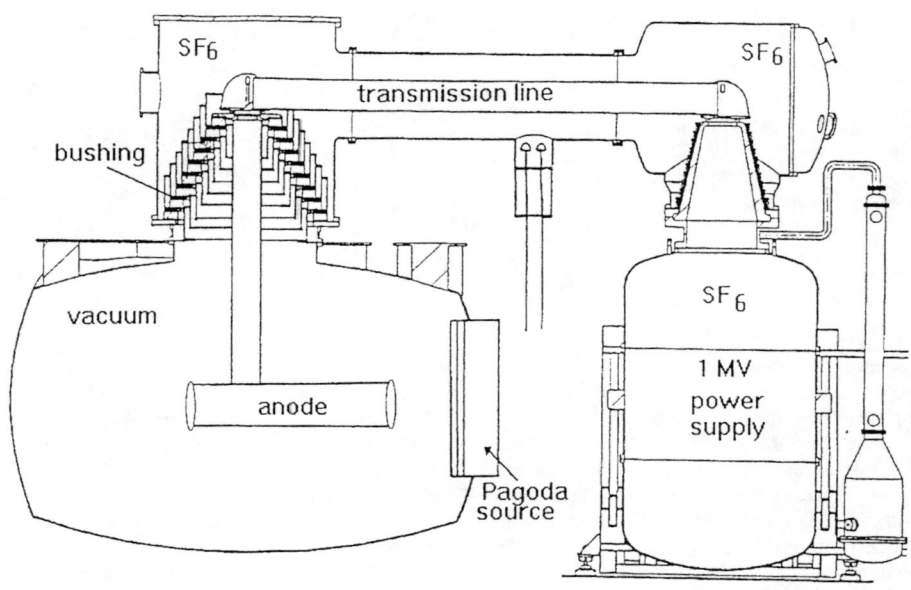

FIGURE 1. SINGAP experimental set-up .

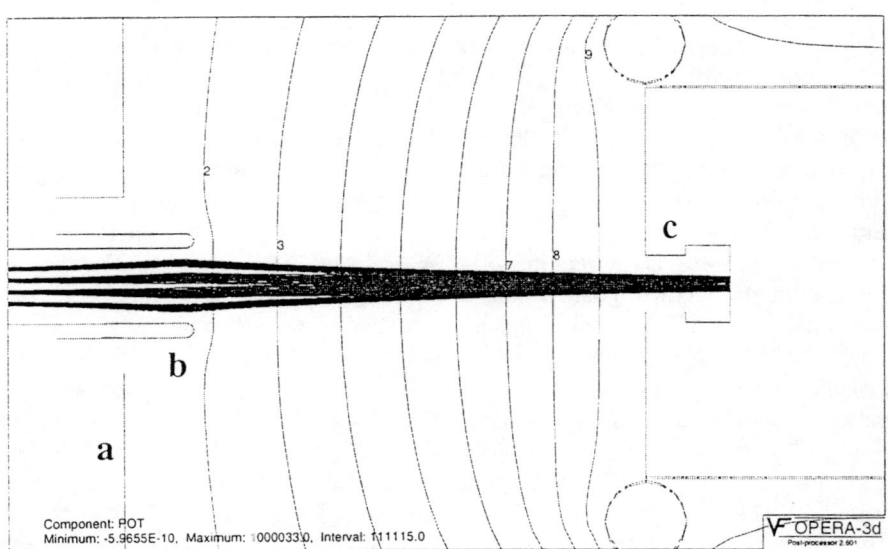

FIGURE 2. SINGAP lens concept . a- cathode at ground potential -b- preaccel electrode (30-50 kV)- c- anode electrode (up to 1 MV) . Beamlets converge by the lens formed by these two electrodes, equipotential surfaces are shown .

In this paper we describe the set-up of the 1 MV, 100 mA D⁻ proof of principle experiment, and present the results obtained from the measurements and 3D beam optics calculations.

2 APPARATUS

2.1 Experimental Set Up

The experimental set up is shown in Fig. 1. An electrostatic potential representation of the SINGAP lens concept is given in Fig. 2. Multiple beamlets are extracted from the Pagoda source [6] at earth potential. The beamlets are pre-accelerated up to 50 keV in a multi-aperture 3-grid structure, then post-accelerated to full energy (eventually up to 1 MeV) across a single gap and simultaneously merged into the single aperture of the cylindrical postaccelerator electrode which is held at a positive high potential :"the anode". The anode is suspended in vacuum from the 1 MV insulator - bushing. The nominal SF_6 pressure in the high voltage transmission line and the 1 MV Cockroft - Walton power supply is 4 bar absolute.

The bushing consists of 9 insulating rings (fibre glass reinforced epoxy) bolted to intermediate metal flanges and stacked on top of each other. These flanges are electrically connected in series through a resistor chain to define and equally distribute the applied voltage. To protect the insulating material from particle bombardment and to minimise the electric field stresses at the metal - insulator junction, concentric metal shields are connected to the intermediate flanges on the vacuum and the SF_6 sides of the bushing.

Stripping collisions between the accelerated D⁻ and the background gas mean that the anode also acts as a (partial) gas neutraliser, converting a fairly small fraction of the accelerated D⁻ into atoms.

A full description of the experimental set-up is given in [7].

2.2 Diagnostics

Two diagnostics tools were applied to measure the composite (ion + neutral) or neutral beam profiles (Fig. 3).

2.2.1 Composite Beam Profile Measurement

At the end of the anode (2.63 m from the pre-accelerator), the beam is dumped onto an inertial unidirectional carbon fibre composite target consisting of 2 plates of Mitsubishi MFC-1A material installed side by side. This material has quasi 1D thermal properties, with the high conductivity essentially only in the beam direction. This enables the beam power density profile to be deduced from temperature measurements using an infrared camera viewing the rear face of the target. The spatial resolution at the target is ~2 mm [8].

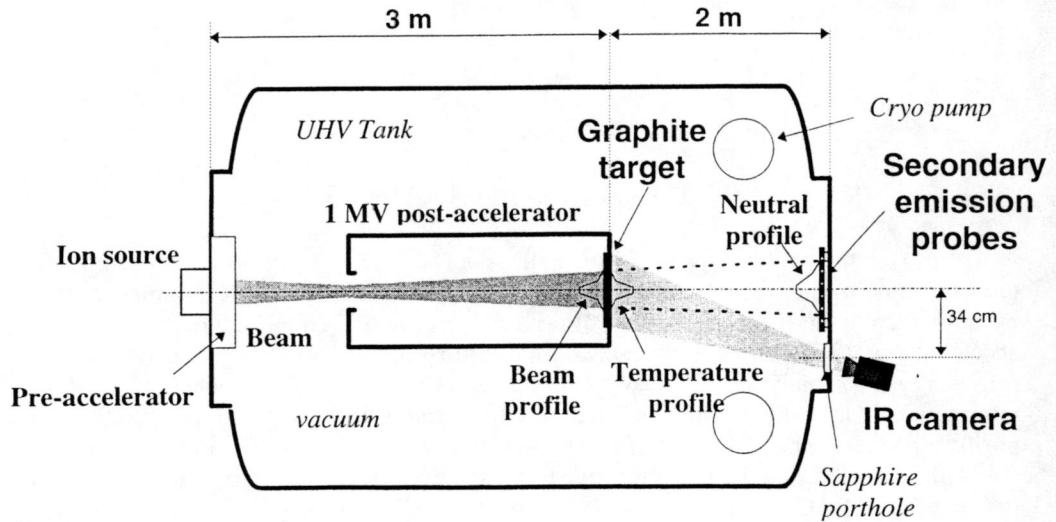

FIGURE 3. Beam diagnostics.

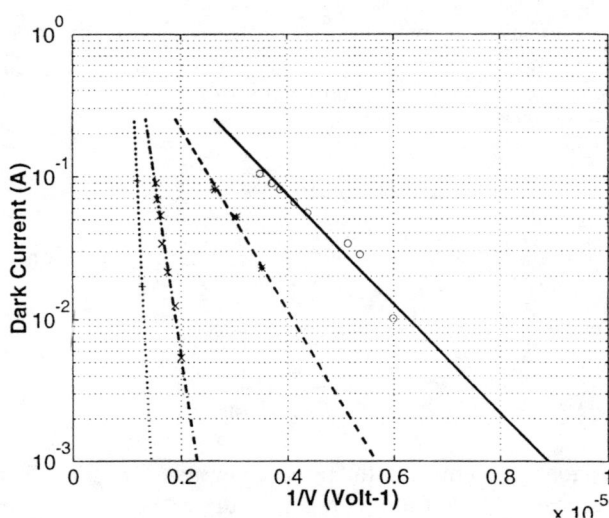

FIGURE 4. Dark current variation with voltage and pressure (first measurements) : [o] 2.10^{-8} mbar, [•] 10^{-6} mbar, [x] 5.10^{-5} mbar, [+] 10^{-4} mbar.

2.2.2 Neutral Beam Profile Measurement

A 1 mm wide slit between the 2 CFC graphite plates allows fast particles to leave the anode and these can be collected on an array of secondary electron emission detectors (12 metal strips aligned parallel to the slit) at 4.63 m distance from the pre-accelerator. Under the experimental conditions used, essentially only D^0 (~5°/∞ of the D^- beam) reaches the array and the distribution of the secondary electron emission current across the different probes is representative of the D^0 beam profile (see 3.3 below).

2.2.3 Aperture Configuration

Two extraction aperture configurations have been used, circular and rectangular. A quasi circular configuration was tried first since a cylindrically symmetric system is most easily adapted to the SINGAP concept. Subsequently a quasi rectangular (ribbon) configuration was used which bears more relevance to the ITER beam geometry.

3 RESULTS

3.1 Voltage Holding

3.1.1 High Voltage Conditioning

Application of high voltage to "unconditioned" in-vessel components is typically accompanied by the generation of "dark" currents, especially at low pressures, which show field emission like behaviour. However, the emission level is found to be orders of magnitude larger than expected from the applied fields and the work function of the cathode, the stainless steel tank walls in this case, which have an area ~60 m². The dark current can be decreased by increasing the operating pressure (Fig. 4). "Conditioning", i.e. the repetitive application of, and gradual increase in, the high voltage is also found to reduce the dark current, even at low pressures, to values small compared to the 100 mA capability of the power supply.

Breakdown-free application of 1.04 MV was achieved after about 35 min accumulated voltage on-time at 10^{-4} mbar operating pressure (helium).

The number of breakdowns encountered during conditioning is very small. Visual light emission is routinely observed in areas of elevated electric field stresses (e.g. interior of the bushing, and the stress rings of the anode) by means of video cameras. Although the electrostatic energy stored in the stray capacitance of the structure is comparatively high (≤300 J) there is no evidence that breakdowns have caused permanent damage to components.

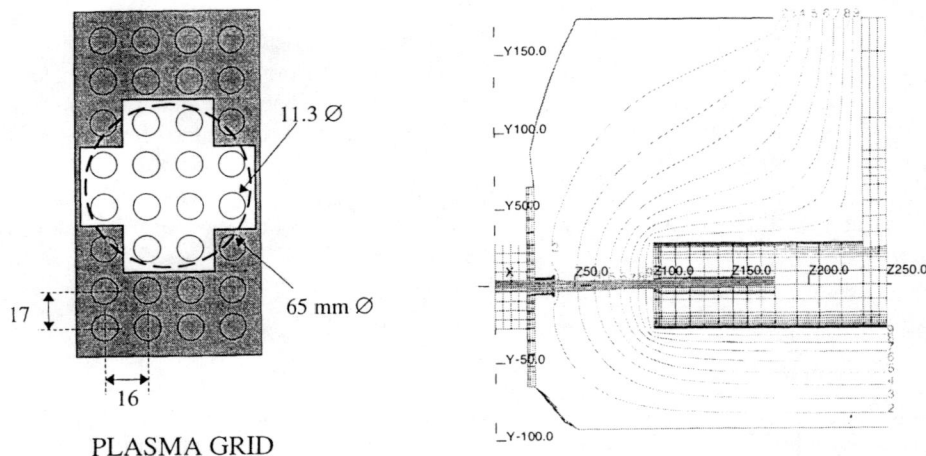

FIGURE 5. Circular configuration of the plasma grid and of the acceleration system. In the vertical cut the 3D calculated beam trajectories are shown for 600 keV, 100 mA D⁻ and 50 kV preaccel voltage.

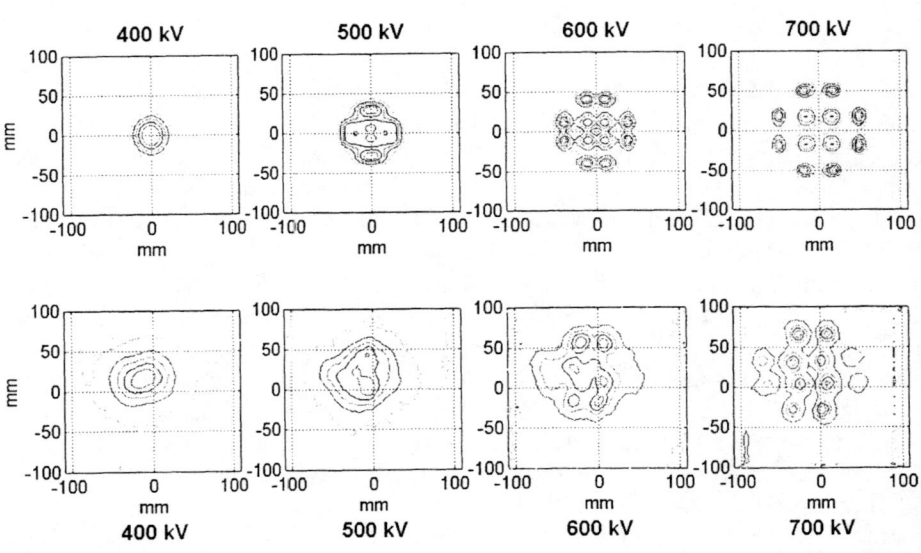

FIGURE 6. Comparison of calculated (top row) and measured (bottom row) 30 mA H⁻ beam profiles.

3.1.2 Insulator - Bushing

Initial results were presented in [9, 10]. We review them briefly here and complete them with the latest results. A few months after the start of the experiments the (top) insulator ring, closest to the 1 MV feed-through, was found perforated in two places as a consequence of carbonisation inside the insulating material. This is ascribed to a manufacturing procedure creating voids inside the epoxy - glass fibre composite as they were not moulded under vacuum. To repair the damage the carbonised material was excavated and the cavities filled with commercially available epoxy resin. The top ring was then tested successfully to voltages >120 kV. Unfortunately soon after operation recommenced and voltages above 700 kV were attempted, new degradation of the voltage holding became apparent. From application of high voltage to individual rings it was shown that the 2 top rings were faulty. In order to carry on with the experimental programme it was decided to limit the applied voltage to below 700 kV and to continue with the measurements. For this reason, the most recent series of experiments was conducted with an applied voltage ≤700 kV.

3.2 Beam Acceleration

In the SINGAP experiment co-extracted electrons are efficiently trapped on the extraction grid developed at Cadarache [11]. Electron leak rates as low as 0.5% have been measured by extracting electrons from an argon discharge in the source, i.e. in the absence of any negative ions.

The first series of experiments was carried out using a circular configuration in which 12 apertures of 1 cm^2 extraction area were disposed in a quasi-circular array of 65 mm diameter. A near axis-symmetric positive lens action on the pre-accelerated beam was obtained by an additional cylindrical electrode at the exit of the pre-acceleration grid assembly (Fig. 5). The anode had a circular aperture of 80 mm diameter, the post-acceleration gap was 0.63 m. For simplicity, the experiments were started with H_2 without caesium seeding of the ion source, and H⁻ beams of up to 860 keV, 40 mA were produced.

Fig. 6 shows a series of experiments [8] in which the accelerated H⁻ current was kept constant around 30 mA and the beam energy was varied between 400 and 700 keV. At 400 keV the beam profile at 2.63 m is near Gaussian with a footprint of nearly the same size as the initial quasi-circle of 65 mm diameter. If the acceleration voltage is gradually increased the beam profile gets larger, develops discrete structures and finally, at 700 kV, splits up into individual beamlets. Such a behaviour has been predicted by the 3D trajectory [12] calculations (Fig. 6 top row). The phenomenon is explained by two effects : the decreasing influence of space charge and the growing strength of the positive lens at the exit of the pre-accelerator with increasing voltage. For voltages high enough the lens transforms the pre-acceleration object plane into an inverse geometric image of about twice the size (the beamlets trajectories cross over each other in both the vertical and horizontal planes). It is worth noting that over the whole range of voltages applied the beam power intercepted by the graphite target is well above 90%. The horizontal displacement of alternate rows with respect to each other found in the measurement (Fig. 6 bottom row) is due to the alternating sign of

PLASMA GRID

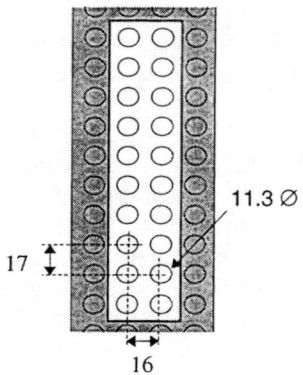 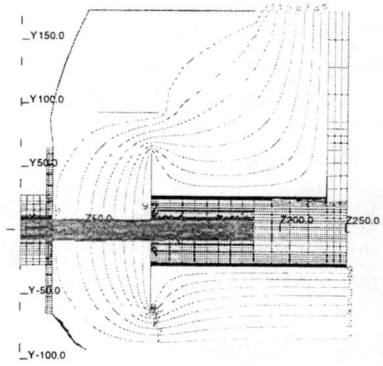

FIGURE 7. Rectangular configuration of the plasma grid and of the acceleration system. In the vertical cut the 3D calculated beam trajectories are shown for 500 keV, 100 mA D⁻ and 50 kV preaccel voltage.

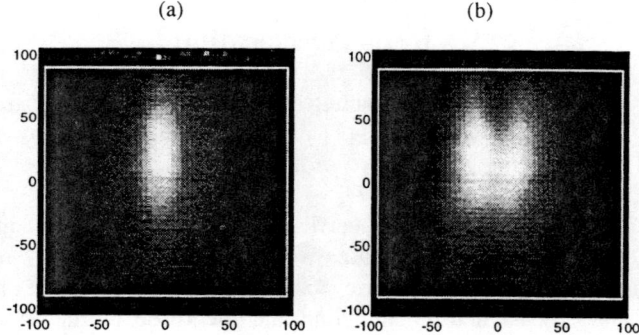

FIGURE 8. Beam profiles in the rectangular configuration for 65 mA D⁻, 50 kV preaccel voltage and two beam energy values: (a) 445 keV, (b) 560 keV.

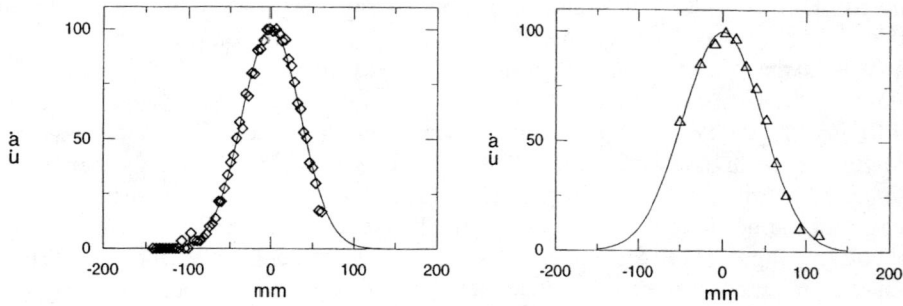

FIGURE 9. In the circular configuration, 25 mA-260 keV D⁻ beam, 37.5 kV preaccel voltage; on the left, particle profile on the carbon target; on the right, neutral profile on the secondary emission array.

the field from the magnets in the extraction electrode of the preaccelerator. This effect has not been included in the simulation.

More recently the working gas was changed to deuterium and, to increase the D⁻ current, caesium was added to the discharge of the ion source ; 105 mA D⁻ beams of ≤ 630 keV have then been produced.

In the latest set of experiments the electrode configuration was changed to obtain a vertical ribbon shaped beam. To achieve this, the plasma grid was masked to extract 20 beamlets from two columns of ten apertures (Fig. 7). The cylindrical extension on the pre-accelerator used for the circular configuration was removed, which resulted in an increase of the post-acceleration gap to 0.76 m, and the SINGAP aperture was changed to a rectangular shape, 86 mm x 300 mm (width x height), with rounded corners. In this configuration an asymmetric positive lens is formed at the exit of the pre-accelerator by the grid support structure, which consists of two parallel vertical plates, and by the lower part of the vacuum tank plus an additional plate above the beam (both at ground potential) - see Fig. 7.

Fig. 8 shows infrared images equivalent to Fig. 6 for two different values of post-acceleration voltage, 445 kV and 560 kV, both for a D⁻ beam of 65 mA. At the lower energy (Fig. 8 (a)) a single ribbon shaped beam profile is measured on the graphite target. The profile at the higher energy becomes separated into a double profile caused by a left-right inversion of the image of the aperture columns. As the positive lens is much weaker in the vertical direction no cross-over occurs upstream of the graphite target, but a clear compression in this direction is observed (Fig. 8 (b)).

3.3 Neutral Beam Analysis

The majority of the secondary electrons generated on the secondary emission detector array are assumed to come from fast D^0 leaving the anode through the 1 mm slit in the graphite target. Contributions from other species are considered to be negligible for the following reasons : Electrons and negative ions will in principle be decelerated to zero energy and be widely dispersed by their own space charge. The amount of D^+ ions produced during "neutralisation" in the thin gas target employed is estimated at only 2% of the D^0 fraction [13]. Additionally the D^+ emerging from the slit, which will be accelerated up to twice the D⁻ energy, will have their transverse profile broadened due to the electrostatic field configuration between the anode and the array.

Neutral beam profiles have been obtained in both extraction configurations, circular and rectangular. Fig. 9 shows an example of a vertical profile of the composite beam in the circular configuration measured on the graphite target for 260 kV, 25 mA D⁻ and the corresponding neutral beam profile detected on the secondary emission array.

For the rectangular configuration the slit in the graphite target is oriented horizontally. Fig. 10 shows (horizontal) beam profiles equivalent to Fig. 9 for the (vertical) ribbon shaped deuterium beams of 430 kV, 100 mA. The "cleavage" in the neutral profile arises because the beam is composed of two columns. The corresponding double Gaussian fit of the profile is indicated in dashed lines (Fig. 10). A more profound analysis of the horizontal ribbon profile taking into account the effect of the field from the magnets in the extraction grid is given in [14].

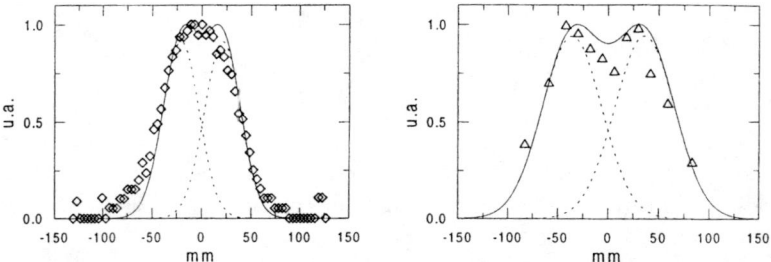

FIGURE 10. In the rectangular configuration, 100 mA-430 keV D⁻, 50 kV preaccel voltage; on the left, particle profile on the carbon target; on the right, neutral profile on the secondary emission array.

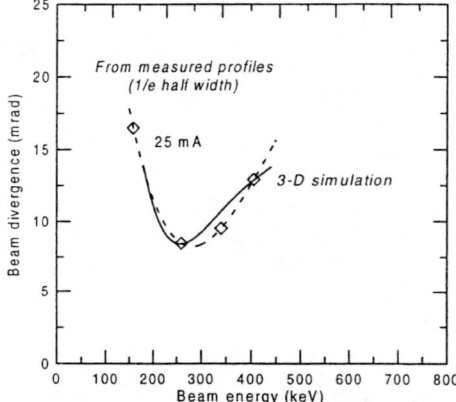

FIGURE 11. In the circular configuration, squares show experimental measures of the beam divergence of 25 mA D⁻, full line show 3D simulation result under the hypothesis of 30 mrad beamlet divergence.

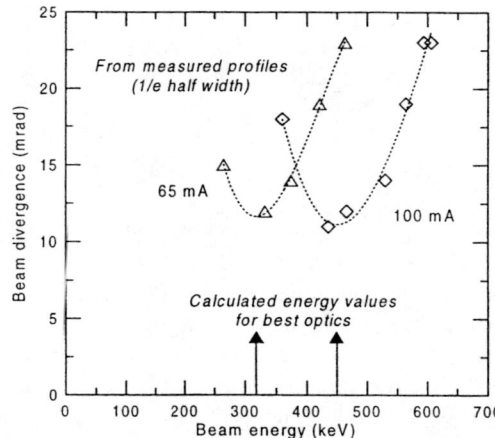

FIGURE 12. In the rectangular configuration, measured beam divergence for 65 mA and 100 mA D⁻, 48 kV and 50 kV preaccel voltage respectively.

3.4 Beam Quality and Perveance

Assuming that the negative ion beam is space charge neutralised inside the anode, the difference between the D^- and D^0 beam profiles should be negligible, and a measure for the "divergence", δ, of the SINGAP beam envelope may be obtained from the profile width at the graphite target and at the secondary emission detector:

$$\delta = \arctan\frac{\Delta W}{L} \tag{1}$$

where ΔW is the difference between the 1/e half widths of the two profiles and L the distance between the two profile monitors. It should be noted that this set up only yields correct results for the case with the focal point upstream of the first (graphite) target. Errors due to the uncertainty in the position of the focal point are estimated as $\leq 25\%$ for the circular and $\leq 10\%$ for the rectangular configuration.

Fig. 11 shows the variation of the envelope divergence, measured in the vertical direction (circular configuration), as a function of energy (perveance curve) for a 25 mA D^- beam. Within the errors of the measurement a minimum envelope divergence of about 10 mrad is obtained at 260 kV. The profiles for this situation of a most parallel beam are shown in Fig. 9. To get agreement between the measurement and the 3D trajectory simulation a beamlet divergence of 30 mrad at the exit of the pre-accelerator had to be assumed in the calculations. This rather high value is thought to arise principally from the non uniform illumination of the extraction array. Additional degradation of the divergence is also expected because the field from the magnets in the extractor results in a horizontal steering and translation of the ion trajectories, which moves the beamlets into areas of the apertures where the electrostatic lens will produce more aberrations. Both the horizontal steering by the magnetic field and the non uniform illumination of the apertures can be clearly seen in Fig. 6.

Perveance curves for D^- beams of 65 and 100 mA in the rectangular configuration, the "divergence" being measured in the horizontal direction, are shown in Fig. 12. It can bee seen that the minimum divergence appears to be about 40% higher than for the circular configuration. This is thought to be due to the influence of the field from the magnets in the extractor on the ion trajectories, which acts in the horizontal direction, influencing strongly the horizontal profiles.

4 CONCLUSION

The results of the experiment appear to be extremely promising. It has been shown that the SINGAP concept can produce high energy negative ion beams with good quality.

Voltage holding at 1 MV and beams up to 860 keV have been obtained with comparatively little effort. Rapid progress towards 1 MeV beams should be possible after solutions are found to the problems with the high voltage insulator - bushing.

These problems are entirely technological and due to manufacturing techniques and, probably, to the choice of insulating material (epoxy).

The results regarding power transmission and SINGAP beam envelope divergence obtained at only about half the design energy and in an incompletely optimised system give hope that a divergence of ~5 mrad (as required for the ITER injectors) may well be achieved at full energy and with some improvements in the pre-accelerated beam. For the understanding of the SINGAP concept and the assessment of the measurements the 3D trajectory code has been useful.

ACKNOWLEDGEMENTS

The authors are greatly indebted to G. Delogu, S. Dutheil, P. Heister, P. Van Coillie, R. Yattou, R. Brugnetti and R. Volpe for their technical assistance.

REFERENCES

1. Hemsworth, R. S. et al., *"Neutral Beams for ITER"*, 16th SOFE, Urbana-Champaign, Il (Oct 1995)
2. Berkner, K. H. et al., *Nucl Fusion* **15** (1975) 249
3. Kuriyama, M. et al., *"Negative Ion Based NBI System for JT60-U"*, 16th SOFE, Urbana-Champaign, Il (Oct 1995)
4. Fumelli, M. et al., *"Proposal for a 1 MeV, 0.1 A, D.C. D⁻ Beam Acceleration Experiment at Cadarache"*, 6th Symp Prod and Neutr of Negative Ions, Brookhaven, N.Y. (1992)
5. Inoue, T. et al., *"Design Study of a Prototype Accelerator and MeV Facility for Demonstration of 1 MeV - 1 A Negative Ion Beam Production"*, JAERI-TECH **94-007** (1994)
6. Fumelli, M. et al., *"Negative Ion Production with the PAGODA Source"*, Fus Eng and Design **26** (1995) 463-472
7. Massmann, P. et al., *"The Cadarache Negative Ion Experiments"*, 16th SOFE, Urbana-Champaign, Il (Oct 1995) 475
8. Massmann, P. et al., *"SINGAP Acceleration of High Energy Negative Ion Beams"*, 16th IAEA Fusion Energy Conference, Montreal (Oct 1996)
9. Bucalossi, J. et al., *"Results of the Cadarache 1 MeV D⁻ SINGAP Experiment"*, 19th SOFT, Lisbon (1996) 685
10. Simonin, A. et al., *"Cadarache 1 MeV Negative Ion Accelerator Development for Application in Thermonuclear Fusion Research"*, P.A.C., Vancouver (Mai 1997)
11. Simonin, A. et al., .*"Extraction system development and stray electron study for high intensity negative ion based NBI"*, 19th SOFT, Lisbon (1996) ...
12. Vector Fields Limited, 24 Bankside, Kidlington, Oxford OX5 1JE ENGLAND
13. Barnett, C. F. (editor), *Atomic Data for Fusion*, ORNL **6086**
14. Bucalossi, J., *PhD Thesis,* Paris VI (Oct 1997)

Beam Transmission in the ITER Neutral Beam Injectors

A. Krylov[1], E. Di Pietro[1], E. Dlugach[2], T. Inoue[1],
M. Hanada[3], R. Hemsworth[4], K. Miyamoto[3], A. Panasenkov[2]

[1]ITER Joint Central Team, Naka Joint Work Site,
[2]RF ITER Home Team, [3]Japan ITER Home Team, [4]EU ITER Home Team

The ITER neutral beam (NB) injection system will consist of three injectors, each delivering to the plasma up to 16.7 MW of 1 MeV D^0 injected through the quasi-tangential main horizontal port with an opening in the blanket structure 50 cm wide × 80 cm high. The beam footprint in the vicinity of the ITER magnetic axis should cover possible operating positions of the axis, 1.3 m < Z < 1.6 m, where Z is the vertical distance, measured from the equatorial plane of ITER. The three NB injectors will be located outside the cryostat inside a common enclosure, the NB cell. Each of the injectors contains, see Fig. 1, a single D^- source, connected to a 1 MeV multi-grid, multi- aperture, electrostatic accelerator[1]. The accelerated ions pass through a D_2-gas neutraliser, and then enter an electrostatic residual ion dump (RID). Emerging from the neutraliser will be a beam consisting of 60% D^0 and almost equal parts of D^-, D^+. The RID deflects the charged particles in the beam onto water-cooled copper surfaces, leaving the neutral beam to continue to the torus. A water-cooled beam dump, the calorimeter, can be moved into the beam path to allow conditioning of the injector independently of the tokamak. Each module also contains a cryopump, to remove the gas flowing from the neutraliser and ion source into the module, and a fast shutter. The function of the latter is to prevent gas flowing from the tokamak to the injector at plasma start up and termination. The duct from the injector to the tokamak will be flexible in order to allow for dimensional changes of the tokamak and to avoid excessive force on the connection between the duct and the torus. The ITER stray magnetic field, which affects the negative ion beam extraction, acceleration and transport, is reduced inside of the injector vacuum vessel by a combination of passive iron layers and active compensation coils. Nuclear shielding around the duct and vessel will reduce to acceptable levels nuclear radiation from the tokamak.

The ITER NB scheme was chosen considering the limited available length (for the injector in the NB cell and for the beamline components in the NB vessel), the optimization of beam losses during acceleration and transport and beam source performance characteristics. A particular feature is the relatively short neutraliser, which is decoupled from the accelerator. It is 3 m long and subdivided into five vertically elongated 80 mm wide channels. Stripping loss in the multi-grid accelerator is a crucial issue, it is calculated to be 47% for 0.3 Pa ion source operation filling pressure[2]. Decoupling of the neutraliser improves the source pumping and the gas pressure profile in the accelerator, reducing stripping loss of the D^- ions before the acceleration is completed. Subdividing the neutraliser enables the necessary gas target thickness to

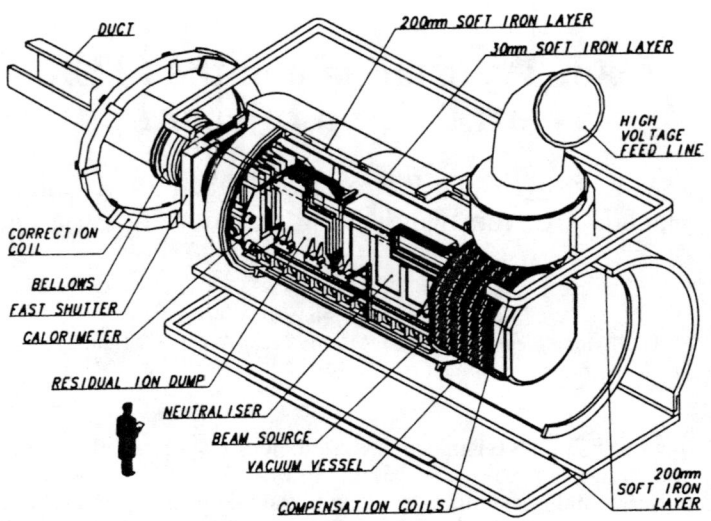

Figure 1. The ITER neutral beam injector

be created with significantly lower gas flow to the neutraliser than would be the case for a simple open neutraliser and so reduces the reionisation loss of D^0 downstream from the neutraliser as well as stripping loss in the accelerator. A disadvantage is that narrow channels restrict the horizontal acceptance of the injector and increase geometrical beam loss.

The RID is also designed with 5 channels matching the neutraliser ones to accommodate dumped D^- and D^+ beam power within the available 1.5 m axial length. Downstream from the RID with the calorimeter moved aside the D^0 beam drifts along a 10 m duct to the torus. The duct section is minimized particularly at the exit, to be compatible with the port and blanket structure and to reduce neutron radiation from the tokamak.

Geometrical Beam Transmission

The beam is to be steered to achieve the necessary beam footprint in the plasma, and acceptable beam transport. The beam is formed by 1300 small beams (beamlets), which are extracted and accelerated separately. For design purpose it is assumed that the beamlets consist of a core with 5 mrad divergence and a 15% halo fraction with 15 mrad divergence. Vertical "beam columns", matching the neutraliser channel are formed by an array of 65 × 5 (vertical × horizontal) beamlets. Space is left between the beam columns to provide enough space for channels walls in the neutraliser and RID. The direction of each beamlet axis in each channel in the horizontal plane is chosen separately to maximize beamlet transmission to the torus.

For the best transmission some of the beamlets must be steered up to 11 mrad. This horizontal beamlet steering is provided partially by offset of the exit aperture axis relative to the beamlet axis. The thin lens theory (with the experimentally determined coefficient) gives [3,4] $\phi = 1.5 \times \delta$, where the angle ϕ is in mrad, and the axis shift δ, is in mm, for the case of 200 kV voltage and a 50 mm gap between the grounded and the preceding grid.

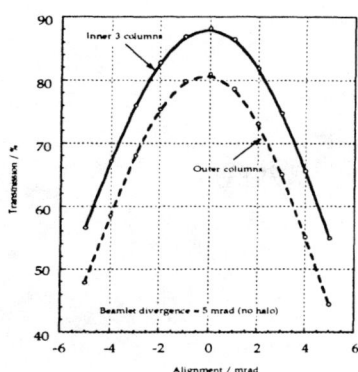

Figure 2. Column beam transmission as a function of horizontal angular misalignment of the whole column beam about the optimum.

Experimental measurements demonstrate that this technique can provide up to 6 mrad beamlet steering.

Additional electrostatic deflection for beam columns is foreseen at the accelerator exit. The outer two beam columns should be deflected towards the beam axis direction by 7 mrad, the next two inner columns - by 3 mrad, and the central column - by 0 mrad. In this case maximum necessary displacement steering for beamlets does not exceed 6 mrad. The electrostatic steering is provided by biased plates at either side of each beam column. Steering plates have 200 mm length and 88 mm gap between them for beam column path. The -16 kV bias voltage results in 10 mrad beam deflection while the calculated divergence growth does not exceed 0.5 mrad.

The calculated beam column transmission and its sensitivity to misalignment is given in Fig. 2. For the outer two beam columns the best transmission is about 80%, the three inner beam columns have better transmission. For good transmission the horizontal misalignment should not exceed ±2 mrad.

The vertical inclination of grid segments is used to create the desired vertical beam profile. Each beam segment constitutes one fifth of each grid (beamlets in 13 rows and 20 columns). The segments should be inclined by (from upper segment to bottom) - 36, -22, -7, +12 and +31 mrad. Within each beam segment beamlets are parallel in a vertical plane. The beam power deposition limit of 100 kW/m^2 on the duct exit and the 0.80 m exit height restrict the vertical misalignment by value of ±4 mrad.

The calculated beam footprints for the best horizontal and vertical alignment are shown in Fig.3 (one half of each footprint is only shown).

Influence of the Magnetic Field

The stray magnetic field from ITER will be up to 0.1 T in the NB cell. Negative ions emerging from the accelerator will be deflected by the magnetic field. Inside of the neutraliser additional distortion of the angular and space distribution of each beamlet takes place due to deflection prior to neutralisation. Calculations show that to keep the beam axis deflection in the range ±1 mrad, the integral of the vertical field component along beam trajectory must be $-0.0002 < \int_0^X B_z dx < 0.0002$, T×m. A similar limitation is found for the magnetic field in the neutraliser. The ITER stray magnetic field in the NB area before and after shielding are shown in Fig.4. The shielded magnetic field is well below the above mentioned limit for Z = 0, see Fig.5. For the upper and lower regions of the beam the integrals are close to the limit and have opposite signs. This effect results in "twisting" the beam column, where upper part is deflected right direction, and the bottom part left. This produces not only additional beam loss but, more important, the increase in the local power density on the leading edges of the neutraliser channel,

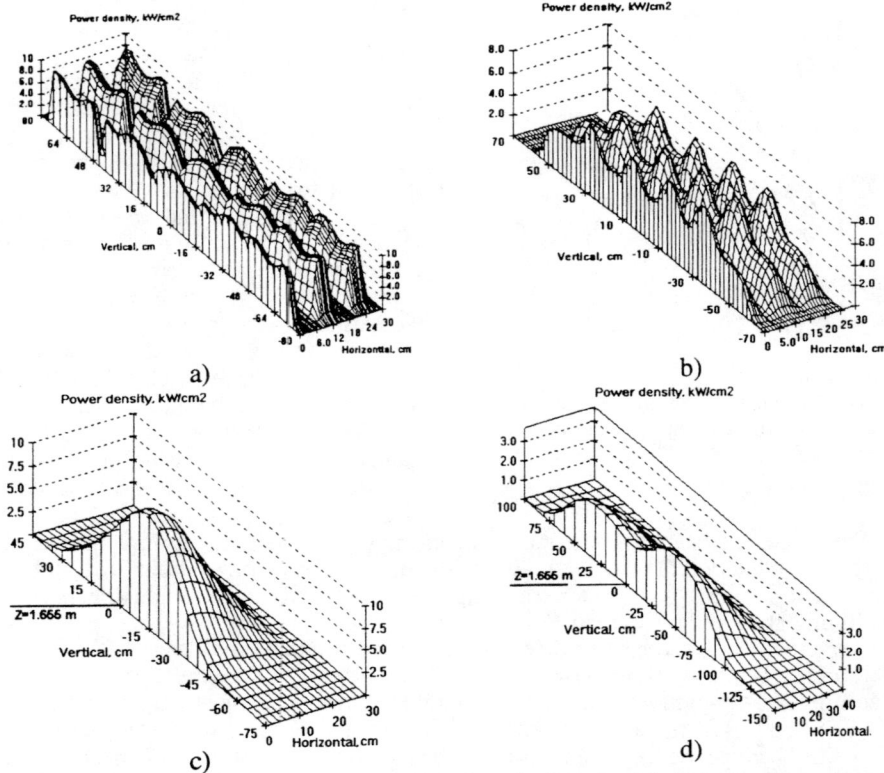

Figure 3. Evolution of the beam footprint (normal to the beam) at a) the neutraliser entrance, b) calorimeter entrance, c) the ITER magnetic axis, and d) the ITER far wall. Line Z=1.655 corresponds to the height of the beam source center in the ITER coordinate system.

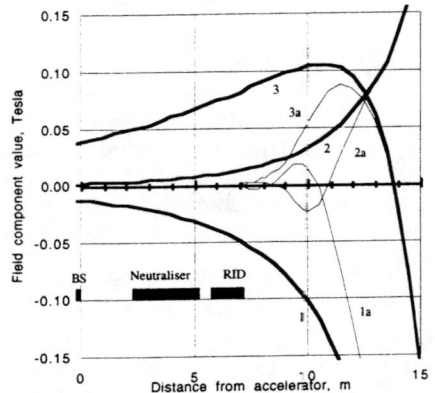

Figure 4. The stray and shielded (lettered "a") magnetic field along beam path (Z = 1.655 m) for standard (end of burn) ITER operation.
1,1a - B_x; 2,2a - B_y; 3,3a - B_z components

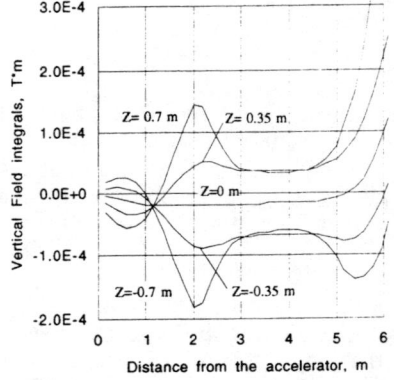

Figure 5. Values of $\int_0^X B_z dx$ calculated along beamlet trajectories at different heights Z relative to the beam axis

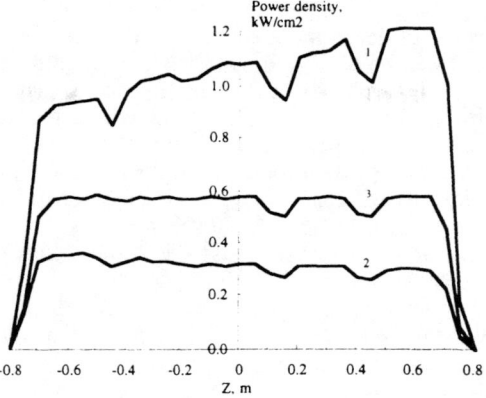

Figure 6. Power deposition on the leading edges of the central neutraliser channel. 1 (left edge) and 2 (right) - with magnetic field and -2 mrad initial misalignment; 3 - perfect alignment, no magnetic field.

see Fig.6. The -2 mrad misalignment gives different power density along the left and right edges. The structure of the curves reflects the different vertical aiming for the beam segments.

Reionisation

Downstream from the neutraliser some part of the D^0 beam is reionised in collisions with gas molecules along the beam path. The distribution of the D_2 pressure in the beam path was calculated by the Monte-Carlo method. Three gas sources were taken into account: gas flow from the beam source, from the neutraliser and from the RID due to recombination of the fast D^+ and D^- dumped there. The result is shown in Fig.7. The accumulated reionisation starting from the neutraliser exit is shown also. Up to 5% of D^0 emerging the neutraliser will be lost due to reionisation, of which about half occurs before the exit from the RID. This half of the 1 MeV D^+ ions produced by reionisation is

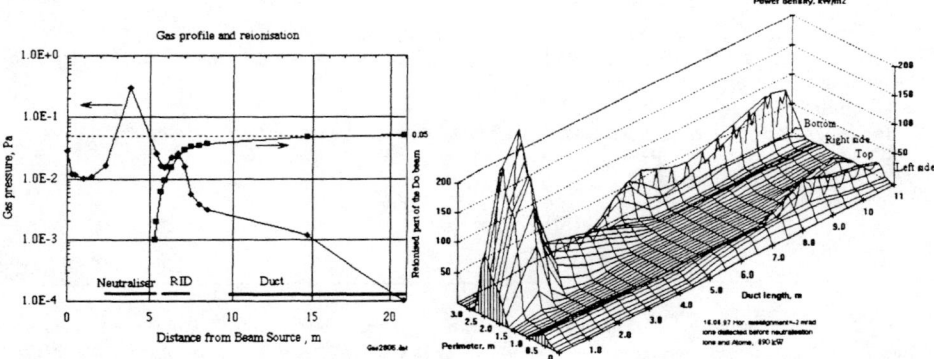

Figure 7. Gas density profile and reionisation in the NB injector beam line

Figure 8. Power density distribution in the NB duct from direct interception and reionisation. Horizontal misalignment -2 mrad.

203

affected by the magnetic field and by the electrical field in the RID. Most of these ions strike the RID plates, the calorimeter or the entrance to the duct. The other half produced downstream from the RID will be deflected by the magnetic field onto duct walls. Power deposition in the duct, which is shown in the Fig. 8, results mainly from reionisation, and it is deposited mainly at the first part of the duct where the gas pressure is high. Input from direct interception defines the power load in the duct close to the exit. Ions in this area have a Larmor radius less then the duct width and do not produce focusing of the reionised heat at the exit. At the exit, where the duct is formed by blanket modules, power deposition is below the limit value of 100 kW/m^2.

Gas release in the duct can, in principle, cause the "beam blocking" effect. However the analysis carried out shows that the ITER NB injector is far from conditions when beam blocking starts. In comparison with injectors based on positive ion beams the ITER NB has relatively small beam current (16.7 eq.A) and smaller reionisation cross section due to higher particle energy.

The following table demonstrates nominal power flow in the ITER NB injector.

Table The ITER NB power distribution

Parameter	Value
D- current density, mA/cm2	18.6
Accelerated beam power, MW	37.2
Geometrical Beam Transmission, %	79.2
D^O fraction	0.60
Reionisation, %	5.0
Power to the neutraliser, MW	5.3
Total power to the RID, MW	14.2
D^O beam power, exiting RID, MW	18.1
Total power to the duct, MW	1.0
Injected D^O beam power, MW	**16.7**

REFERENCES

[1] R.S. Hemsworth et al., *16th Symp. On Fusion Engineering*, Champaign, Sep.30 -Oct 5, 1995.
[2] For details see paper of M. Hanada, et al., "Stripping Loss and Power Loading of the Acceleration Grid", presented on this Symposium.
[3] Y. Okumura et. al., RSI 51/4, 471 (1980).
[4] T. Inoue, et al., *6th Intern. Symp. On Prod. and Neutr. of Negative Ions and Beams*, NY, Nov. 1992.

This report is an account of work undertaken within the frame work of the ITER EDA Agreement. The views of authors do not necessarily reflect those of the ITER Director, the Parties to the ITER EDA Agreement, or the International Atomic Energy Agency.

Radiation Induced Conductivity and Voltage Holding Characteristics of Insulation Gas for the ITER NBI

Y. Fujiwara, M. Hanada, T. Inoue, K. Miyamoto,
N. Miyamoto, Y. Ohara, Y. Okumura, and K. Watanabe

Japan Atomic Energy Research Institute
801-1, Mukoyama, Naka-machi, Naka-gun, Ibaraki-ken, 311-01, Japan

Abstract. An irradiation experiment on insulation gas was carried out to evaluate ionization current and voltage holding characteristics with the gamma-rays from Co-60. The gases of air, SF_6, C_2F_6, CO_2, and mixture of air and SF_6 were investigated using parallel-disk electrodes at the absorbed dose rate of up to 0.45 Gy/s. Ionization current increased linearly with gap length between electrodes, gas pressure, absorbed dose rate, and molecular weight. An experimental formula to estimate ionization current was obtained on the basis of experimental results. Degradation of breakdown voltage with gamma-rays was at most about 10 %; however, the degree of the degradation did not depend strongly on absorbed dose rate.

I. INTRODUCTION

Insulation gas like SF_6 has been used in high-energy accelerators and in high-voltage power devices for high-voltage insulation. The use of insulation gas allows one to design equipment easily, resulting in a reduction in cost as well as size. For this reason, gas insulation will be attractive in high-energy neutral beam injection (NBI) systems for next generation fusion devices or reactors. In the engineering design of NBI for International Thermonuclear Experimental Reactor (ITER), in fact, gas insulation are being examined for 1-MV insulation between a beam source and a beam source vessel. However, in the use for fusion reactors, there will happen a problem unique to fusion reactors. Since fusion reactions will produce high-energy neutrons which will generate radiation such as gamma-rays and X-rays, insulation gas will be irradiated and ionized. In July 1996, Dr. Eric Hodgson reported on results of an irradiation experiment on insulation gases of Helium, N_2, air, CO_2, and SF_6. The results indicated that ionization current of insulation gas would be very high at ITER condition, leading to much dissipation of acceleration power inside the vessel [1].

On the design of the ITER NBI, it has become important and urgent to clarify Radiation Induced Conductivity (RIC) or ionization current and voltage holding characteristics of insulation gases, so that an irradiation experiment was carried out at Japan Atomic Energy Research Institute (JAERI). The gases of air, CO_2, C_2F_6, SF_6, and mixtures of air and SF_6 were investigated using the gamma-rays from cobalt-60. In the present paper, experimental results for ionization current and voltage holding capability are reported after a description of an experimental setup. Next an experimental formula is obtained to estimate ionization current flowing through insulation gas around the beam source during the operation of ITER.

II. EXPERIMENTAL SETUP

A gamma-ray irradiation experiment was performed at Co-60 irradiation facility at JAERI. The facility has two kind of sources of gamma-rays: One is $4.1*10^{15}$ Bq and the other is $0.22*10^{15}$ Bq. The sources are 134 mm in outer diameter and 405 mm in height. The absorbed dose rate is defined as that for air; the absorbed dose rate in Gy/s is given by multiplying exposure in C/kg/s by averaged energy needed to produce a pair of electron and positive ion (or W-value) of air in eV, which is 33.8 eV for the gamma-rays from Co-60.

Figure 1 shows an experimental apparatus for measuring ionization current and breakdown voltage. This is composed of a test chamber, electrical cables, and a power supply. The test chamber consists of a circular pyrex-glass, upper and lower flanges, a bourdon-tube pressure gauge, a linear motion introducer, a gas supply line, and parallel-

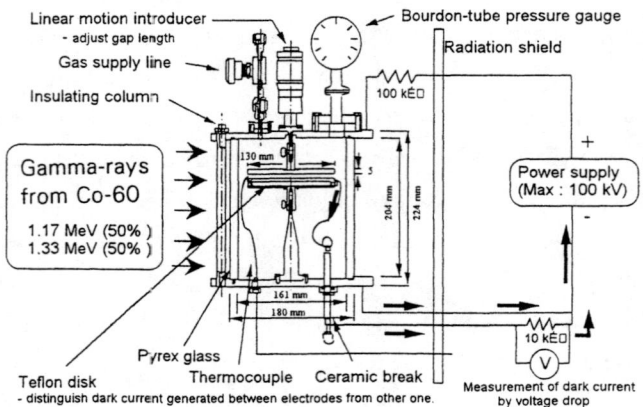

FIGURE 1. Experimental apparatus for measuring ionization current and breakdown voltage: the apparatus is pumped to vacuum and then is filled with test gas through the gas line.

disk electrodes made of stainless steel. The pyrex-glass has an internal diameter of 161 mm, an external diameter of 180 mm, and a height of 204 mm. The volume of the test chamber is about 4 litter. The bourdon-tube pressure gauge shows the pressure of test gas filled in the chamber. The linear motion introducer enables one to change the gap length between the parallel-disk electrodes at the range of 0 mm to 20 mm. After the test chamber is pumped to vacuum, it is filled with test gas through the gas line. The disk electrodes are 130 mm in diameter and 5 mm in thickness. A cathode electrode is insulated by a 8-mm-thick teflon disk. An electrical cable connected to the cathode electrode is insulated by a ceramic break to distinguish ionization current generated between the electrodes from that generated at other space. Capacity of the power supply is 100 kV, 3 mA at direct current. Ionization current is evaluated as a voltage drop in a resister of 10 k ohm.

III. EXPERIMENTAL RESULT

A. Applied Voltage Dependence

Figure 2 shows the dependence of ionization current for some gases against voltage applied between the gap of electrodes. Gas pressure is 600 mb and absorbed dose rate is 0.45 Gy/s. Table 1 is a summary of averaged relative dielectric strength normalized by the dielectric strength of SF_6. It was found that although the breakdown voltage of SF_6 was highest, the ionization current of SF_6 was highest. On the other hand, the breakdown voltage of air was lowest, while the ionization current of air proved to be lowest.

TABLE 1. Averaged relative dielectric strength: each dielectric strength is normalized by that of SF_6.

Gas	Relative dielectric strength
SF_6	1
C_2F_6	0.81
Air (80%) + SF_6 (20%)	0.75
Air (90%) + SF_6 (10%)	0.69
Air (95%) + SF_6 (5%)	0.63
CO_2	0.51
Air	0.49

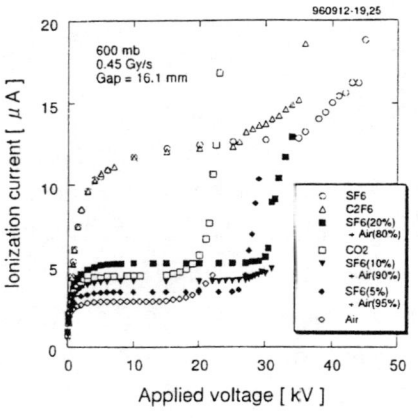 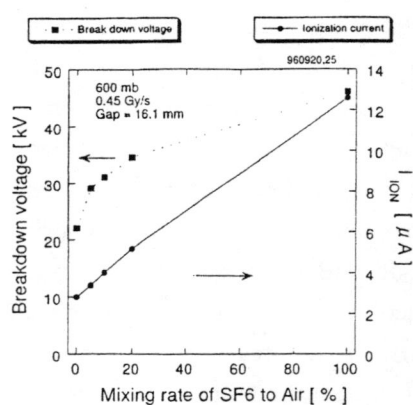

FIGURE 2. Ionization current of test gases.   **FIGURE 3.** Breakdown voltage and ionization current of mixed gas.

The dependence of ionization current against applied voltage is well similar to that at conventional ionization chambers. It can be seen that the curve of ionization current is composed of three parts, from region I to region III. Region I is recombination region. Applied voltage, namely, electric field is low, so that both positive ions and electrons move at the low speed. Since charged particles at the low speed are easy to recombine, they are difficult to reach the electrodes; ionization current is low. As the applied voltage increases, the speed of charged particles increases, resulting in the decrease in the ratio of recombination. Region II is saturation region. The ionization current becomes saturated, or the ionization current is almost independent of the applied voltage in this region, where the number of charged particles collected at the electrodes would be equal to the number of charged particles produced by the gamma-rays. Region III is secondary ionization region. With the applied voltage, the speed of electrons becomes so high that they cause secondary ionization, which are followed by electron avalanche and breakdown.

To decrease ionization current, nevertheless, keeping high-voltage capability, mixing of air and SF_6 gases was studied. Experimental results for mixture of air and SF_6 gases are shown in Fig. 3. Left side of vertical axis is breakdown voltage and right side is a new term of " I_{ION} ". The term of I_{ION} is, for convenience, defined as the ionization current at a half of breakdown voltage; it is practically regarded as saturation current. The graph

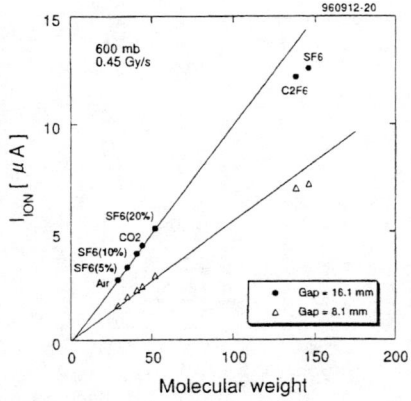
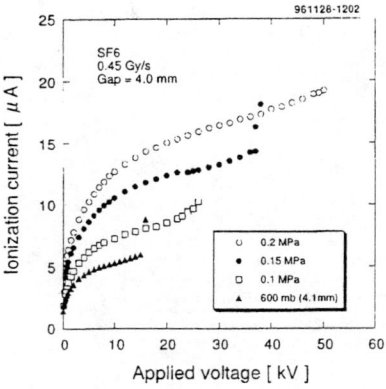

FIGURE 4. I_{ION} as a function of molecular weight.

FIGURE 5. Ionization current of SF6 as a function of applied voltage.

shows that a small quantity of SF6 made breakdown voltage increase, and that I_{ION} was proportional to volume rate of SF6 gas. Breakdown voltage of the mixture of 95 % air and 5% SF6 gases was 63 % of that of pure SF6. On the other hand, I_{ION} was 30 % of that of pure SF6. This indicates that the mixture of air and a little SF6 gases has low ionization current and high capability of voltage holding. Therefore, mixing gas will be effective from the view point of suppressing ionization current.

Figure 4 is a plot of I_{ION} versus molecular weight. Circle is I_{ION} for a gap length of 16.1 mm. Triangle is I_{ION} for a gap length of 8.1 mm. Molecular weight of mixing gases is defined as averaged molecular weight which is the sum of molecular weight multiplied by ratio of partial pressure. This figure shows that I_{ION}, namely, saturation current was proportional to molecular weight.

B. Pressure Dependence of SF6 Gas

Ionization current of SF6 gas was measured with gas pressure changed. Figure 5 shows the dependence of ionization current of SF6 against voltage applied at the gap length of 4.0 mm. Gas pressure was at 0.59 (600 mb), 0.1, 0.15 and 0.2 MPa. It should be noted that only the gap length at 0.59 MPa (600 mb) is not 4.0 mm but 4.1 mm. Figure 6 shows the pressure dependence of I_{ION} of SF6 gas. I_{ION} proved to increase linearly with gas pressure.

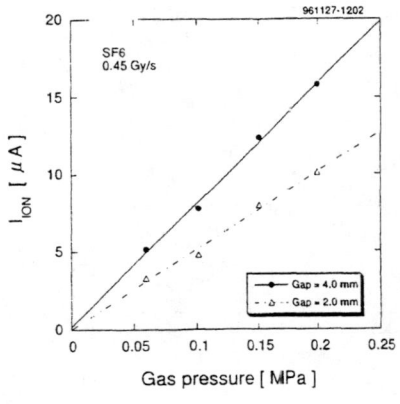
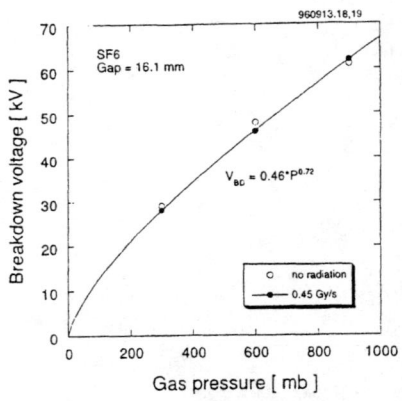

FIGURE 6. I_{ION} as a function of gas pressure.

FIGURE 7. Breakdown voltage as a function of gas pressure.

Breakdown voltage of SF6 are shown as a function of gas pressure in Fig. 7. Open sign is the breakdown voltage without gamma-rays and closed sign is that with gamma-rays. Breakdown voltage increased with gas pressure to the 0.72 in both conditions.

C. Gap Length Dependence

Figure 8 shows I_{ION} as a function of the gap length. I_{ION} increased linearly with the gap length. Linear relationship between I_{ION} and the gap length implies that there will be linear relationship between I_{ION} and the volume of gas. The figure also indicates that I_{ION} is not zero at the gap length of 0 mm. This might be attributed to the effect of a space between electrodes and pyrex glass; charged particles generated in the space also might be collected by the electrodes.

Breakdown voltage of gases is plotted as a function of the gap length in Fig. 9. Open sign is breakdown voltage without gamma-rays and closed sign is that with gamma-rays. Breakdown voltage increased with the gap length to the 0.73 in both cases.

D. Absorbed Dose Rate Dependence of SF6 Gas

Changing absorbed dose rate is equivalent to changing the distance from the source of

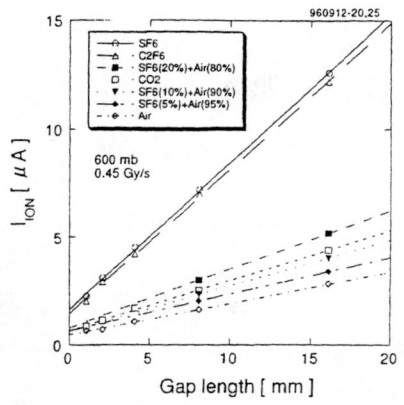
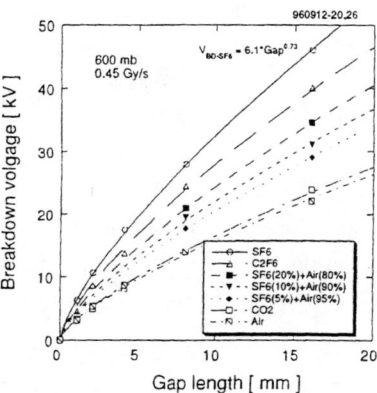

FIGURE 8. I_{ION} as a function of gap length between electrodes.

FIGURE 9. Breakdown voltage as a function of gap length.

gamma-rays to the center of electrodes. Figure 10 shows the linear relationship between I_{ION} and absorbed dose rate.

Break down voltage as a function of absorbed dose rate is shown in Fig. 11. As compared with breakdown voltage without gamma-rays, the decrease in breakdown voltage with gamma-rays was about 10 %. However, there was not clear relationship linking absorbed dose rate and the degree of the decrease in breakdown voltage.

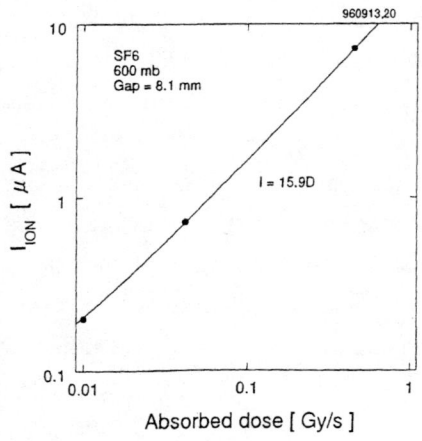
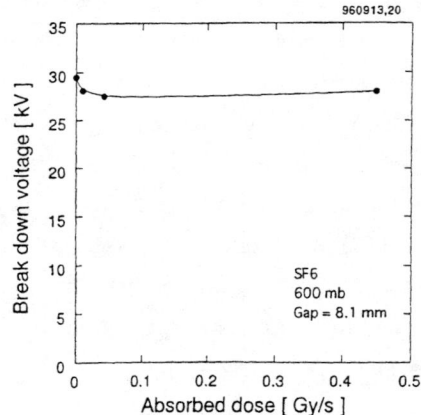

FIGURE 10. I_{ION} as a function of absorbed dose rate.

FIGURE 11. Breakdown voltage as a function of absorbed dose rate.

IV. DISCUSSION

A. Ionization Current

Although electrons are prone to be attached to SF_6 molecules, experiment demonstrated that ionization current of SF_6 gas had been high. This might be attributable to the electron-energy dependence of the electron-attachment cross section of a SF_6 molecule. It is true that the cross section has the highest peak at the electron-energy of nearly zero eV, but it is very small at higher region of electron energy [3]. Since the energy of electrons generated by the interactions with gamma-rays is almost much high, the electrons will not be attached to SF_6 molecules; therefore, the electrons can migrate in the gas.

The observation of the linear relationship between I_{ION} and molecular weight is of interest. It is inferred to be attributed to the fact that a dominant interaction is compton scattering at the energy of Co-60 gamma-rays, and that the cross section of compton scattering increases proportionally with atomic number [4]; ionization current might be proportional to the molecular weight at same gas pressure, same temperature and same volume.

If other nuclear reaction except for compton scattering is dominant, it is unknown whether the results with Co-60 gamma-rays are applicable. This is because both the energy and the nuclide dependence of the cross sections for other nuclear reaction is different from those for Co-60 gamma-rays. In ITER, a neutronics simulation has revealed that the space around the beam source would be mainly exposed to neutrons [2]. The interaction of neutrons with matter is very complicated: neutrons will be scattered by nuclei elastically or inelastically, will be captured and followed by gamma-ray emission, will induce (n, p) and (n, alpha) reactions, or will cause fission. Secondary particles generated by the neutrons, such as scattered neutrons, gamma-rays, protons, alpha particles, and fission products can cause next nuclear reactions. Unlike the gamma-rays from Co-60, the ionization processes resulting from neutrons are so complicated that the experimental results with the gamma-rays might not be applicable to the case of neutrons. More accurate estimation of ionization current will require irradiation experiments with neutrons.

B. Breakdown Voltage

Breakdown voltage was also measured with the parallel-disk electrodes. It is true that it is suitable for measuring ionization current, but it might not be proper for measuring breakdown voltage. Electrodes like the Rogowski electrode will be needed for accurate measurement of breakdown voltage. However, much information such as relative dielectric strength, gap length dependence, pressure dependence and absorbed dose dependence with gamma-rays is thought to be useful. But, of course, more detailed studies by the use of the proper electrodes like the Rogowski electrode will be necessary.

C. Estimation of Ionization Current at the ITER NBI

An experimental formula, which can estimate saturation current, is obtained on the basis of above experimental results. Saturation current, whose dimension is ampere, generated by Co-60 gamma-rays is

$$I_{total} [A] = fgas * M * V * P * D \text{-----} (1)$$

where the fgas is constant obtained from the experiment, which was 0.00143. M is molecular weight. V is the volume of gas: dimension is m^3. P is the pressure of gas: dimension is atmospheric pressure. D is absorbed dose rate in case of air: dimension is Gy/s. It should be noted that electric field between electrodes is assumed to be high enough to collect all charged particles generated.

In the ITER NBI, the volume of the space between the beam source and its vessel will be about 28.7 m^3. Table 2 shows required gas pressure around the beam source. At the each pressure, dielectric capability of each gas is as same as that of CO_2 at 20 atm. Then, the total ionization current can be estimated by assuming the absorbed dose rate. Some neutronics simulations have been conducted to evaluate the absorbed dose rate so far, but they have not shown good agreement. There are uncertainties of two order of magnitudes, about 0.02 Gy/s to about 2 Gy/s. Supposing the absorbed dose rate of 2 Gy/s, the total ionization currents are estimated, as shown in Table 3. In case of SF_6, total ionization current is estimated to be 88 A. In case of mixing gas of 5% SF_6 and 95% air, it is estimated to be 41 A. It is clear that those value will not be acceptable from the view point of capacity of both acceleration power supply and cooling system. However, of course, these values depend on the accuracy of the neutronics simulations; therefore, it will be significant issue to evaluate the absorbed dose rate more precisely.

V. SUMMARY

An gamma-rays irradiation experiments was done at Co-60 irradiation facility at JAERI. Ionization currents and voltage holding characteristics were measured for gases of air, SF_6, C_2F_6, CO_2, and mixture of air and SF_6. Ionization current increased linearly with gap length between electrodes, gas pressure, absorbed dose rate, and molecular weight. The experimental formula to estimate ionization current was attained on the basis of the experimental results. Decrease of breakdown voltage with gamma-rays was at most about 10 %. However, it was found that the degree of the degradation of breakdown voltage did not depend strongly on absorbed dose rate.

ACKNOWLEDGMENTS

We would like to thank Mr. H. Nagayama and Mr. T. Haneta of Co-60 facility at JAERI for technical assistance throughout the experiment. We would also thank Dr. R. Hemsworth of ITER JCT and other members of NBI Heating Laboratory for their valuable discussions and comments. We would acknowledge Dr. M. Ohta and Dr. P. L. Mondino for their support and encouragement.

TABLE 2. Required gas pressure for the ITER NBI: dielectric capability of each gas is as same as that of CO_2 at 20 atm.

Gas	Pressure [atm]
SF6	7.9
C2F6	10.6
Air (80%) + SF6 (20%)	11.8
Air (90%) + SF6 (10%)	13.2
Air (95%) + SF6 (5%)	15
Air	21.2
CO2	20

TABLE 3. Total ionization current at the ITER NBI: ionization current is estimated from the experimental formula.

Gas	Current [A]
SF6	88
C2F6	116
Air (80%) + SF6 (20%)	56
Air (90%) + SF6 (10%)	46
Air (95%) + SF6 (5%)	41
Air	50
CO2	75

REFERENCE

1. E.Hodgson : Insulating Gas Characterization, The 1st technical meeting, Cadarache 2-5 July 1996
2. T.Inoue : Radiation calculations for the ITER NBI, The 2nd technical meeting, Naka 17-20 February 1997
3. M. Hayashi and T.Nimura : J. Phys., D17, 2215 (1984)
4. N.Tsoulfanidis : Measurement and Detection of Radiation, Hemisphere Publication Corporation (1983)

Surface Charging of Insulated Materials by Negative Ion Beam Bombardment

Junzo Ishikawa, Hiroshi Tsuji and Yasuhito Gotoh

*Department of Electronic Science and Engineering,
Kyoto University, Sakyo-ku, Kyoto 606-01, Japan*

Abstract. When a negative ion beam bombards an insulated material such as an isolated electrode or an insulator, a surface voltage is induced by the surface charging and will affect the ion beam trajectories during beam transport. The mechanism of the surface charging of an insulated electrode is related with its secondary electron emission factor and energy distribution. The charging voltage in this case is as low as several volts and its polarity is positive. On the other hand, in case of an insulator, the polarity is negative and its value is several volts because an electric dipole layer is generated just on the surface due to negative ion beam bombardment. These low surface charging voltages lead an advantage in negative ion beam applications as well as in negative ion beam transport.

INTRODUCTION

In positive ion beam transport systems, surface charging voltages of insulated materials due to ion bombardment are quite large, and sometimes they make the beam transport difficult. The surface charging voltage of an insulated material is growing during positive ion bombardment because both incoming positive charge of positive ion and outgoing negative charge of secondary electrons increase positive charging of the material surface, and it saturates at an acceleration voltage of ion beam.

On the other hand, in the case of negative ion bombardment, incoming charge is negative and outgoing charge is also negative. It is, therefore, expected that the charge balance on the material surface could be easily obtained and the surface charging voltage is relatively low. Evaluation of this charging voltage is vital to design negative ion beam transport systems and for negative ion applications.

In this paper, the mechanism and experimental data of surface charging voltage of insulated materials such as isolated electrode and insulator due to negative ion beam bombardment are discussed together with secondary electron emission factor and energy distribution.

SECONDARY ELECTRON EMISSION BY NEGATIVE ION BOMBARDMENT

The secondary electron emission factors as a function of ion velocity when negative ions of C^- and C^+ bombard the surface of platinum metal and SiO_2 film, is shown in Fig.1(1). In SiO_2 film, thermally grown silicon dioxide films (120 nm thick) on silicon substrate was used, and the measurement was performed with an order of the pA ion current to avoid surface charging of the films.

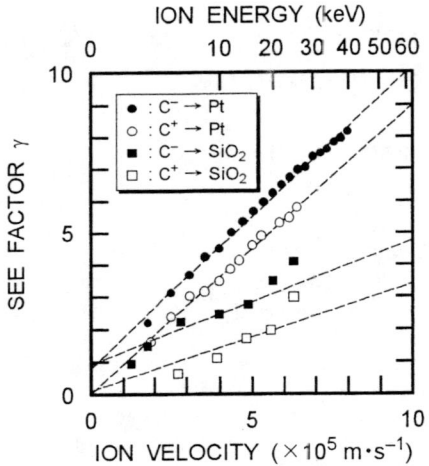
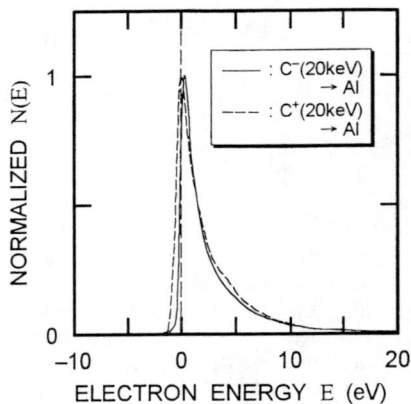

FIGURE 1. Secondary electron emission factor as a function of ion velocity.

FIGURE 2. Normalized energy distribution of secondary electrons.

Secondary emission is mainly caused by kinetic emission in the experimental energy range because such characteristics are in proportion to ion velocity. The negative-ion-induced secondary electron emission factor is larger by about 1 than the positive-ion-induced emission. It is thought that a negative ion easily releases an extra-electron when colliding with surface atoms. Thus, the negative-ion-induced secondary electron emission factor is usually above unity when the energy of negative ion beam is larger than several keV.

The normalized energy distributions of secondary electrons by negative- and positive-ion impacts are indicated in Fig. 2. The distribution by negative-ion impact is similar to that of positive-ion impact, which peaks at 1 - 2 eV and the rapidly decreases as the energy increases in the range of 2 eV to several tens eV. The shape of the distribution slightly depends on the material but is almost independent of ion energy.

CHARGING VOLTAGE OF ISOLATED ELECTRODE

In the isolated electrode, it is thought that the voltage of the electrode is uniform in space since electrons quickly diffuse in conductive material. When negative ions bombard the isolated electrode, the voltage of the electrode initially increases positively because the secondary electron emission factor is more than unity. The positive voltage pulls back the low-energy secondary electrons with time. At the steady state, the voltage of the electrode becomes saturated under the condition of charge balance where the outgoing charge of secondary electrons having an energy higher than the voltage of the electrode, coincides with the incoming charge of negative ions, as illustrated in Fig. 3(2-5). This condition is expressed by the following equation.

$$\gamma \int_{eV_c}^{E_{max}} N(E)dE = 1, \qquad (1)$$

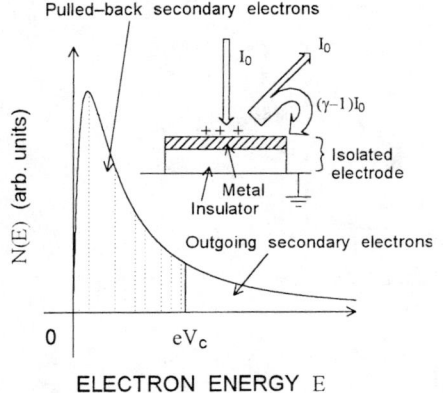
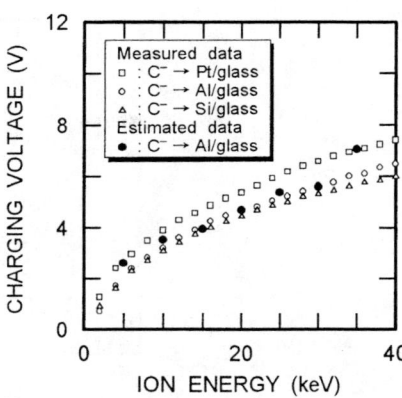

FIGURE 3. Schematic diagram illustrating the charge equilibrium of an isolated electrode during negative ion bombardment.

FIGURE 4. Charging voltage of the isolated electrodes as a function of ion energy when carbon negative ions bombarded isolated Pt, Al and Si. Calculated values using eq.(1) are also indicated by black circles.

where γ is the secondary electron emission factor, and $N(E)$ is the energy distribution function of secondary electrons. Therefore, the charging voltage of an isolated electrode can be calculated from the emission factor and the energy distribution function of secondary electrons.

A comparison with the charging voltage of direct measurement by a high-impedance voltmeter, and the calculated value by using the above equation is shown in Fig. 4. These are in good agreement, thus, the charging model for an isolated electrode is accurate as shown in Fig. 3. From a simple analysis, it is found that the charging voltage is proportional to the square root of negative-ion energy.

In Fig. 4, the charging voltages of various material electrodes are also indicated. The charging voltage of isolated electrode due to negative ion bombardment is as low as several positive volts.

CHARGING VOLTAGE OF INSULATOR

In insulators, charging voltage cannot be measured directly. However, it can be estimated from the peak energy shift of the electron distribution function of secondary electrons from the insulator surface(2-5). The estimated charging voltage of quartz glass and photoresist surfaces as a function of ion energy in carbon negative ion bombardment are indicated in Fig. 5. These voltages are as low as several negative volts, which differ in polarity from the isolated electrode case.

The charging mechanism for insulators is thought that the secondary electron emission factor is limited to unity due to a small negative potential which is formed by an electric double layer generated in the vicinity of the insulating surface in negative-ion impact. The electric double layer consists of the negative charge layer on the surface, and the positive charge layer near the ion projected range. By using this model, the surface charging voltage of quartz glass whose secondary electron emission

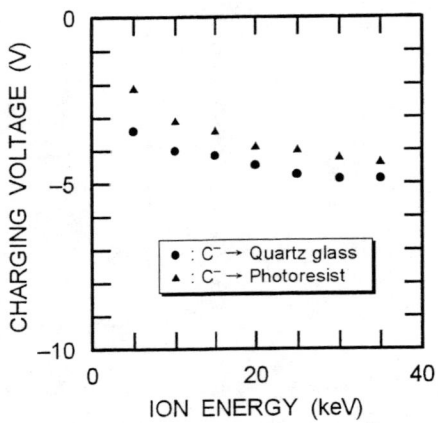

FIGURE 5. Charging voltage of quartz glass and photoresist due to carbon negative ion bombardment as a function of ion energy.

factor is assumed to be 2 - 4, is calculated as several negative volts.

POSSIBILITIES OF NEGATIVE ION APPLICATION

To estimate a yield rates of ULSI under the fabrication process with negative-ion implantation, a yield rate of the test element group (TEG) device with negative-ion implantation was measured. From the experiment, almost no damage would be expected in the ULSI fabrication process with negative-ion implantation.

When micron-sized powder particles are implanted with positive ions, they are scattered by the Coulomb repulsion force due to surface charging, and sufficient ion implantation is difficult to achieve. With negative ions, however, no scattering was observed.

CONCLUSIONS

Surface charging voltages due to negative ion beam bombardment on insulated materials are within ± 10 volts. Thus, negative ion has a great advantage in its beam transport. This advantage can be used for negative ion applications such as ULSI fabrication and scatter-less ion implantation into micron-sized powders.

REFERENCES

1. Toyota. Y., Tsuji. H., Gotoh. Y. and Ishikawa. J., Jpn. J. Appl. Phys. **34**, 6487 (1995).
2. Ishikawa. J., Tsuji. H., Toyota. Y., Gotoh. Y., Matsuda. K., Tanjo. M. and Sakai. S., Nucl. Instrum. & Methods **B96,** 7 (1995).
3. Ishikawa. J., *Mat. Res. Soc. Symp. Proc.,* Vol.354, p.99 (MRS,Pittsburgh,1995).
4. Ishikawa. J., Surface and Coating Technology **65**, 64 (1994).
5. Ishikawa. J., Rev. Sci. Instrum. **65**, 1290 (1994).

Secondary Electron Emission from Insulators*

H.J. Hopman and J. Verhoeven

FOM-INSTITUTE OF ATOMIC AND MOLECULAR PHYSICS
Kruislaan 407, Nl-1098 SJ Amsterdam, The Netherlands

Abstract. A modified LEED/Auger diagnostic allows the measurement of the coefficient of secondary electron emission, δ, under electron bombardment. To measure currents on insulators, the system is pulsed. Problems due to charging are alleviated by alternating between pulses in which the target charges negatively ($\delta < 1$) or positively ($\delta > 1$).
A measurement of δ requires a total primary beam charge of $\approx$ - 0.2 nC. Results obtained on alumina and synthetic diamond will be presented.

* Work supported by Philips Research, Eindhoven, The Netherlands.

INTRODUCTION

Fig. 1 presents examples of a total yield measurement. Given is the ratio δ, of the current of secondary electrons (SE) leaving the sample and that of the primary electrons (PE) impacting on the material, as a function of the energy of the primary electrons E_p. The general behaviour is explained in terms of the mean free paths in the material [1]. For a primary energy $E_p = E_{max}$, $\delta(E_p)$ reaches a maximum value when the mean free path of the PE, λ_{pe}, and the mean free path of the SE, λ_{se}, are roughly equal. For $E_p < E_{max}$, we have $\lambda_{pe} < \lambda_{se}$. With increasing primary energy, more SE are created, which all have a similar escape probability. So δ increases linearly with E_p. For $E_p > E_{max}$, we have $\lambda_{pe} > \lambda_{se}$. Now δ decreases with increasing primary energy, because the number of electrons created in the escape layer of thickness $\approx 4\lambda_{se}$ decreases with the decrease of the cross section for production of SE.

Irradiation of insulators results in accumulation of charges. For large primary energies, when $\lambda_{pe} > \lambda_{se}$, one may expect the formation of a (not necessarily symmetric) double layer: the keV PE come to rest some 100 nm below the surface, the SE leave behind a positive charge in the escape layer at the surface (some 10 nm thick). Averaged over the total volume, there is a net positive charge when more than one SE per incoming PE leaves the insulator ($\delta > 1$). This positive charge results in a positive potential V_s at the surface. The majority of SE have an energy of a few eV. Then, a small positive surface potential, preventing the escape from the target of low energy SE, results in large errors in the measurement of δ. On the other hand, when $\lambda_{pe} < \lambda_{se}$, one should not expect a separation of negative and positive charges.

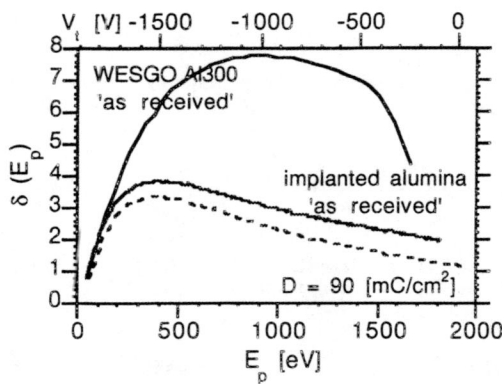

FIGURE 1. Secondary electron emission curves $\delta(E_p)$ measured on 1 mm thick disks of alumina. E_p is varied by sweeping V_t (top axis) from +90 to -1910 V. The full curves present data taken on virginal spots of a reference sample and an implanted sample. The dotted curve gives data taken on a spot irradiated with a 2 keV beam up to a dose D = 90 mC/cm^2.

Work is performed in a UHV system (p ≤ 10^{-9} mbar), equipped with a LEED/Auger system. The electron beam (φ = 1 mm) is incident normally to the surfaces studied. The beam is pulsed for 5 µs at a 100 Hz repetition frequency. The sample is clamped to a molybdenum support, which can be biased at an adjustable target potential V_t. The impact kinetic energy of the PE on the sample surface is E_p = - e (V_b - V_t), where V_b is the gun voltage. To vary E_p, V_t is swept from +90 to -1910 V at constant gun voltage. Measured is the current I_t through the sample to ground. The coefficient of secondary electron emission δ is given by δ = 1 - (I_t/I_p). The primary beam current I_p is obtained from the current at a positively biased target (V_t = + 90 V), for which the escape of the SE is prevented ($\delta = 0$).

Net charging of the insulator is reduced by alternating between a pulse with a negative target bias and an adequate number of pulses with a positive bias [2]. Care is taken that the net injected charge is negative. The result is an error in the value of E_p. The trick aims at trading a large error in δ_{max} for a small one in E_p.

REFERENCE AND IMPLANTED ALUMINA

CERN are running a project to improve the performance of high voltage insulators. By treating alumina surfaces with Farmsum Associates' proprietary "smart 2 low" high dose ion implantation process at AEA Technology, Harwell, UK, it proved possible to reduce the break down frequency [3]. Below we report on two effects that may influence the breakdown behaviour: reduced secondary emission and enhanced surface conduction for implanted alumina.

The upper curve in Fig. 1 gives the $\delta(E_p)$ function obtained on a virginal spot of a reference sample of Al300 Wesgo alumina. It has $\delta_{max} = 7.8$ at $E_{max} \approx 870$ eV. Probing the sample at 20 different positions, we find $\delta_{max} = 7.6 \pm 0.4$. The break in the slope of the curve at $E_p \approx 1500$ eV indicates a perturbation due to charging effects. For $V_t > -500$ V, a fraction of the SE is trapped in the potential well formed by the double layer. Repeating the measurement 10 times on the same spot (electron dose D ≈ 0.3 μC/cm^2), δ_{max} decreased to 7.15. This value represents an initial saturation value. At higher dose, D ≈ 6 mC/cm^2, the emission decreased further to $\delta_{max} = 6.65$.

The middle curve of Fig. 1 presents the secondary electron yield of implanted alumina. In this graph $\delta_{max} \approx 3.75$ and $E_{max} \approx 330$ eV. The curve is not perturbed by charging. Probing the sample at 10 different position, we find $\delta_{max} = 3.80 \pm 0.05$. This value is comparable to the one for glass ($\delta_{max} = 3.2$, $E_{max} \approx 290$ eV). Bombarding the target with a 2.7 keV beam to a dose dose D ≈ 90 mC/cm^2, δ_{max} decreased $\approx 10\%$ (dotted curve in Fig. 1).

In agreement with experiments done by other groups [4], it is found that the implanted alumina sample has a conducting surface layer. It follows from the fact that with a DC beam $I_t \neq 0$. Further, the conductivity inhibits the formation of a double layer. Measurements do not show the break in the slope at $E_p \approx 1500$ eV.

SYNTHETIC DIAMOND

Both, natural and synthetic diamond are wide band gap insulators with a negative electron affinity, meaning that the conduction band lies above the vacuum level. In principle, this makes diamond an interesting converter material for a D$^-$ source. Recent measurements resulted in conversion efficiencies (H$^+$ to H$^-$) of 5% [5]. However, our measurements demonstrate an extreme sensitivity of the value of δ on surface preparation. This suggests that appropriate conditioning of a diamond converter may lead to appreciably higher conversion efficiencies than 5%.

Measurements have been done on coarse crystalline (X-tal size ≈ 1 μm) CVD diamond films (≈ 10 μm thick) deposited on a Mo substrate [6]. The samples as received showed a position dependence with $3.0 < \delta_{max} < 5.0$. The observed value of E_{max} [eV] $\approx 770 \pm 60$. Fig. 2 presents a δ-curve measured at virgin positions of high δ_{max}. All samples conduct a DC beam current. There are no clear signs of charging of the samples.

Irradiating a sample by a DC beam ($I_p = 45$ nA, $E_p = 1910$ eV, $\delta > 1$), it is seen that the positive target current I_t ($\delta > 1$) exponentially approaches a limiting value. The characteristic growth time $\tau \approx 100$ s, corresponds to a characteristic dose $D_c \approx 0.5$ mC/cm^2. The high yield curve in Fig. 2 gives the δ-curve measured after reaching saturation. The increase is local at the irradiated spot only. The high yield did not deteriorate over a period of 140 hours in UHV. The nearly linear relation of δ with E_{max} strongly resembles data published by Shih [7].

The exponential time dependence of the secondary emission suggests a one-step process like charge trapping or particle desorption from the surface as the cause. The small characteristic dose $D \approx 1$ mC/cm^2 makes particle desorption unlikely. Irradiation of insulators results in the formation of a double layer. The associated electric field causes the SE to drift toward the surface, probably with small (sub eV) drift energies. Only in materials displaying a negative electron affinity, such electrons escape the solid on arriving at the surface. Therefore, this effect is absent in alumina.

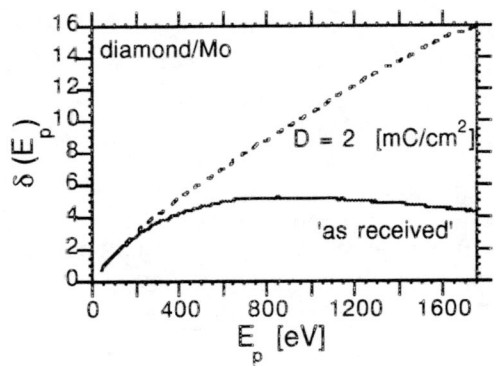

FIGURE 2. Secondary emission curves for synthetic diamond, measured before and after irradiation of a spot with an electron dose D = 2 mC/cm^2.

ACKNOWLEDGEMENTS

The authors are indebted to Dr Peter K. Bachmann, Philips Res. labs. Aachen, Germany, for providing the diamond samples, and to Mr Bill Taylor of Farmsum Ass., Hartlepool UK, for the alumina samples. They thank Hans Alberda, Idsart Attema, and Hans Zeylemaker, of the FOM-Institute AMOLF, for technical support.

REFERENCES

1. H. Bruining, "*The physics and applications of secondary electron emision*", London, Pergamon Press (1954).
2. J.J. Scholtz, Philips Electron Optics, Eindhoven, The Netherlands, private comm.
3. B.J. Goddard et al., Proc. 17-th Int. *Symp. on Discharges and Electrical Insulation in Vacuum*, Berkeley, July 1996, ISBN 0-7803-2906-6.
4. V.V. Lopatin, A.V. Kabyshev, L.S. Bushnev, *Phys. Stat. Sol.* **116** (1989) K69.
5. P. Wurz, R. Schletti, M.R. Aelig, *Surf. Sci.* **373** (1997) 56.
6. P.K. Bachmann, W. van Enckefort, *Diamond Rel. Mater.* **1** (1992) 1021.
7. A. Shih, J. Yater, P. Pehrsson, J. Butler, C. Hor, R. Abrams, *J. Appl. Phys.* **82** (Aug. 1997).

APPLICATIONS AND SYSTEMS

UPDATE OF NEUTRAL BEAMS FOR ITER

R. S. Hemsworth[1], E Di Pietro[1], T. Inoue[2], A. Krylov[2], Y. Okumura[3], M. Tanii[5], and M. Watson[5]

1 Association EURATOM - C. E. A., C. E. Cadarache, 13108 St. Paul-lez-Durance, France

2 ITER Joint Central Team, Naka, Japan

3 JAERI, Naka, Japan

4 Nissin Co. Ltd.., Kyoto, Japan

5 JET Joint Undertaking, Abingdon, UK

Abstract : The design of the neutral beam injection heating and current drive system for the International Tokamak Experimental Reactor is based on the neutralisation of accelerated D⁻ ions. This paper describes changes in the design which have occurred since it was presented at the first of these symposia (1) and outlines the negative ion beam source proposed as the basis of the ITER diagnostic neutral beam injector. The proposed NBI system consists of three negative ion based neutral injectors, delivering a total of 50 MW of 1 MeV D^0 to the ITER plasma for pulse lengths of >10,000 s. The diagnostic beam system consists of a single beamline delivering a 5 MW modulated beam of 125 keV H^0 to the ITER plasma.

NBI HEATING AND CURRENT DRIVE SYSTEM

The design of the injectors for ITER is dominated by four factors: The basic specification, 50 MW, 1 MeV D^0, ≥10,000 s, and the need to induce rotation and/or drive current in the plasma; the ITER "geography", which limits severely the available space envelope, the extensive stray magnetic field of ITER; and the hostile radiation environment. To induce rotation and/or to produce current drive requires that the beams must be injected at an angle such that they are quasi-tangential to the magnetic axis, hence they must be injected through the large horizontal ports on the equatorial plane of ITER. As ITER is to be located below ground, this means that they have to be situated between the biological shield and the wall of the "service" gallery which is inboard of the pit wall. This restricts the overall length of the beamline and beam source to the order of 10 m. Other ITER sub-systems are situated above and below the injectors, which restricts their height and prevents vertical stacking of the injectors at any given port.

As the neutralisation of 1 MeV D^+ is too low to allow this to be considered for the production of 1 MeV D^0 beams, the neutralisation of 1 MeV D^- will be used, the efficiency of which is ≈60% with a simple gas target. Detailed calculations of beam neutralisation, reionization and transmission have led to the general specification given in Table 1. The present structure of the injectors can be seen in Figs. 1 and 2.

TABLE 1   GENERAL SPECIFICATION OF THE ITER NEUTRAL BEAM SYSTEM

Total power to ITER	50 MW	Energy	1 MeV
Species	D^0	Tangency radius	6.5 m
Number of ports	3	Number of injectors	3
Power to ITER per injector	16.7 MW	Pulse length	>1000 s
Number of sources per injector	1	Accelerated D^- per beam source	≈40 A
Geometric beam transmission	>85%	High divergent fraction of beam (aberrations etc.)	15%
Neutralisation efficiency	60%	Reionization loss	≈7%

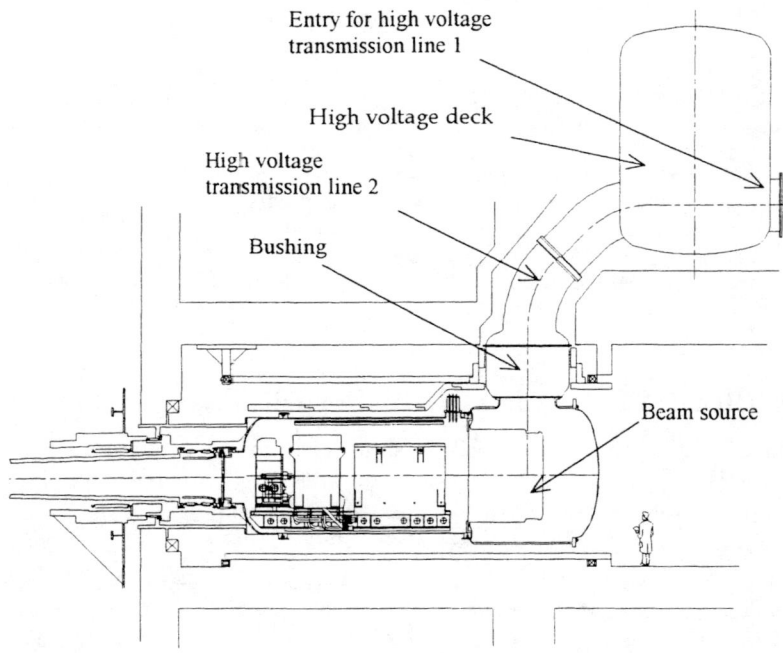

FIGURE 1. Elevation section through an ITER injector

The "High Voltage Deck" shown in Fig.1 contains the high current auxiliary supplies for the arc, plasma grid filter, filaments, plasma grid bias and the extraction grid. All high potentials (1000, 800, 600, and 200 kV) and the ac power for the high voltage deck are supplied via high voltage transmission line 1.

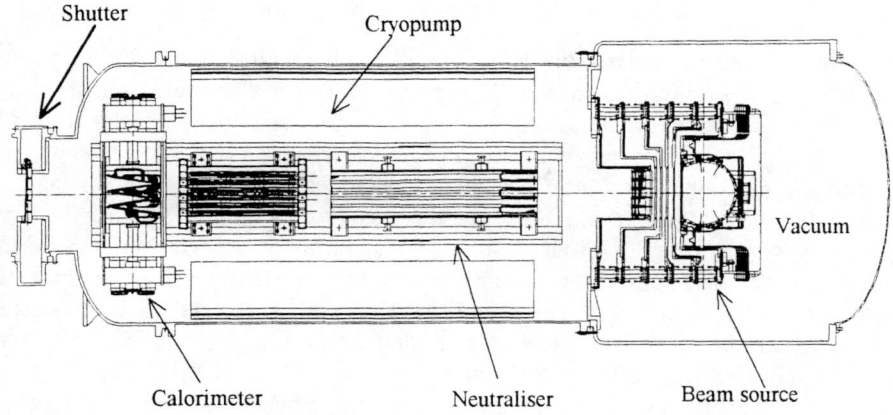

FIGURE 2. Plan section through an ITER injector

DESIGN CHANGES

Changes, and proposed changes, from the beamline design reported at the first of these joint meetings (1) are given below.

Absolute Valve Removal

It was proposed previously to have an absolute valve between the injector and the torus, which would have used all metal seals. The body of the valve would be part of the first confinement barrier, hence it would have had to be capable of withstanding the 0.5 MPa internal pressure that could be generated by an accident in the torus. The main role of the valve would have been to allow maintenance of the injector without venting the torus. It has been decided that this is insufficient reason for this valve considering that:

- The overall length of the injectors is fixed by the ITER "geography", so that the use of the valve reduced the space available for the beamline itself.
- Remote maintenance, such as replacement of the metal seal, would have been very difficult and time consuming considering the location of the valve.
- Large quantities of dust are expected to be generated within the torus, which could be carbon and/or tungsten. As the exclusion of this dust from the valve cannot be guaranteed, the reliability of the valve is uncertain.
- The cost of the valve was high.

R&D on the valve seal, in the absence and presence of carbon and tungsten dust, is still continuing as an ITER task in the EU, which, if successful, could allow the use of such a valve to be reconsidered.

Relocation of the Fast Shutter

The shutter is used to prevent the ingress of T_2 into the injectors prior to plasma initiation, and at the termination of the ITER pulse, for either a normal termination or a termination provoked by a disruption. It will also be used to avoid gas from the injectors flowing to the torus when the injector cryopumps are regenerated. To perform these functions the shutter has to be positioned at the end of the beamline. Previously the shutter was located just inside the beamline vacuum vessel with the drive mechanism passing through the wall of the vessel, an arrangement which minimised the impact on the overall length of the system. Given that the shutter will open and close at least twice during every ITER pulse, and that the "seal" is metallic, it is considered that it will require maintenance several times during the 20 year life of ITER. To avoid having to remove all the beamline components in order to maintain the shutter, it has been moved to a position immediately downstream of the beamline vessel in a separate casing (see Figs. 1 and 2). Maintenance will be performed in the ITER hot cell after cutting of the line up and downstream of the casing and transporting the shutter to the hot cell in a sealed cask.

Calorimeter Design

The calorimeter design has now advanced significantly. It consists of 5 vertical panels which are combined to have one group of 3 in the form of an "N" an a group of 2 in the form of a "V". When the beams are being injected into ITER, these sets of panels are located either side of the beam path. For commissioning of the injectors, the panels are moved horizontally to intercept the 5 columns of beam emerging from the 5 channels of the residual ion dump. The panels are made from arrays of interlocking, vertically disposed, swirl tubes which have a cross section which is essentially rectangular. A schematic section of the calorimeter is shown as Fig. 3.

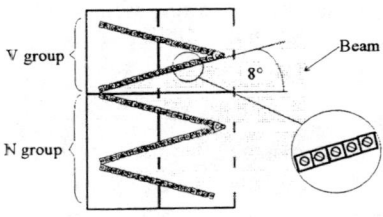

FIGURE 3. Schematic section through the calorimeter

Vacuum Insulated Source Option

The beam source may now be "vacuum insulated", as shown in Figs. 1 and 2. The multi-grid electrostatic accelerator and the ion source are held at -1 MV and the negative ions are extracted from the ion source and accelerated up to ground potential. In the gas insulated design the ion source enclosure and the insulator column formed the vacuum boundary and high pressure gas provided insulation between these and the pressure vessel walls at ground potential. The insulator column consists of 5 sets of alumina ceramic columns, 290 mm tall, 2.7 m internal diameter, interspaced with metal flanges, onto which the plasma grid and the accelerator grid supports are mounted. In order to minimise mechanical displacement and stress between the flange and the ceramic due to temperature variations, titanium alloy, which has a low thermal expansion coefficient, is chosen for the material of the flanges.

Experiments carried out in Europe and Japan (2, 3, 4) have shown that the conductivity of all the insulating gases so far considered increases considerably in the presence of radiation, and calculations have shown that the currents flowing in the gas surrounding the ITER beam source could be many amperes (5). This is not acceptable as it would result in considerable power loss, hence a considerable reduction in the overall efficiency of the injector, and possibly excessive heating of the gas, resulting in an unacceptable increase in the gas pressure. To avoid this a vacuum insulated beam source (VIBS) is now being designed.

In the VIBS there is no insulating gas, and vacuum provides the insulation between the beam source and the beam source vessel wall, which forms the vacuum boundary. Power to the beam source is provided via high voltage transmission line 2 (see Fig. 1) and the bushing at the junction of the transmission line and the beam source vessel is part of the vacuum boundary. The VIBS is illustrated in Fig. 4.

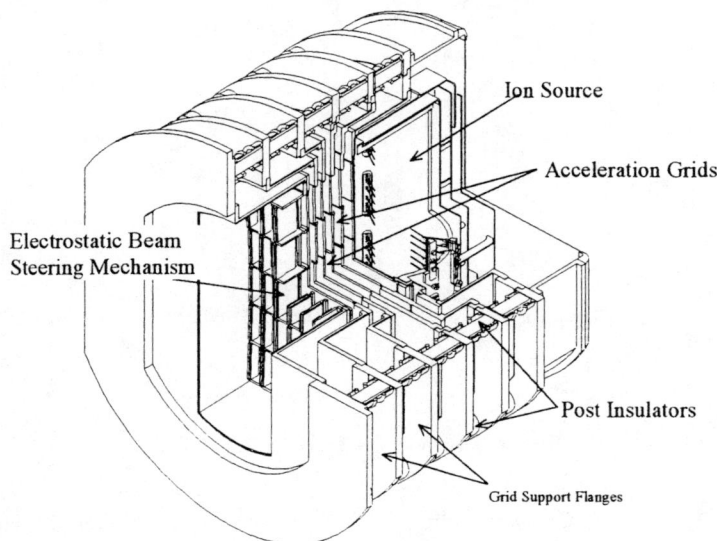

FIGURE 4. The vacuum insulated beam source

As there is vacuum both inside and outside of the VIBS, the insulators of the accelerator no longer form part of the vacuum boundary, and the insulator column of the gas insulated version has been replaced by a set of "post insulators". Additionally the metal flanges on which the plasma grid and the accelerator grid supports are mounted can be made from stainless steel instead of titanium alloy, and as the ion source enclosure no longer needs to support the pressure of the insulating gas, this becomes a rather simple cover which defines the electrostatic potentials.

As mentioned above, the bushing at the junction of high voltage transmission line 2 and the beam source vessel is part of the vacuum boundary. The basic structure of this bushing is similar to that of the insulator column of the gas insulated beam source, i.e. it consists of 5 sets of alumina ceramic columns interspaced with titanium alloy flanges. However, the size of the insulators is smaller than those of the gas insulated beam source, the height is 210 mm, and the internal diameter is 1.8 m, which is expected to lead to a reduction in both the manufacturing difficulty and the cost compared to the insulator column for the gas insulated source. An additional advantage compared to the gas insulated source insulator column is that the insulators are in compression only. The bushing for the VIBS is shown as Fig. 5.

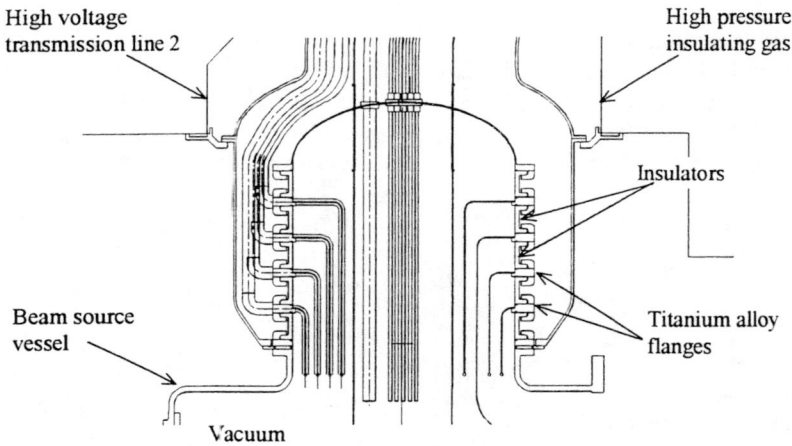

FIGURE 5. The bushing for the vacuum insulated beam source

Change in the Vertical Beam Aiming

One of the aims of ITER is to operate in so called "advanced scenarios" in which the current in the plasma is driven non-inductively. The role of neutral beam driven current in these scenarios is to drive current at the magnetic axis of ITER. Extensive studies of possible equilibria in ITER have now been carried out, and the magnetic axis position not the same for all the equilibria of interest. Additionally, changes in the design of the ITER shield blanket have led to an increase in the height of the axis of the opening for the beams to a position above the magnetic axis of the standard plasma scenario, and a decrease in the height of the opening. The vertical aiming of the beams has been changed to accommodate, as far as is possible, changes in the position of the ITER magnetic axis for the different equilibria, and to reduce the height of the beam footprint at the shield blanket to ensure good transmission to the plasma.

For reasons previously explained (1), only geometrical steering of the beam is used in the vertical plane, which is achieved by tilting each of the 5 segments making up each grid so that they are "aimed" in the appropriate direction. Fig. 6 shows the new aiming for the different segments. The beam envelope aims downwards through the shield blanket, the beam height is small at the shield blanket, and the beam is somewhat spread in the vertical direction at the magnetic axis. With this arrangement the beams can drive adequate current at the magnetic axis for the majority of the plasma equilibria now considered for ITER.

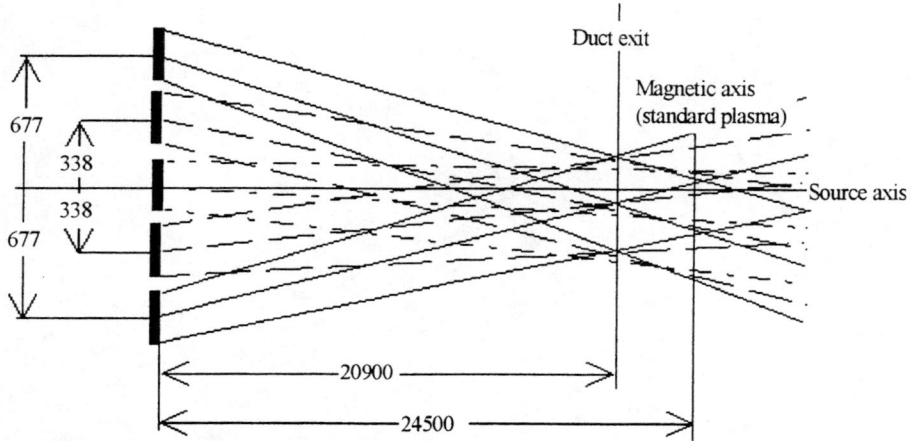

FIGURE 6. New vertical aiming of the beam. Dimensions in mm. Starting from the top, the centrelines of the segments cross the magnetic axis of the standard ITER plasma +20, -50, -170, -190 and -210 mm respectively above (positive values) or below (negative values) the horizontal axis of the source.

DIAGNOSTIC NEUTRAL BEAM SOURCE

A diagnostic neutral beam system is proposed for ITER which will be used to measure the thermal helium density profile in real time. The aim is to produce singly ionised helium via charge exchange with the beam, i.e. :

$$H^0 + He^{++} \Rightarrow He^+ + H^+$$

The basic specification of the injector is given in Table 2.

TABLE 2   GENERAL SPECIFICATION OF THE ITER DIAGNOSTIC INJECTOR

Beam Species	H^0	Measurement period	Throughout the ITER pulse
Beam energy	125 keV	Measurement frequency	Every 10 to 20 s
Beam power	5 MW	Pulse modulation	5 Hz, 100%

and to deduce the original He^{++} concentration via the measurement of the radiation from the subsequent excitation of the He^+ by the plasma electrons. The beam energy has been optimised to produce the maximum signal considering the rapid decrease in the charge exchange cross section at above 100 keV, and the beam penetration into the plasma, which requires energies of >100 keV to penetrate deep into the ITER plasma. To increase the measurement sensitivity, the beam is to be modulated at about 5 Hz, and to be on for ≈5 s. The measurements will be made throughout the ITER pulse, every 10 to 20 s.

The design of the diagnostic neutral beam is still evolving, but the basic features of the present outline design can be seen in Fig. 7.

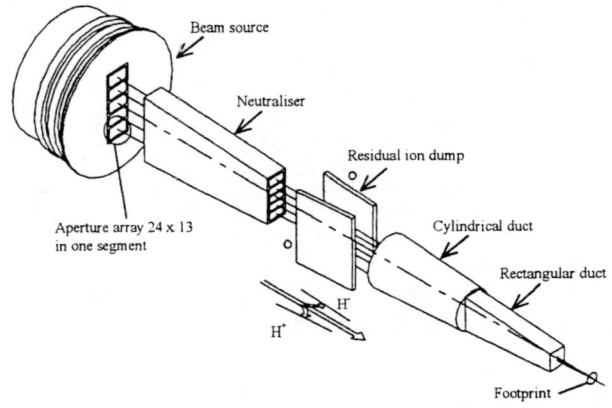

FIGURE 7. Schematic of the proposed diagnostic neutral beam

The injector will be based on the neutralisation of an accelerated H^- beam, and, to avoid costly R&D, the beam source is to be based on the source being designed and developed for the main heating systems. The negative ion source will be identical to that of the main heating beams, and the structure and insulator for the extractor and accelerator.

The aperture pattern will be 24 columns x 13 rows on each of the 5 segments making up each grid, and there will be only one stage of acceleration. The beam footprint is required to be small, some 200 mm x 200 mm at 20 m from the accelerator. The accelerated H^- density is to be 360 A/m^2, and the total accelerated current will be about 87 A. A section through the proposed source is shown as Fig.8.

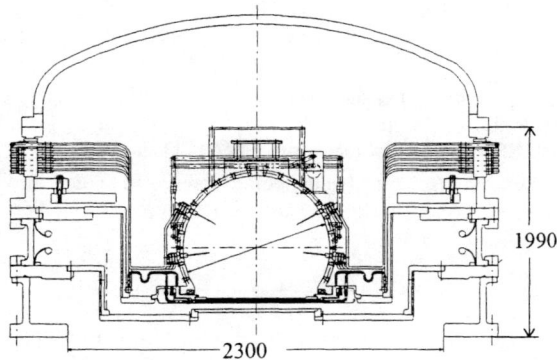

FIGURE 8 Section through the beam source for the diagnostic neutral beam. Dimensions in mm.

CONCLUSIONS

The use of neutral beams to heat ITER raises many new challenges in that not only does the required 1 MeV beam energy necessitate a negative ion based system, but the injectors will have to operate in an environment with high magnetic fields, severe space constraints and high radiation levels. The overall design of the injectors has advanced considerably, however, results from R&D initiated by ITER, more detailed studies of component reliability and advances in the study of the remote maintenance have all led to changes in the injector design, which are described above.

A diagnostic neutral beam is proposed which will deliver a 5 MW beam of 125 keV H^0 to the ITER plasma. An outline design has been studied which is based on the neutralisation of a 125 keV H^- beam. To avoid additional R&D, the beam source required is based on the ion source being developed for the heating beams, plus a single stage accelerator using an insulator and grid support system developed for the heating beams.

REFERENCES

1 Hemsworth R S et al., *Production and Neutralisation of Negative Ions and Beams*, Proc. Seventh International Symposium, Production and Application of Light Negative Ions, Sixth European Workshop, AIP Conference Proceedings 380, 1995, pp 504-517.

2 Hodgson, E. R., "Insulator Gas Characterisation", presented at the First EU-Japan Workshop on Neutral Beam Injectors, Naka, Japan, Feb. 1997. (Work carried out as part of the ITER R&D task N53 TT 07.)

3   Fujiwara, Y. et al, "Radiation Effects On Insulating Gases", presented at the First EU-Japan Workshop on Neutral Beam Injectors, Naka, Japan, 1997. (Work carried out as part of the ITER R&D task N53 TT 08.)

4   Hodgson, E. R., "Insulators - Insulating Materials - Insulating Gases", , presented at the Third ITER NBI Design Review Meeting, ITER Joint Work Site, Naka, Japan, Sept. 1997. Part of the ITER R&D task N53 TT 05, final report submitted to ITER Oct. 1997.

5   Inoue, T, et al, "Vacuum Insulated Beam Source", presented at the Third ITER NBI Design Review Meeting, ITER Joint Work Site, Naka, Japan, Sept. 1997. (Work carried out as part of the ITER R&D task N53 TT 08.)

This report is an account of work undertaken within the framework of the ITER EDA Agreement. The views of the authors do not necessarily reflect those of the ITER Director, the Parties to the ITER EDA Agreement, or the International Atomic Energy Agency.

NEGATIVE ION BEAM SOURCE DEVELOPMENT IN EUROPE

R. S. Hemsworth[1], J. Bucalossi[1], C. Desgranges[1], P. Frank[2], M. Fumelli[1], B. Heinemann[2], E. R. Hodgson[4], C. Jacquot[1], W. Krause[2], P. Massmann[1], C. Michaut[3], W. Ott[2], H.-P. Penningsfeld[2], D. Riz[3], A. Simonin[1], E. Speth[2], A. Stäbler[2], R. Trainham[1], and O. Volmer[2]

[1] *Association EURATOM - C. E. A., C. E. A. Cadarache, 13108 St Paul-lez-Durance, France*
[2] *Max Plank Institut für Plasmaphysik, Garching bei München, Germany*
[3] *L. P. M. I., Ecole Polytechnique, 91128 Palaiseau, France*
[4] *CIEMAT, Madrid, Spain*

Abstract : The development in Europe of negative ion sources and accelerators is carried out principally at Cadarache, France and at Garching, Germany. Other work is in progress at the Ecole Polytechnique, France, and CIEMAT, Spain. The main thrust of the programme so far has been driven by the requirements for the 1 MeV D^0 injection on ITER. It is now being proposed to develop a new negative ion source and accelerator for use on present and future European fusion experiments. This paper summarises the present and proposed future programmes in Europe.

THE PRESENT PROGRAMME AND PRESENT STATUS

European development of negative ion sources and accelerators is carried out principally at Cadarache, France and at Garching, Germany. Other work is in progress at the Ecole Polytechnique, France, and CIEMAT, Spain. Dublin City University supplies diagnostic support for the joint IPP/Cadarache programme. The main parts of the development programme in Europe are given below along with the objectives.

MeV Accelerator Development

This is the development of the SINGAP (<u>Sin</u>gle <u>Gap</u>, <u>Sin</u>gle <u>Ap</u>erture) accelerator concept at Cadarache for use on the ITER 1 MeV injection system. This is a novel form of electrostatic accelerator in which the negative ions are extracted and pre-accelerated to ≅50 keV via a 3 grid multi aperture system, and then accelerated to 1 MeV across a single gap. For the ITER accelerator, the beamlets from the pre-accelerator are arranged in 25 groups of 4 columns and 13 rows of apertures, and each group passes through a single rectangular aperture in the final grid (1, 2).

This concept offers several advantages compared to more conventional multi-aperture, multi-gap accelerator (MAMuG) which is presently the reference design of the ITER injectors: There are fewer acceleration stages, therefore the power and the high voltage transmission line are simplified as the number of intermediate potentials is substantially reduced. Additionally, the large single main acceleration gap allows better pumping, hence potentially lower "stripping" loss in the accelerator, which could be as low as 20% compared with 44% for the MAMuG accelerator (3).

The SINGAP concept is rather successful, as is described more fully in (4). The production of negative ion beams appears to be extremely easy with essentially no breakdowns occurring across the main acceleration gap, and no degradation in voltage holding being observed in the presence of the beam. Progress to date in reaching higher energies has been limited by the voltage holding of the bushing at the end of the HV transmission line. Following the initial success of sustaining >1 MV, two of the 9 insulators making up the bushing failed due to faults originating from the manufacturing process used, and the voltage limit of the remaining 7 insulator is about 700 kV. The bushing is now disassembled and each insulator is being examined. The two defective insulators will be replaced in October '97 and the bushing re-assembled.

Voltage holding across the accelerator of >1 MV has been demonstrated, and high quality H⁻ and D⁻ beams have been produced, 40 mA at ≈860 keV and 106 mA at 630 keV respectively. Importantly the measured beam optics corresponds well with the predictions using a 3D code, as, for example, shown by Fig. 1, which gives confidence in predictions for a full accelerator geometry for ITER.

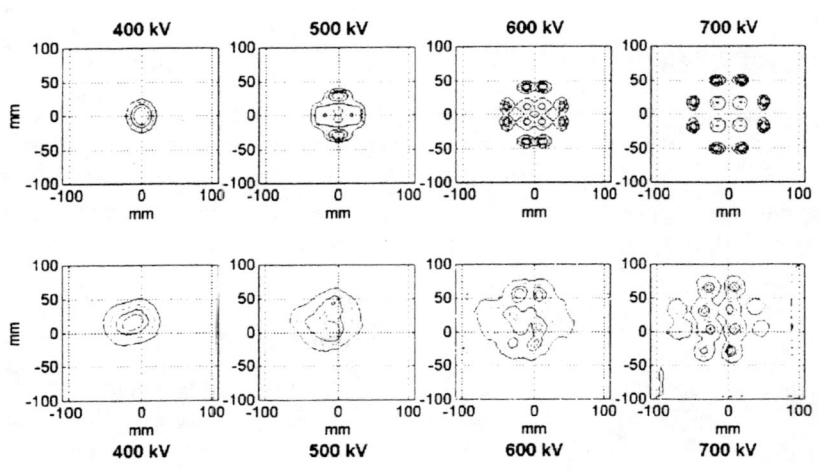

FIGURE 1. Measured (bottom) and calculated (top) beam profiles from SINGAP.

The main objective of the work with SINGAP is now to demonstrate reliable acceleration of D⁻ to 1 MeV at the ITER design current density (200 A/m²) with beam optics acceptable for the ITER injector.

Pre-accelerator Development

This a detailed study of a single beamlet experiment at the Ecole Polytechnique in collaboration with Cadarache, with the aim of improving understanding of the pre-accelerator beam optics. Experiments started in May '97 and so far the emittance of a 3 mA/cm² H⁻ beam at 30 keV, extracted from a "pure volume" source has been measured and good agreement is found with the emittance predicted by the 2D code of the Ecole Polytechnique (5).

High Frequency Driven Negative Ion Source Development

The development of a high frequency (rf) driven negative ion source is being carried out at IPP Garching in collaboration with CEA Cadarache. The aim of the programme is to establish the design basis for an rf driven source suitable for neutral beam injectors. The first prototype source, with metal walls forming a Faraday screen for an externally mounted multi-turn antenna, has been tested in H_2. Operating in "pure volume" mode, up to 4.5 mA/cm² of H⁻ was extracted, but at rather high pressure, 1.6 Pa. The addition of caesium to that source enhanced the yield (compared to the pure volume mode) at 0.6 Pa by a factor 4 (6).

D⁻ Arc Discharge Negative Ion Source Development

This is a collaboration between Cadarache, France and JAERI on the testing and development of a "model" ITER negative ion source. The ion source was designed and built by JAERI and it is tested and operated in D_2 at Cadarache. The objectives are to reduce the operating pressure, and the extracted electron current, and to assess the pulse length capability of the ion source and the plasma grid.

Current densities in excess of those required by ITER (200 A/m²) have been obtained at 0.35 Pa, i.e. nearly the desired operating pressure of 0.3 Pa, with the co-extracted electron current below the accelerated D⁻ current (7). An example of a 5 s pulse at >200 A:m² is shown in Fig. 2.

Extension of the pulse length to 1000 s is now the main objective of this work. So far D⁻ beams with 19 mA/cm² have been accelerated for 20 s and H⁻ beams of similar current densities for pulse lengths up to 60 s (7).

Insulator Development

The insulator used in the accelerator of high energy beam sources (>800 keV) has been identified as a critical item. Cadarache, in conjunction with European industry is carrying out R&D on the technology required to manufacture ceramic insulators of the size required by ITER (1.8 m to 2.7 m internal diameter). Porcelain appears promising in that several manufacturers are confident of producing insulators of the required size, and the radiation tolerance of porcelain has been measured (see Table 1). A porcelain insulator is now being be built for testing on the SINGAP MV test bed.

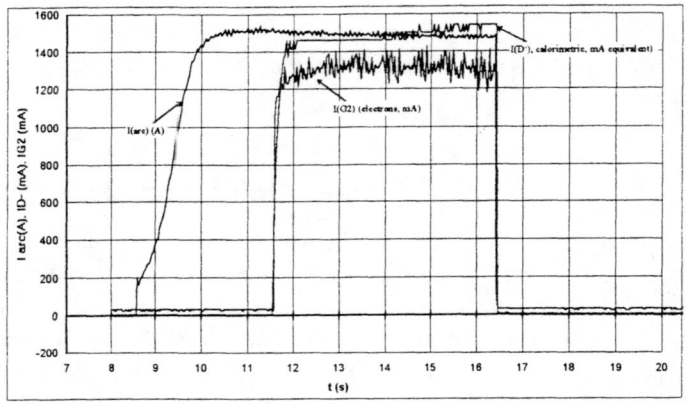

FIGURE 2. 5s pulse with >200 A/m2 D⁻ accelerated from the Kamaboko III source
p = 0.35 Pa, arc power = 80 kW, average $J(D^-) = 220$ A/m^2

Radiation Induced Conductivity in Solid Insulators

For the beam sources of the injectors for ITER and future fusion reactors, radiation induced conductivity (RIC) in the insulators limits the choice of insulating material. ITER has chosen to use high density alumina for the insulators used in the neutral beam injectors as this was the only insulating material identified as having negligible induced conductivity. Unfortunately high density alumina is expensive and it is difficult to produce insulators of the required dimensions in this material. The work in Europe, at CIEMAT in Spain, is aimed at measuring RIC in potential alternative materials. This has now been done for various types of porcelain and one type of epoxy resin, as summarised in Table 1 (8). Given the predicted radiation levels (9, 10), these measurements, show that the RIC is sufficiently low that either porcelain or epoxy can be considered for the ITER beam sources. Porcelain is particularly interesting for use in the beam source and/or the bushing at the end of the ITER high voltage transmission line as its vacuum properties and resistance to arcs are similar to those for alumina, but it is easier to manufacture the large diameter insulators required by ITER in porcelain than in alumina. Epoxy insulators will be used in the ITER NBI high voltage transmission line, and measurements such as these will enable a sensible choice of epoxy resin to be made.

Table 1   ELECTRICAL CONDUCTIVITY OF INSULATORS UNDER RADIATION

Sample	Conductivity at 285 K (S/m)	Conductivity at 400 K (S/m)	Conductivity increase with 0.5 Gy/s (S/m)
Fibre Glass Reinforced Epoxy			
Permaglas RE 230	5×10^{-13} *4.5×10^{-12}	6.5×10^{-12}	1.3×10^{-11}
Porcelain			
c120a			1.6×10^{-10}
c120b	$(3.5 \pm 0.5) \times 10^{-12}$	6.5×10^{-9}	7.4×10^{-12}
c120c	1.5×10^{-11} <limit*	1.5×10^{-8} *1×10^{-8}	2.8×10^{-11}
c110	$(1.5 \pm 0.5) \times 10^{-11}$	7×10^{-7}	<limits
c221	8×10^{-13}	1.0×10^{-12}	5.5×10^{-11}

Background at 285 K: $0.5 - 2.5 \times 10^{-12}$ S/m (sample dependent)
* Denotes the observed difference from heating to cooling
± Denotes the error margin of the measurement

Radiation Induced Conductivity in Gases

Experiments from CIEMAT, Spain, reported last year, indicated that RIC in the insulating gas surrounding the ITER beam source could lead to currents of several amperes flowing in the gas (11). Experiments since appear to validate the extrapolations (8, 11, 12) and, as the resulting power loss (several megawatts) would not be acceptable, ITER is now considering a vacuum insulated beam source (10, 13).

FUTURE PROGRAMME

The main elements of the ongoing programme described briefly above will continue with a high priority being given to the production of 1 MeV D⁻ beams and the understanding of the beam optics. However, it is clear from experience, both in Europe and Japan (14), that emphasis must be placed also on producing a reliable 1 MV insulator.

Whilst most of the successful magnetic confinement fusion experiments operating in Europe already use positive ion based high power neutral beam systems, upgrades to those machines and future European machines will require neutral beams of substantially higher energies, which can only be efficiently produced from accelerated negative ions. It is proposed to develop a "common" negative ion beam source, i.e. ion source, extractor and accelerator, which could form the basis of future neutral beam injectors in Europe. Potential "users" have been considered, and the conclusions are summarised in table 2.

Preliminary studies have shown that good transmission can be obtained in all cases using an accelerator 360 mm wide by 800 mm high. After taking into account beam losses (geometrical transmission and reionization) and the efficiency of conversion to neutrals (a gas target is assumed), it is concluded that 20 A of D⁻ accelerated to 350 keV would satisfy the requirements of all the machines listed.

Table 2 POSSIBLE USES OF HIGH ENERGY NEUTRAL BEAMS IN EUROPE

EXPERIMENT	REQUIREMENT	ENERGY[1] keV	POWER MW	PULSE s
ASDEX-U	Off axis current drive for advanced scenarios	250 to 350	2 * 2.5	10
W 7X	Heating of high density plasmas by balanced injection	250 to 350	2 * 2.5	10
Tore Supra	Central current drive for advanced scenarios	350 to 750	3.0	>100
JET	High power "central heating"	300 to 500	10	20

[1]The limits on the beam energy given above arise, at the high end, from shinethrough considerations, and, at the lower end, either from poor beam penetration into the plasma (JET, Tore Supra, W 7X), or low current drive efficiency (Tore Supra, ASDEX-U).

It is proposed that the development of such a beam source should be carried out jointly by the neutral beam development groups of Cadarache and IPP Garching. Present work is oriented towards costing of the project and an examination of how to share the work between the laboratories.

CONCLUSIONS

Europe is making a worthwhile contribution towards developing negative ion sources for ITER in collaboration with JAERI, and work has started in Europe on the development of an rf driven ion source. The development of the SINGAP accelerator for high energy negative ion acceleration is extremely promising, and should be considered for use on ITER as it offers a substantial simplification in the design of the beam source, the 1 MV power supply and the high voltage transmission line. In the near future the development programme will concentrate on achieving the objectives set by ITER, principally the production of 1 MeV D⁻ beams with the SINGAP accelerator. The proposed development of a long pulse 350 keV D⁻/H⁻ beam source for use on many European fusion experiments represents a significant challenge, which, if successful, would eventually greatly enhance the experience base for the use of negative ion based neutral beam injector systems.

REFERENCES

1. Hemsworth R. S. et al., *Production and Neutralisation of Negative Ions and Beams, Proc. Seventh International Symposium, Production and Application of Light Negative Ions, Sixth European Workshop*, AIP Conference Proceedings 380, 1995, pp 504-517..
2. Hemsworth, R. S. et al "Update of Neutral Beams for ITER", presented at this symposium, 1997.
3. Hanada, M., "Stripping Loss Upgrade", presented at the Third ITER NBI Design Review Meeting, ITER Joint Work Site, Naka, Japan, Sept. 1997.
4. Desgranges, C, et al, "Results of the 1 MV SINGAP Experiment", presented at this symposium, 1997.
5. Michaut, C., et al, "Measurement of Negative Ion Beam Emittance", presented at this symposium, 1997.
6. Frank, P., et al, "A Large Area RF Source for Negative Hydrogen Ions", presented at this symposium 1997.
7. Trainham, R., et al, "Kamaboko Long Pulse Results", presented at this symposium 1997.
8. Hodgson, E. R., "Insulators - Insulating Materials - Insulating Gases", , presented at the Third ITER NBI Design Review Meeting, ITER Joint Work Site, Naka, Japan, Sept. 1997. (Work carried out as part of the ITER R&D task N53 TT 07.)
9. Nightingale, M. P. S., " ITER NBI Neutronics", presented at the Second ITER Technical Meeting on NBI, ITER Joint Work Site, Naka, Japan, Feb. 1997. (Work carried out as part of the ITER R&D task N53 TT 07.)
10. Inoue, T., et al, "Vacuum Insulated Beam Source", presented at the Third ITER NBI Design Review Meeting, ITER Joint Work Site, Naka, Japan, Sept. 1997.
11. Fujiwara, Y. et al, "Radiation Effects On Insulating Gases", presented at the First EU-Japan Workshop on Neutral Beam Injectors, Naka, Japan, 1997. (Work carried out as part of the ITER R&D task N53 TT 08.)
12. Hodgson, E. R., "Insulator Gas Characterisation", presented at the First EU-Japan Workshop on Neutral Beam Injectors, Naka, Japan, 1997. (Work carried out as part of the ITER R&D task N53 TT 07.)
13. Hemsworth, R. S. et al, "Update of Neutral Beams for ITER", presented at this symposium 1997.
14. Watanabe, K., et al, "Multi-stage, Multi-aperture Electrostatic Accelerator for H⁻ Beams", ", presented at this symposium 1997.

SEPARATION OF BEAM AND ELECTRONS IN THE SPALLATION NEUTRON SOURCE H-ION SOURCE

J. H. Whealton and R. J. Raridon
Oak Ridge National Laboratory

K. N. Leung
Lawrence Berkeley National Laboratory

The Spallation Neutron Source (SNS) requires an ion source producing an H-beam with a peak current of 35mA at a 6.2 percent duty factor. For the design of this ion source, extracted electrons must be transported and dumped without adversely affecting the H-beam optics. Two issues are considered: (1) electron containment transport and controlled removal; and (2) first-order H-beam steering. For electron containment, various magnetic, geometric and electrode biasing configurations are analyzed. A kinetic description for the negative ions and electrons is employed with self-consistent fields obtained from a steady-state solution to Poisson's equation. Guiding center electron trajectories are used when the gyroradius is sufficiently small. The magnetic fields used to control the transport of the electrons and the asymmetric sheath produced by the gyrating electrons steer the ion beam. Scenarios for correcting this steering by split acceleration and focusing electrodes will be considered in some detail.

Extraction of negative ions from volume sources is usually attended by a large unwanted flux of extracted electrons, which cause heat loading and sometimes affect the beam directly or indirectly. Direct effects on the beam include enhanced space charge. But most deleterious effects of these electrons on the ions include concomitant effects of the necessary magnetic field normally present in the extraction region. The ambient magnetic fields cause a drift motion of the electrons across the extraction aperture leading to an accumulation of electrons on one side of the aperture, which in turn could cause a steering of the negative ion beam (opposite to the steering of the beam due to the magnetic field itself) and perhaps an azmuthal density asymmetry in the extracted negative ion beam.

Two facets of the electron issue will be addressed in this paper. The correction to steering of the ion beam by imposing a dipole field is done in a 3-D geometry having some similarity to accelerator designs being considered for the SNS. The second consideration will be a plasma electrode trapping of these electrons before they get to high energy.

Research sponsored by the LDRD Program of Oak Ridge National Laboratory, managed by Lockheed Martin Energy Research Corp. for the U. S. Department of Energy under contract number DE-AC05-96OR22464.

To offset beam steering due to the magnetic field, we will consider a split accelerator (second) electrode such as to provide an average electric field restoring the beam. The object of this preliminary study is to see if the RMS emittance of the beam can be brought back to its symmetric value (without magnetic field). The conventional model of negative ion extraction is used [1] where the positive ions are represented by an equilibrium Boltzmann distribution and the negative charges are represented by a kinetic description (collisionless Boltzman equation) with the charge balance in the presheath assumed to be symmetric with the case of positive ion extraction.

To make this comparison, we examine Fig. 1, which is the baseline symmetric case. Figure 1c shows the conventional view of a slice through the accelerator in a plane perpendicular to the applied magnetic field (although in this case the magnetic field is zero. The electrodes including a LEBT are shown with the entrance to the RFQ appearing at the right-hand side. An end view is shown in Fig. 1a, which shows a more or less axially symmetric beam. An emittance, or phase space occupation, distribution is shown in Fig. 1b in that transverse plane corresponding to Fig. 1c. This emittance diagram is approximately symmetric about the origin as expected.

Placing a magnetic field (assumed uniform for this study) of 1700 gauss over a region from the left hand side of Fig. 2c to about twice the extent of the plasma electrode thickness yields the result of Fig. 2. Here we can see that the beam is not cylindrically symmetric, almost impinging on the third electrode as Fig. 2c shows. (The aperture displacement steering method [2] exacerbates this defect.) The end view, Fig. 2a, and the emittance plot clearly show the asymmetry and the RMS transverse emittance is about twice that of Fig. 1 without a magnetic field (as shown in Fig. 4).

Placing a gap in the acceleration electrode, indicated in Fig. 3a (the asymmetric potential can be seen in Fig. 3c) tends to resteer the beam as seen especially in the emittance plot (Fig. 3b). The RMS emittance decreases by an application of a 2-5 KV potential across this gap, as shown in Fig. 4, reducing the emittance to almost the value without the magnetic field, which proves the principle that this technique not only resteers the beam but reduced the emittance growth caused by the magnetic field. The technique of aperture displacement [2] often used in multiaperture positive ion sources may be equivalent to the split electrode technique described here with respect to steering (first order). Since the beam in the aperture displacement technique gets even closer to the second electrode one might expect the emittance to grow even more than the zero resteering case (with magnetic field). However, the split electrode technique sees a decrease in the RMS emittance from the zero resteering case since the beam is further from the second electrode than either the zero displacement case, or especially the aperture displacement correction technique [2].

Next consider a split (longitudinally) plasma electrode to see the possible trapping effectiveness in a special case (2D slot geometry). A typical case is shown in Fig. 5 with both extracted electrons (going up) and negative ions. If we split the plasma electrode as shown in Fig. 6 and bias the split part positively with respect to the plasma side to attract electrons and assume no secondary electrons are created we see Fig. 6b. In this case, a significant fraction of the electrons are attracted towards the split electrode, at a couple hundred volts instead of being extracted to the acceleration electrode (50 KV). The effectiveness of the bias on the split electrode in trapping electrons is shown in Fig. 7 for this geometry and conditions. We conclude from this that, if materials with very low, secondary electron emissions can be found, and the primary ion optics are not adversely affected, then this technique might be useful.

References:

[1] Computer modelling of negative ion beam formation, J. H. Whealton, M. A. Bell, R. J. Raridon, K. E. Rothe, and P. M. Ryan, J. Appl. Phys. 64, 6210 (1988).

[2] L. D. Stewart, J. Kim, and S. Matsuda, Rev. Sci. Instrum., 46, 1193 (1975).

Acknowledgements:

The authors wish to thank M. A. Akerman, and J. B. Green for their technical advice and J. M. Shover and E. D. Stratman for their assistance.

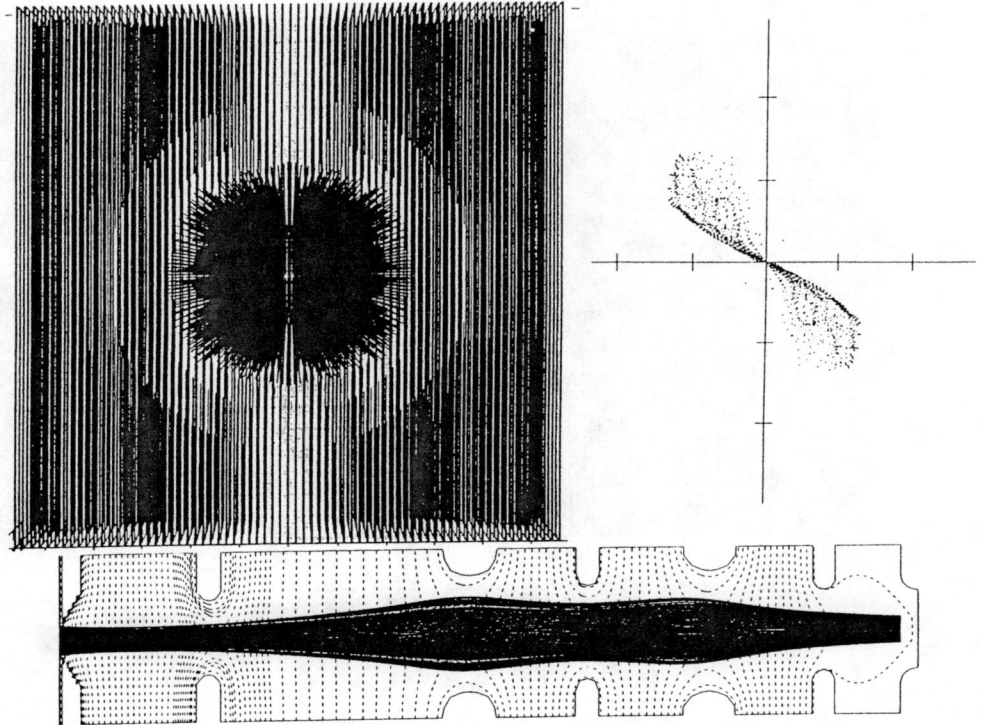

Figure 1. The baseline symmetric case. Fig. 1c shows the conventional view of a slice through the accelerator in a plane perpendicular to the applied magnetic field (although in this case the magnetic field is zero. The electrodes including a LEBT are shown with the entrance to the RFQ appearing at the right-hand side. An end view is shown in Fig. 1a, which shows a more or less axially symmetric beam. An emittance, or phase space occupation, distribution is shown in Fig. 1b in the transverse place corresponding to Fig.. 1c.

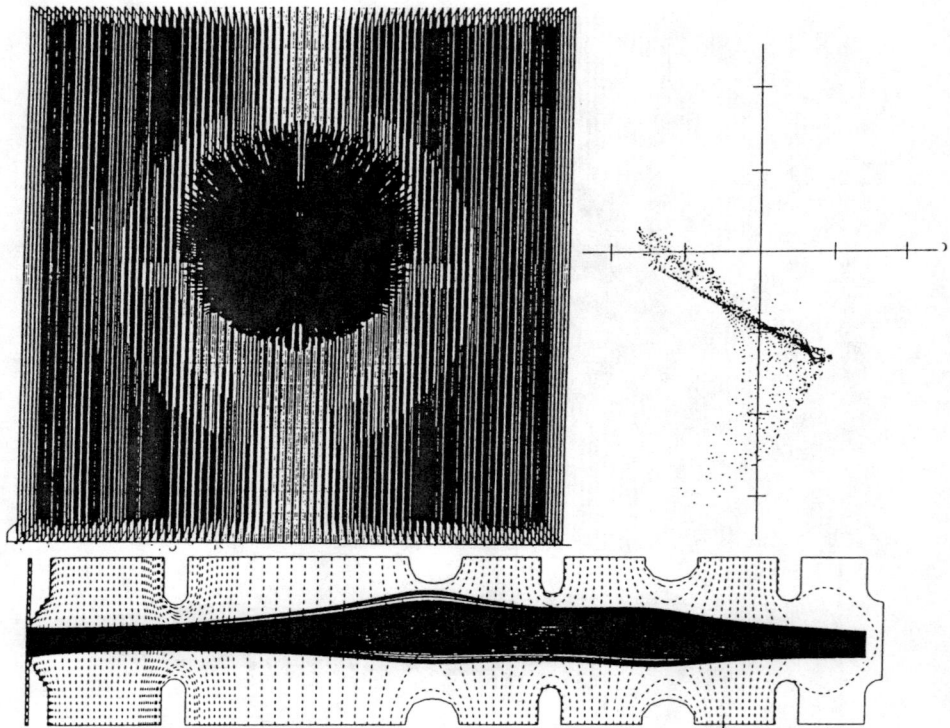

Figure 2. Same as Fig. 1 but including a magnetic field (assumed uniform for this study) of 1700 gauss over a region from the left-hand side of Fig. 2c to about twice the extent of the plasma electrode thickness. Here we can see that the beam is not cylindrically symmetric; almost impinging on the third electrode as Fig. 2c shows.

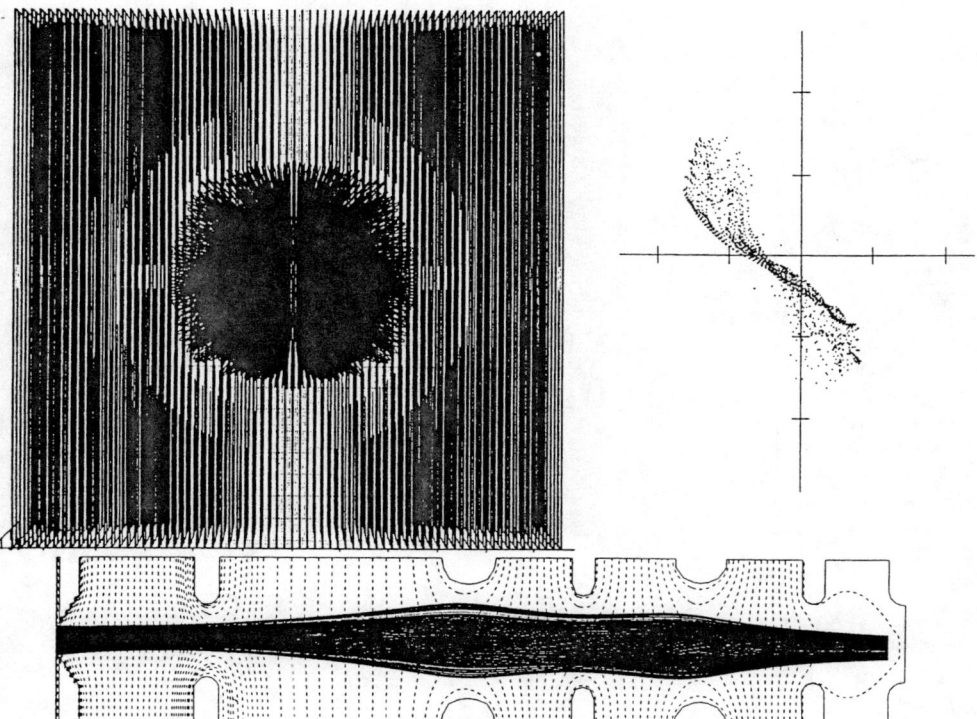

Figure 3. Same as Fig.. 2 but including a gap in the acceleration electrode, indicated in Fig. 3a (the asymmetric potential can be seen in Fig. 3c). Tends to resteer the beam as seen especially in the emittance plot (Fig. 3b).

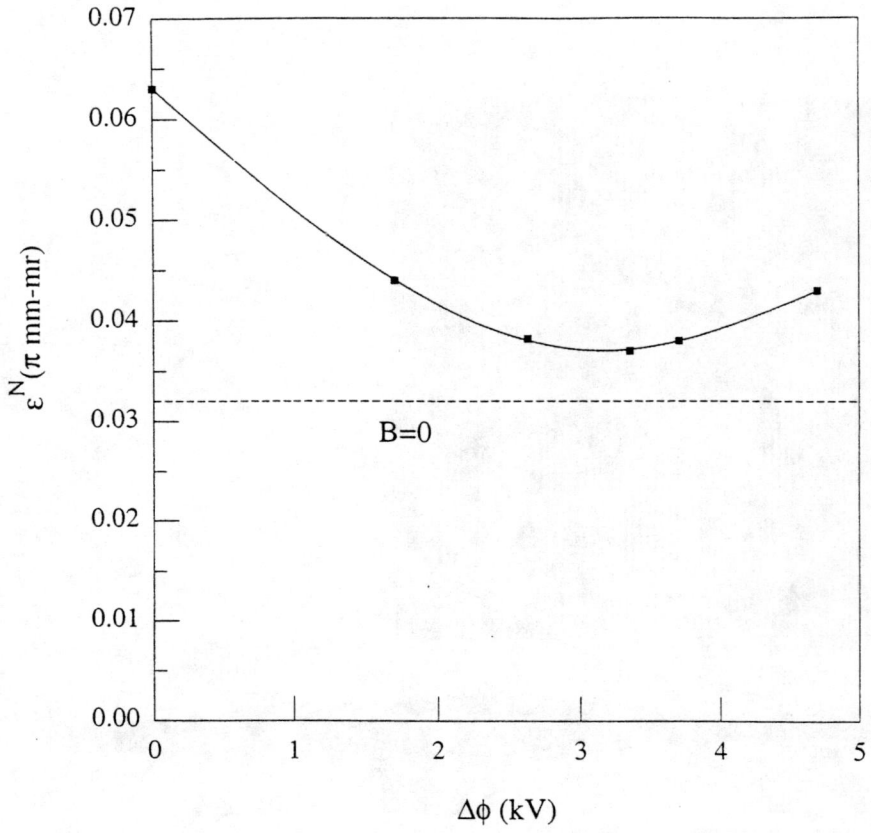

Figure 4. Beam RMS emittance as a function of gap voltage also showing the emittance without a magnetic field.

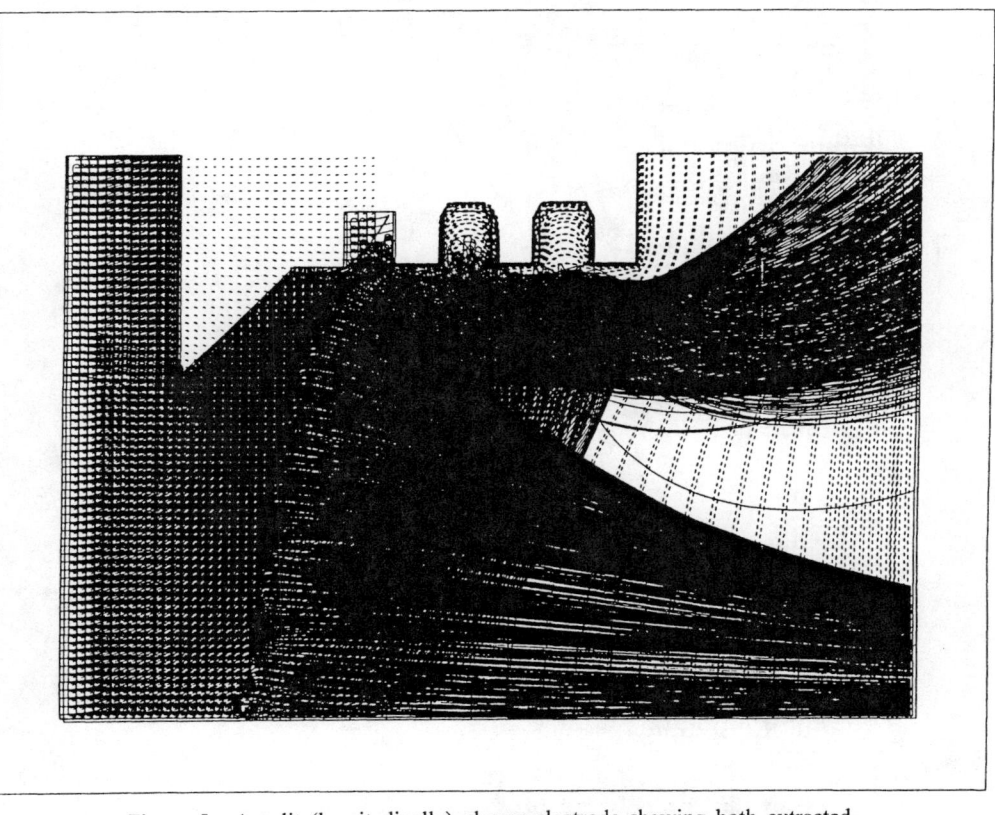

Figure 5. A split (longitudinally) plasma electrode showing both extracted electrons (going up) and negative ions.

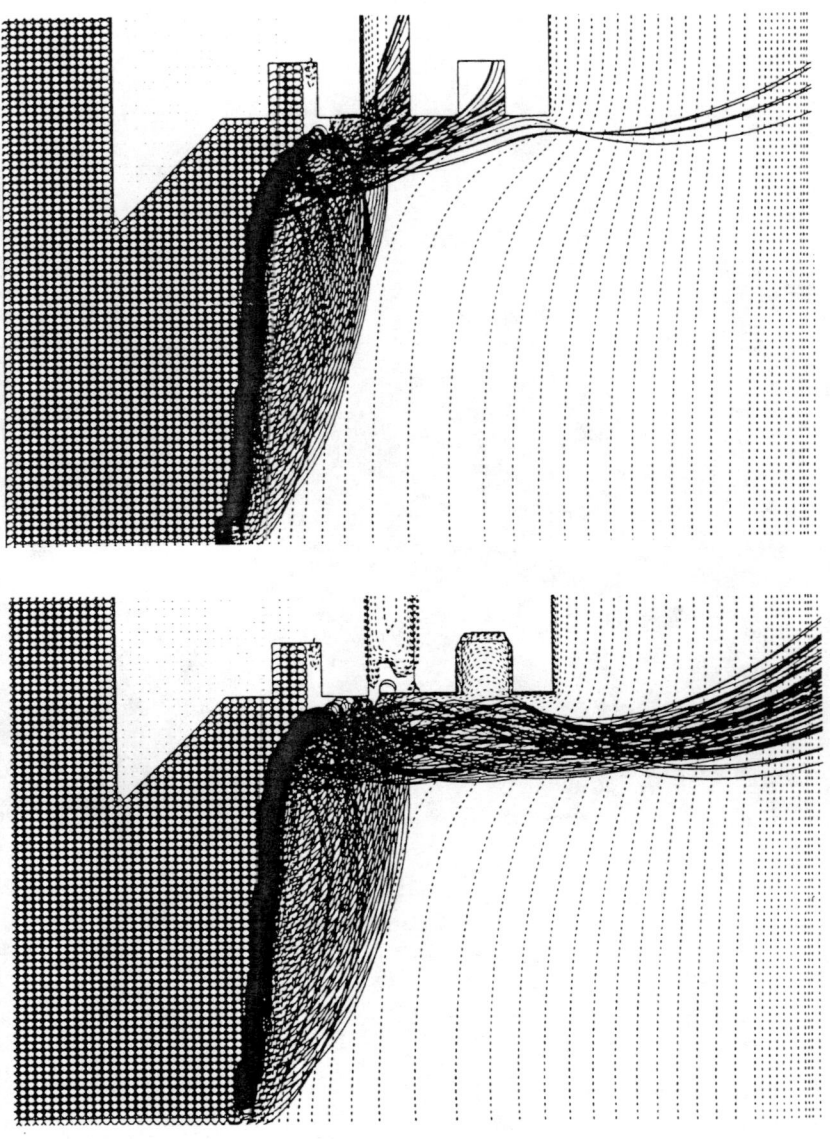

Figure 6. Electron trajectories as a function of split electrode bias.

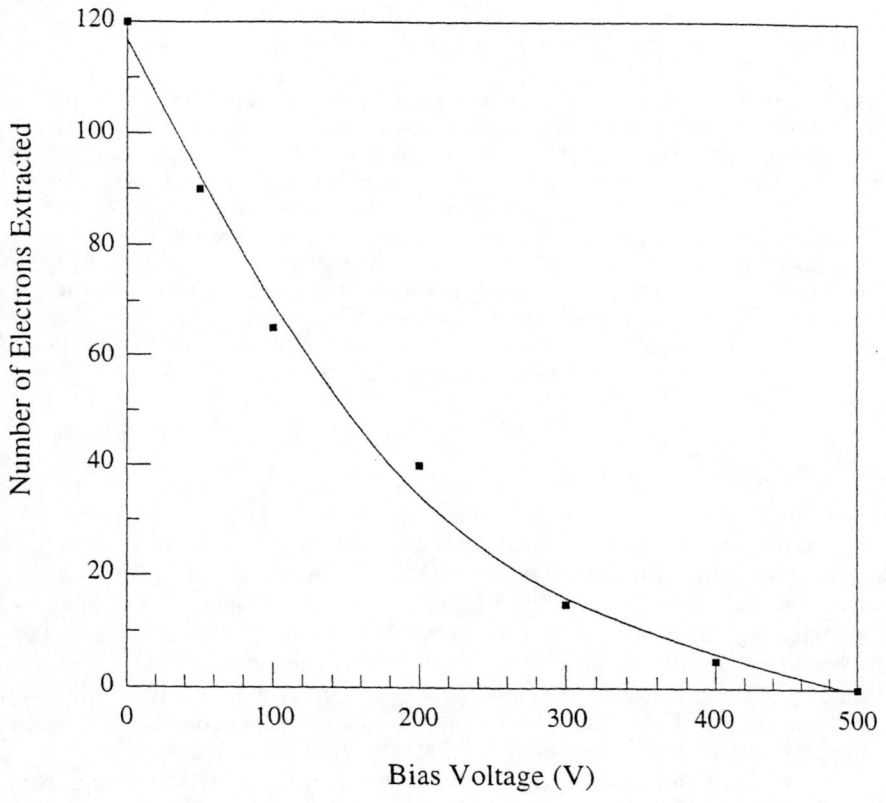

Figure 7. Electron transmission as a function of the bias on the split electrode.

Historical Perspective of the H⁻ Ion Source Symposia

C. W. Schmidt

Fermilab National Accelerator Laboratory
MS 307, P.O. Box 500, Batavia, IL 60510

Abstract. The International Symposium on the Production and Neutralization of Negative Hydrogen Ions and Beams is entering its third decade of providing a forum for the H⁻ ion source community. The first meeting was held at Brookhaven National Laboratory in 1977 and has returned there every three years to 1995. This is the eighth meeting in this series and for the first time is in Europe, hosted by CEA/Center de Cadarache. Since this Symposium is meeting in Europe many new people have had an opportunity to attend and many of these are of a younger generation. On the 20th anniversary of the First Symposium it seems fitting that a historical review should be given. The Symposium meetings and its archiving of information has been a valuable asset to this community. I hope to briefly describe the early H⁻ source work and provide some insight into the success of the H⁻ source effort.

I. INTRODUCTION

In 1951, at the University of California, Berkeley, a 2-MV generator was proposed to produce 4-MeV protons by injecting and stripping H⁻ ions in the HV terminal. The note describing this proposal also proposed the use of H⁻ ions for charge-exchange injection into synchrotrons.[1] At this time, H⁻ beams were obtained from the charge exchange of proton beams so only a few hundred microamperes could be expected. Nevertheless, the "swindletron", as it was called at Berkeley, became the Tandem Van de Graaff. Throughout the 60's H⁻ ions were obtained by charge exchange of proton beams and many sources along this line were developed[2]. Uses for H⁻ ions such as charge-exchange extraction from cyclotrons[3] and charge-exchange injection into synchrotrons were done[4-6].

In 1971 a Symposium on Ion Sources and Formation of Ion Beams was held at Brookhaven[7]. H⁻ charge-exchange ion sources were represented with papers from Los Alamos, Oxford University and the University of Wisconsin using donor gases of hydrogen, lithium and cesium. Kansas State University used a diode source for direct extraction of negative ions and Toshiba Research Center of Japan obtained H⁻ ions by direct extraction from a duoplasmatron. Currents of tens of microamperes to a few milliamperes were possible. Intense charge-exchange sources using a cesium canal to make H⁻ ions and H neutrals for fusion application were also mentioned[8].

In 1974 the Second Symposium on Ion Sources and Formation of Ion Beams was held at Berkeley[9]. A major contribution to this meeting was a paper from Novosibirsk on "Surface-Plasma Source of Negative Ions". This was the first report of cesiated surface-plasma sources in the west although several papers had been published in Russia in 1972 and 1973. A Brookhaven report states that "first experiments (at Novosibirsk) with cesium were done in 1971!"[11] A comprehensive review of H⁻ source work to that time was presented[12], and papers on direct-extraction of negative ions from duoplasmatrons, major charge-exchange sources, etc. were given by the University of Washington, Los Alamos, Argonne, Brookhaven, and the Institut de Physique Nucleaire.

For the next several years H⁻ sources of many types were under development for accelerators and for fusion consideration. Novosibirsk and Brookhaven were actively studying the cesiated surface-plasma sources; large charge-exchange sources were under development at Livermore; and charge-exchange injection into the Argonne ZGS was successful[13] using a proton source in tandem with a hydrogen charge-exchange cell to obtain ~25 mA of H⁻ beam.

II. THE H⁻ SYMPOSIA

With the rapid rise in H⁻ source work, Brookhaven choose to host a Symposium on the Production and Neutralization of Negative Hydrogen Ions and Beams, September 26-30, 1977[14]. More than 50 people attended the meeting with 42 papers presented. The predominant consideration was the cesiated surface-plasma source. Several sources were already working or under development for accelerators. Fermilab was using a surface-plasma magnetron (planatron) source for injection of H⁻ ions onto the Linac, and Los Alamos was developing a high-duty-factor surface-plasma Penning source. Brookhaven, Berkeley, Oak Ridge, Novosibirsk, and others were considering these sources for high-power fusion use. Charge-exchange sources were still very much in the running especially for high-power fusion applications. For both the surface-plasma and the charge-exchange sources considerable effort was spent on discussing fundamental processes. Information for collision cross-sections, surface effects, space charge phenomena, etc. were necessary and represented work in progress or needed.

In the midst of all this was a paper on the "Production of Negative Hydrogen Ions in a Low Pressure Hydrogen Plasma," from Ecole Polytechnique. This paper, expounding the possibility of significant H⁻ ions within the volume of a plasma was rather shocking. There were cries of, "How could this be!"

Three years later, 1980, ion source people convened again at Brookhaven for the 2nd International Symposium on the Production and Neutralization of Negative Hydrogen Ions and Beams[15]. This time the symposium was titled "international" although international cooperation and participation had existed earlier. This meeting concentrated heavily on fundamental processes for both surface-plasma and volume-production sources. Yes, the production of H⁻ ions in the plasma volume had been further studied since the last symposium and it stood up. Now it needed to be explained and the formation of high vibrational states seemed to be part of the answer. Numerous ideas and measurements for surface-plasma and volume production came from Livermore, Berkeley, Oak Ridge, Ecole Polytechnique, FOM-Institute for Atomic Physics (Amsterdam), University of Pittsburgh, Stevens Institute of Technology, University of Ulster (Northern Ireland), Wesleyan University, Instituto de Fisica (Mexico), the U.S. Air Force, SRI International, and others.

Ion sources with significant improvement and of new designs were presented. Berkeley had developed the converter source, which was also being studied at Culham; Novosibirsk change to a semi-planatron; Brookhaven was working on ampere surface-plasma sources; Oak Ridge had a surface-ionization source; and Nagoya University was working toward high-current surface-plasma sources. Smaller H⁻ sources for accelerators were doing well at Brookhaven, Fermilab, and Los Alamos.

Charge-exchange sources were presented by Livermore and Centre d'Etudes Nucleaires (Grenoble) but they were on their way out; and cluster sources were under development at the Institut fur Kernverfahrenstechnik (Karlsruhe).

The needed for multi-megawatt fusion injectors intensified research on beam lines, accelerators, neutralizers, etc.

The 3rd International Symposium on the Production and Neutralization of Negative Hydrogen Ions and Beams[16] (1983) concentrated heavily on fundamental processes for both volume and surface sources. A full day was given to each. Fundamental principles for surface-plasma sources were looking rather good and agreeing with source data, but volume processes were still uncertain. This was followed by many talks on sources themselves: sources for accelerators of good intensity and duty-factor, sources anticipating ampere and multi-ampere beams for fusion, sources based on surface effects and others based on volume principles. The magnetic filter field had been realized and it seemed to make possible a large-aperture scalable volume source for fusion. There was also many ideas on high-current accelerating and neutralizing systems in anticipation of fusion applications which was the ultimate goal for much of this effort. What fusion really needed was energetic neutral deuterium beams and much had to be done to address this matter.

At the 4th International Symposium[17] (1986) volume sources took the lead. Fundamental processes were concerned primarily with volume effects. Volume source development adopted the idea of a bucket or multi-cusp source with a magnetic-filter field to pass the low energy electrons into the extraction region. Still, these sources had low current densities, but they could apparently be scaled up to give hope for a high-current source for fusion. There were some sources of the surface-plasma and others of the volume type for accelerators. The SDI (Strategic Defense Initiative - "Star Wars") was interested in intense long-pulse H⁻ sources while down-to-earth accelerators were more modest and could use what was being produced. Improvements of earlier sources were occurring and new ideas based on previous concepts were being tried. Fusion continued to be interested in neutral beams so large accelerator systems and neutralizers were a part of the program. There was also some discussion on other negative ions and polarized ions.

Significant at this meeting was the introduction of cesium into a volume H⁻ source at the Kurchatov Institute. Cesium seeding immediately gave a large increase in the extracted H⁻ ions. This prompted several theories as to what might be going on, which continues to this day.

The Symposia have continued on a three year basis. At times the future seemed in doubt, especially when the fusion program was severely cut in the U.S. and when SDI ended; but, for accelerators H⁻ ions have become a necessity and the need for ever more intensity and duty-factor has pushed the H⁻ source community. Still, in other parts of the world, Japan and Europe, fusion remained and in spite of RF heating as a competitor, there was some impetus to build large sources.

At its peak, about the fourth or fifth symposium, approximately 100 participants attended, representing about a dozen countries. The atmosphere was always cordial, the presentations good, the discussion lively and the conference activities excellent. Over the several meetings there were many activities, ranging from a participants soccer match on the Brookhaven field to attending the N.Y. Metropolitan Opera.

III. CONCLUSIONS

These Symposia have undoubtedly been significant to the H⁻ ion source program. It has been important to bring people together in an atmosphere where all ideas could be presented and considered, and so it did. I has also been important in archiving the information and progress of this field as noted by the many references cited in these proceedings. Unfortunately the first two symposia proceedings had limited printing and are difficult to get.

The ion source community has had great success in developing H⁻ sources for numerous applications and in making new hardware work. This has been our forte. Understanding the fundamental processes, however, has developed at a somewhat slower pace. This has been due in part to the difficulty of the problem, and the lack or disparity of the information. Different investigators often found different effects. It seems that no sooner was one idea resolved than two more came up. For the most part, we were so busy building sources that we seldom had time to study them in detail, and given that they worked, we had to get on with using them.

IV. FINAL REMARKS

This review has covered the early years prior to the Brookhaven H⁻ Symposium and has gone essentially through the first decade, the first four sessions, of the Symposia. With this, the 8th Symposium, the second decade has come to a close and in another ten years someone can hopefully look back and review the second decade. The first decade was highly productive and I think it is clear that the second decade is as well.

During the present panel session on fundamental processes and ion sources there was some discussion about the early Symposia and their significance. Since this meeting was being held for the first time in Europe have been many new people attending this meeting, many of whom are young, so there appeared to be a need and purpose for this review.

In giving this review I have avoided persons names. It is clear that many people are important to the development of H⁻ ion sources and since it would be impossible to recognize everyone I have chosen not to give names. Still, many people will know who the players are and if not, one can go to the proceedings to find some of the major people. I use institution names where possible and I apologize if I did not include some places. It should also be pointed out that in addition to this Symposia there were other ion source conferences that paralleled the information presented here. Note worthy among these is the European Workshop on the Production and Application of Light Negative Ions, with which this Symposium has merged. This merger will hopefully allow these combined meetings to be held in different parts of the world thereby reaching a broader audience and continuing to serve the world community.

Everyone who has worked on the H⁻ source problem, from the machinist and the lab technician to the senior scientists and the funding support personnel should feel pleased at a job well done. We have come a long way and it has been fruitful and good. At the first few symposia there were some people who stood tall among the group. Some of them are still amongst us, in the laboratories and at the meetings. Some have had to move to other areas. And some have retired but are not forgotten. The effort has gone on for many years. Unfortunately, some are no longer with us except in memory, but what a pleasant memory it is. On this the 20th anniversary of the first symposium (1977-1997), we should all be pleased with the progress that has been made and with our contribution, whether big or small. We should also take time to remember our colleagues.

REFERENCES

1. L. Alvarez, R.S.I., 22, 705 (1951).
2. K. Ehlers, N.I.M., 32, 309 (1965).
3. K.W. Ehlers, B.F. Gavin and E.L. Hubbard, N.I.M., 22, 87 (1963).
4. G.I. Budker and G.I. Dimov, *International Conf. on High Energy Accel.*, Dubna, USSR, August 21-27, 1963.
5. G.I. Dimov, et al, *USSR National Conf. on Part. Accelerators*, Moscow, October, 9-16, 1968.
6. R.L. Martin, IEEE Trans. on Nucl. Sci., NS-18, No. 3, 953, (June 1971).
7. *Proceedings of the Symposium on Ion Sources and Formation of Ion Beams*, Th. Sluyters, Editor. Brookhaven, Upton, N.Y., BNL 50310, October 19-21, 1971.
8. J.E. Osher, ibid, 137.
9. *Second Symposium on Ion Sources and Formation of Ion Beams.* Berkeley, LBL-3399, October 22-25, 1974.
10. Yu.I. Belchenko, G.I. Dimov and V.G. Dudnikov, ibid, VIII-1-1.
11. K. Prelec, AGS Division H- Technical Note, No 35, 1976.
12. Th. Sluyters, ibid, VIII-2-1.
13. C.W. Potts, IEEE Trans. on Nucl. Sci., NS-24, No. 3, 1385, (June 1977).
14. *Proceedings of the Symposium on the Production and Neutralization of Negative Hydrogen Ions and Beams.* Brookhaven, September 26-30, 1977, BNL 50727.
15. *Proceedings of the Second Symposium on the Production and Neutralization of Negative Hydrogen Ions and Beams.* Brookhaven, October 6-10, 1980, BNL 51304.
16. *Proceedings of the Third Symposium on the Production and Neutralization of Negative Hydrogen Ions and Beams.* Brookhaven, 1983, AIP Conference Proceedings, No. 111.
17. *Proceedings of the Fourth Symposium on the Production and Neutralization of Negative Hydrogen Ions and Beams.* Brookhaven, 1986, AIP Conference Proceedings, No. 158.

Summary and Closing Remarks
Peter Massmann

... "any one of you who now plans to come to the next meeting without a paper, well,
you could just inherit this job (of presenting the Summary)!"
Ken Ehlers, "Brookhaven '86"

"This time it happened to me, and I have no regrets!"
Peter Massmann, "Brookhaven in Giens '97"

1. INTRODUCTION

Last time the "Symposium with the long name" (Ken Ehlers), usually refered to by insiders as "Brookhaven", was combined with the *European Workshop on the Production and Application of Light Negative Ions*, with the virtually unavoidable consequence to bear a double long name. Venues were agreed to alternate between Brookhaven and Europe. This time, at its 20th anniversary (which went not unnoticed as we will see below), the Symposium was held "outside" Brookhaven for the first time. Miraculously, the organizer, *Claude Jacquot*, managed to conserve (or should we say: "enhance without caesium"?) the spirit of "Brookhaven" so well that one cannot help to call it "Brookhaven in Giens".

The participants at this conference will realize that the written version of this Summary is (as announced) not just a transcript of the oral analogue presented immediately after the last session. The main reason is, of course, that such an interesting symposium as ours is extremely hard to bring to an end - which is very encouraging. But, to finish in time, was absolutely necessary since not even the keenest negative ion enthusiast (including the author of this Summary) would have liked to "miss the boat" sailing for the Conference Dinner on the lovely island of *Port Cros*.

In this situation the author found himself "sandwiched" between, on the one hand, the slightly "overboarding" panel discussion on *Neutral Beam Injectors and Other Negative Ion Applications* and, on the other hand, the rock solid deadline imposed by the boat departure. After all, there were so many things to say. But too little time was left to present all the viewgraphs bravely prepared in the spare hours between the sessions throughout the week. A selection had to be made - resulting in a summary of the Summary.

Although the written word is known to be substantially different from a live speach "fired from the hip" nothing documented below will be different in spirit from what has been said verbally in Giens. It is a good thing however, to take full advantage of the medium which hopefully will succeed in giving a bit more structure to the story (to avoid confusion adopting a chronological order and *Claude*'s partition of the sessions) and to do more justice to everybody in the review of the presentations. Like in Giens, the Summary will contain a review of the papers and a section with some concluding remarks - not necessarily conclusive.

CP439, *Production and Neutralization of Negative Ions and Beams:* Eighth International Sysposium/
Production and Application of Light Negative Ions: Seventh European Workshop---A Joint Meeting
edited by Claude Jacquot © 1998 The American Institute of Physics 1-56396-737-5/98/$15.00

2. REVIEW

2.1 Fundamental Processes

It was *Marthe Bacal* who kicked off the lectures with a contribution carrying an important message regarding the rôle of caesium in negative ion sources. The (volume) effect of caesium vapour partial pressure on the plasma parameters was studied in the cylindrical multipole filter source *CAMEMBERT III*. These measurements were done in the centre of the extraction region, far enough away from the walls and the plasma grid to exclude surface production (consistent with *David Riz*'s theory below). It follows that caesium, possibly more often quoted for its surface effects, also gives a substantial enhancement in volume negative ion production, especially pronounced (about a factor 3) at low discharge pressures around 1 mtorr. Simultaneously the plasma (volume) parameters change significantly with caesium, resulting in a decrease of the electron density and temperature and of the plasma potential. In a way, it looks like the increase in H⁻ goes at the expense of the electrons. The message: whatever caesium may do to the (plasma grid) "surface" it does not go without profound changes in the "volume".

Volume processes, in particular the distribution and density of highly vibrationally excited H_2 molecules, were also the subject of *Fiodor Baksht*'s theoretical two chamber "experiment". It was shown that the highly vibrational level population of $H_2^{(v)}$ created in the low voltage caesiated hydrogen discharge of the first chamber may increase significantly after streaming through the cold hydrogen gas in the second chamber (with unheated plane copper walls). The idea is to capitalize on this in obtaining a more efficient H⁻ volume production. One could, of course, go a step further and use a wall material (tantalum, caesiated tungsten, ... etc.) that boosts the negative ion yield by surface ionization.

That also unexpected materials may feature a quite efficient surface production of negative ions became clear in *Michael Gleeson*'s presentation dealing with H⁻ production by scattering of H^+, H_2^+ and H_3^+ from polycrystalline or **H**ighly **O**rientated **P**yrolitic (*HOP*) graphite surfaces. This subject is of certain importance for tokamak divertor physics and plasma (graphite) wall interaction. A remarkably high H⁻ yield of 30% was reported for the poly-graphite (*HOP* graphite only 1%) which is even more amazing considered the comparatively high workfunction of about 5 eV.

In his triple Vlasov model of negative ion extraction from a volume source with a negative plasma electrode bias *John Whealton* showed us once again what a tough problem it is to tackle all the processes involved in a self-consistent manner. Although not complete (e.g. no magnetic field), the model manages to arrive at some plausible explanations. The self-consistently calculated electrostatic potential in the pre-sheath and extraction region turns out to form a saddle point with the consequence that most of the negative ions generated in the volume are not being extracted. In the region where the extracted current is space charge limited we find back many of the beneficial effects of caesium. Not only does caesium increase the volume negative ion production (see *Marthe Bacal*'s contribution above) but, as it gets ionized, it also adds to the slow positive charges with the result that the sheath moves inward which increases the space charge limit. It is also quite plausible that the positive charges of the Cs^+ ions will decrease the fraction of co-extracted electrons and help to reduce sheath instability (in contrast to positive ion extraction the Bohm sheath stability criterion appears not to be generally satisfied for negative ions). Considered the substantial effort put into this work (throughout years) one wonders whether we will ever arrive at a complete model including magnetic fields, conversion on a caesiated surface, plasma grid temperature etc...

2.2 Source Modeling

Unusual (but nevertheless very interesting) in an environment where the majority is trying to enhance D⁻ or H⁻ yields is the work on negative helium ions presented by *Masaki Nishiura*. He⁻ does exist although in (several) metastable states. The experiment at *NIFS* (*N*ational *I*nstitute of *F*usion *S*cience, aimed at α particle diagnostics) produces a He⁻ beam by double charge exchange in a rubidium vapour cell. Conversion efficiencies of the order of 1% have been measured by photo-detachment resulting in beam current densities of about 10 $\mu A/cm^2$.

No-one less than *NIETZSCHE* (*N*egative *I*on *E*xtraction and *T*ransport *(Z)S*imulation *C*ode for *H*ydrog*E*n, of course!) was necessary to help *David Riz* joining together some more pieces in the puzzle of H⁻ / D⁻ extraction from a plasma source. The 3D code calculates the chance to pass the apertures of the plasma grid for negative ions launched either somewhere in the volume or at the surface of the plasma source. Taken into account are destruction processes, elastic collisions, charge exchange and the magnetized plasma parameters (a spatial variation of the electrostatic potential is not included as only probabilities are calculated). A crucial quantity in the model is the collision parameter R_L / λ_{coll} (Larmor radius over collision mean free path) reflecting the important difference between high and low power discharges.

For the high power discharges of interest for neutral injection the results are as follows. To stand a reasonable chance of extraction negative ions produced in volume have to be generated in the vicinity (cm's) of the plasma electrode. If they are produced on a surface only those genretated on the plasma electrode will be able to make their way to extraction. Furthermore, only H^0 / D^0 converted on surfaces seem to yield the "right" isotope effect, n(H⁻) ~ 1.5 n(D⁻). This is supported by experiments which show that the fraction of extracted negative ions is relatively insensitive to the strength of magnetic filter applied.

2.3 Diagnostics

In the first of his two contributions *Hans-Michael Katsch* presented photo-detachment measurements of H⁻ in the afterglow of a pulsed magnetic multipole discharge. Is is well kown for some time that afterglow plasmas can be relatively rich in H⁻ due to the difference in the loss times between $H_2^{(v)}$ (ms) and fast electrons (µs). In resemblance to its magnetic analogue this effect is often called a "temporal filter". Most amazing and until now unexplained is the finding that in these circumstances the maximum H⁻ yield or the $H_2^{(v)}$ density derived thereof does not seem to vary with discharge current in the range 1 - 20 A.

In his second talk *Hans-Michael Katsch* compared various H⁻ diagnostic methods applied to pulsed or RF discharges. Between the measurement techniques using Langmuir probes (directly), mass spectronmetry, photo-detachment or laser absorption the photo-detachment method was judged to give most confidence for not too high negative ion densities. Laser absorption could yield satifactory results at high n$^-$.

2.4 Poster

In her poster *Claire Michaut* showed first results of emittance measurements on a single aperture H$^-$ beam, for the time being, still without caesium seeding of the plasma source. A nice feature of the diagnostics applied is that the emittances of H$^-$, H$^+$ and H^0 may be obtained simultaneously in one experiment. Considered the many open questions regarding negative ion extraction and transport we certainly would wish to see more of such fundamental experiments and results in this domain.

2.5 Source 1

The first of the two (Japanese) high power negative ion sources for neutral beam injection into large scale fusion machines was presented by *Katsuyoshi Tsumori*. The "Nagoya" source has the following parameters: extraction area 120 cm^H x 25 cm^W, extracted current density 40 mA/cm^2 H$^-$, operating pressure about 2 mTorr (caesium seeded) and arc power 220 kW. Four of these source will power the two injectors of the *Large Helical Device* (now in the commissioning phase) designed to deliver 15 MW of 180 keV H^0 beams for up to 10 s with a required divergence <10 mrad.

The features are quite typical not only for this source but also of many negative ion beam sources proposed (e.g. *ITER*) or operating (*JT60-Upgrade*): caesiated plasma discharge, magnetic multipole confinement, external magnetic filter close to the molybdenum plasma electrode biased a few volts positive, permanent magnets and active water cooling in the extraction grid and, last but not least, vertical geometric beam steering. Also common to other sources in this domain is the individual regulation of the filament emission to improve plasma uniformity across the extraction area. With the presented extracted currents of 4 A (at 73 kV for 0.3 s and at 4.5 mtorr) obtained from only the centre section of the 5 section grid structure the experiments are well on their way to approach the design parameters. Difficulties seem still to exist with an electron leakage through the extraction grid, maybe one of the toughest problems in these kind of sources.

The experiments with the second (Japanese) source in this domain reported by *Rusty Trainham* are aimed at a demonstration of long pulse operation up to 1000 s on the MANTIS (*M*ulti *A*mpere *N*ega*T*ive *I*on *S*ource) test bed. For this purpose a so called "*Kamaboko*" source (named after the popular Japanese culinary speciality of typical semi-cylindrical shape) which was supplied in the framework of a collaboration with *JAERI* (*J*apanese *A*tomic *E*nergy *R*esearch *I*nstitute), is being used. Athough of reduced scale (extraction area 70 cm^2) the other parameters are by all means relevant for neutral injection purposes: arc power 80 kW, extracted current 1.4 A D$^-$ (20 mA/cm^2), operating pressure 0.3 Pa. The conceptual features are in principle not very different from what has been mentioned above.

The results are summarized as follows. As anywhere, caesium admixture improves the D$^-$ yield and simultaneously reduces the fraction of co-extracted electrons. Increasing the strength of the magnetic (plasma grid) filter to 1800 Gcm enhances the negative

current ratio $I(D^-) / I(e^-)$ to about 2. A positive bias of 5 V on the plasma grid reduces the electron fraction without effecting the negative ion yield. The optimum yield is obtained at a plasma grid temperature of approximately 300 ^{0}C which (from the work of Taylor - Langmuir) corresponds to a 0.6 - 0.7 monolayer caesium coverage.
Assuming surface production on the plasma grid as the main mechanism it was tried to enhance the D$^-$ generating area by employing a thicker grid (12 mm instead of 6 mm). As a result, the fraction of co-extracted electrons was again reduced but the D$^-$ current went down because of inferior optics.
Longer pulses (at a stable plasma grid temperature of 300 ^{0}C) require a grid design with a thermal barrier between the exposed metal and the cooling channels. Two different designs will be tested in the near future. Until now pulses up to 1 min have been achieved.

In a new country, with a new identity but with the same old name the experiment *DENISE* was presented by *Paul McNeely*. The recomissioning of this experiment represents a real metamorphosis: from the **DE**nse **N**egative **I**on **S**urface **E**xperiment of the early eighties in Amsterdam (filamented source, H$^-$ beam by specular reflection from 2 caesiated W(110) monocrystals) to the **DE**uterium **N**egative **I**on **S**ource **E**xperiment in Dublin for use in the late nineties employing either a filamented or a 13.5 MHz RF source and D$^-$ to be extracted from the discharge with and without caesium. Compared to the old days *DENISE* also gains substantially in diagnostics (new Langmuir probe, new photo-detachment system etc.) and receives a new PC based control and data acquisition system.
The status is such that all major components are ready (except the YAG laser for the photo-detachment measurements). Good functioning of the subsystems has been demonstrated in preliminary low power (600 W) RF discharges. We all should look forward to lots of interesting results, especially at the low discharge pressures below 0.3 Pa required for the *ITER* injectors.

2.6 Source II

As time moves on, the originators of the positive ion RF sources for one of the ASDEX Upgrade injectors have turned to negative ions, as *Davar Feili* reported. Lately they have developed an RF (2.5 MHz) negative ion source of reduced scale (volume about 1 l) featuring a 6 turn quartz sleeve insulated internal antenna, metal walls and multi (8) cusp confinement. Both adding caesium to the discharge and using a magnetic filter have given a distinct enhancement of the negative ion yield. Although stable operation up to 15 min has been obtained there remains the problem of quartz sputtering from the antenna which after a while leads to coating of the metal walls with an insulating layer.

Anybody who did not fancy *BATMAN* until now must no doubt change his mind after *Peter Frank*'s presentation on the **BA**varian **T**est **MA**chine for **N**egative **I**ons. This RF source development is directed towards maintenance free steady state operation as a prefered option for *ITER* neutral beam injection. A volume of about 30 l and up to 100 kW of RF power (1 MHz) makes the extrapolation to *ITER* size more credible. To save on pumps (and also to compare results with those from a filamented source of the same extraction area on the *MANTIS* or *SINGAP* test bed, see below) the extraction area is reduced for the time being to 48 cm^2.

The source features a 6 turn antenna outside the discharge chamber (two versions: vacuum tight Faraday screen or quartz wall with internal Faraday screen) a magnetic filter close to the plasma grid but (except for permanent magnets on the back plate) no magnetic multipole confinement (as yet). Best value of extracted H$^-$ current density so far is 4.5 mA/cm^2 without caesium at elevated pressure of 1.6 Pa with a power input of 3 kW/l. Addition of caesium to the discharge results in a reduced operating pressure (2x with the same yield) and an increase of the H$^-$ yield which is strongest (5x) at the lowest operating pressure of 0.6 Pa achieved until now. Clear goal for the future is to improve the confinement and the source efficiency at low pressures (<0.3 Pa).

Volume H$^-$ production in a helicon (ECR) plasma confined in a magnetic mirror field (B_z = 250 G) was the subject of *Daiyu Hayashi*'s talk. He showed that relatively high plasma densities (10^{11} - 10^{12} cm^{-3}) may be obtained using this method (advantage: no erosive electrodes inside the plasma). with 0.8 kW RF power at discharge pressures between 6 - 10 mtorr. Typical for this type of plasma is that the electron temperature is relatively high in the plasma centre (T_e = 7 eV) and low at the edge (T_e = 1 eV). Not unexpected therefore (from the "classical" mechanism of volume production) that the highest H$^-$ concentration of 5 x 10^{10} cm^{-3} is found (by photo-detachment) at the plasma edge. The negative ion to electron fraction in cw operation is about 0.3. This may be "enriched" to 0.7 in the afterglow of (repetitive) pulsed discharges. Athough these results are quite encouraging extraction from these sheet plasmas and high quality beam formation may not be easy because of the expected high poloidal velocity of the particles.

A completely different "kettle of fish" with current densities well in excess of what is envisaged for neutral injection are the negative ion pulses required for the *E*uropean *S*pallation *S*ource (*ESS*), as *Andreas Maaser* reported. The specifications of the source under development in his group are quite impressive indeed: high brightness 70 mA H$^-$ beam from a single aperture in 1.3 ms pulses with a repetition rate of 50 Hz. This all needs to be combined with low noise and high reliability.

The best values obtained so far applying 22 kW in the arc (volume 0.5 l, pressure 1 to 5 Pa) and caesium vapour injection near the plasma electrode are 40 mA through an aperture of 4.25 mm radius (70 mA/cm^2). It is found that adding caesium increases the H$^-$ yield and simultaneously reduces the electron fraction, both by a factor 3 compared to pure volume production (where the negative ion to electron ratio is about 1/100). The optimum negative ion yield is achieved at a plasma grid temperature of 180 °C. The extracted H$^-$ current density tends to increase with smaller apertures. Although the negative ion yield does not increase linearly with the arc power there is good hope that the envisaged 70 mA may be reached by using the full installed power of 50 kW available from the arc supply.

2.7 Source III

Igor Soloshenko's presentation contained both experiments with a reflex type negative ion source and modeling work on volume and surface processes incorporating no less than 26 reaction equations. The reflex source is capable of producing attractively high plasma densities (10^{13} cm^{-3}) but at pressures (10 - 100 mTorr) which would make an application in neutral injection rather difficult. The best extracted H⁻ current density is an impressive 80 mA/cm^2 at a discharge power density of 500 W/cm^2. The negative ion to electron fraction is 0.2.

The results from the modeling show no increase with caesium in "volume" but up to 5x enhancement on a caesiated surface where the conversion efficiency is taken as 10%. Even at lower conversion rates the surface effect seems to dominate. But attention, the predicted yield in "volume" is about a factor 3 lower than measured!

The effect of caesium added to the discharge of a tandem "volume" source has been modeled by *Osamu Fukumusa*. The model takes into account not only volume processes but also surface conversion (on the plasma electrode) where the efficiency is taken as a function of the caesium coverage (work function). For both volume and surface production an enhancement of a factor 5 is found with caesium at the "optimum" pressure (1 Pa) for maximum extraction (including stripping losses). The strong variation of the negative ion yield with the electron temperature in pure volume mode gets largely moderated with caesium which indicates less destruction. The pressure dependence in "volume" shows a maximum whereas with caesium saturation is found with increasing pressure. Despite the changes due to caesium in "volume" it appears that surface production is dominant.

With the practical view of a good experimentalist *Claude Jacquot* summarized his experience with the "*Kamaboko*" source (30 l, 1.4 A D⁻, 80 kW arc - see *Rusty Trainham*'s contribution above). The experimental results in combination with the work of *David Riz* (see above) led him to believe that surface conversion of D^0 on the plasma electrode must be the dominant effect for the following reasons. The magnetic filter does hardly effect the D⁻ yield (but improves electron suppression) which could mean that D⁻ is mainly produced by conversion of D^0 and not D$^+$. The D⁻ yield goes still up after the admission of caesium has been switched off and no emission of caesium spectral lines from the discharge is detected. This seems hard to explain without a surface effect. From the work of Taylor - Langmuir a stable caesium coverage of 0.7 monolayers is obtained at 300 ^{0}C which is consistent with the minimum workfunction achievable for caesiated molybdenum or tungsten (from the filaments possibly deposited on the plasma electrode). The plasma electrode temperature effect is only present if high enough quantities of caesium (>600 mg) are used which might be necessary ro reach equilibrium in high power discharges.

A general finding with these sources is that the power efficiency seems to scale with size, i.e. about 2 kW/l source volume or 15 W/cm^2 extraction area. Large sources seem to have the advantage that the operating range extends to lower pressures.

Some hint concerning the mobility of caesium was given by means of a photograph of the inside of the source after some time of operation. To be seen were walls free of tungsten exept for sharply defined areas behind the filament stems which were covered with tungsten. These "shadows" are exactly the areas which particles emerging in straight lines from the caesium injection opening would not reach. *Claude*'s conclusion: caesium "does not like" tungsten and is "not very mobile".

That negative ion beams can do much more than "serve" in the fields of high energy accelerators or neutral beams for fusion was shown in a refreshing manner by *Samar Guhary* in his contribution dealing with compact surface plasma sources and new applications. Driven by the stringent requirements in both of the fields mentioned above significant progress has been made in the development of negative ion sources throughout the last decade. In recent times the merits of these sources and beams (e.g. potentially high brightness and the fact that negative ions do not charge a beam exposed surface like positive ions) have been recognized for a whole series of novel applications in areas like *I*on *P*rojection *L*ithography (*IPL*), ion implantation (into insulators), surface treatment, *P*article *I*nduced *X*-ray *E*mission (*PIXE*), *S*econdary *I*on *M*ass *S*pectroscopy (*SIMS*) and, last but not least, spallation neutron sources. Depending on the application the beam energy may range between some 100 keV to several MeV (e.g. PIXE 2 MeV).

Especially the micro beams to be used for electronic chip design put extremely high demands on beam brightness and good optics, i.e. negative ion currents of several 10 µA focused into 3^0 or widths in the 10 nm range. The experimental results from the Penning surface plasma source development in *Guhary*'s group are equally impressive: 170 mA extracted through an aperture of 6 mm diameter, meaning no less than a current density of 600 mA/cm^2.

The panel discussion on *Fundamental Processes and Ion Sources* was moderated by the author of this Summary. The members of the panel (in alphabetic order) were: *Marthe Bacal, Fiodor Baksht, Samar Guhary, Rusty Trainham* and *Katsuyoshi Tsumori*. To get the discussion going the moderator presented a couple of viewgraphs containing a cartoon of a "*classical volume source 'spoiled' by caesium and high power*" (indicating that the keen source developer longing for progress tends to try new things, like putting a known "surface drug" into the "volume", rather than going for a complete understanding of all the fundamentals - until he gets stuck), one graph explaining the essentials in the energy level diagram of surface ionization (broadening and lowering of the affinity level by the image forces at close distances to the surface) and a second graph showing the dependence of the work function with coverage for a caesiated tungsten (110) monocrystal (minimum at 1.45 eV, higher for polycrystals or other metals). Finally, a viewgraph with the typical parameters of "successful" (20 mA/cm^2 D$^-$ or 30 mA/cm^2 H$^-$, $n_-/n_e \approx 2$, $p \leq 0.3$ Pa) high power sources for neutral beam injection was presented to summarize the "facts" (*features*: caesium super drug, tungsten filaments, magnetic multipole confinement, magnetic filter near or at the plasma grid, plasma grid at 300 ^{0}C; *dependences*: $j_- \sim V_{ex}^{3/2}$, $j_- \sim P_{arc}$, $P_{arc} \approx 2$ kW/l or 15 W/cm^2; *working hypothesis*: surface ionization on the plasma grid) and to form a basis for discussion.

From the series of proposed topics the points *"Surface" or "Volume"* and *Diagnostics* attracted the most interest and were lively discussed. The results (by no means conclusive) may be summarized as follows. There is growing evidence that surface conversion is dominant but the mechanisms of production and extraction are far from being fully understood. Furthermore, as we have learned above, adding caesium to the discharge also causes significant changes in the plasma (volume) parameters. The profound wish that to improve understanding there should be more synergy between the source "builders" (who are presssed by a tight time schedule) and the source "fundamentalists" (who's work is very important for long term progress) was expressed by one of the "builders".

Generally it was found that the diagnostics need to be extended and improved. Two suggestions were made in particular. Firstly, a beam emittance measurement with and without caesium in the discharge could yield valuable information about the (change of the) negative ion temperature (transverse energy). Secondly, it was suggested to apply the laser absorption technique (suitable for larger negative ion concentrations and to be imroved, if necessary) in the region close to the plasma grid. This could possibly give some answers regarding the negative ion concentration in front of the grid (with and without caesium) and what it is that "points" the negative ions in the right direction for extraction (charge exchange of H$^-$ with H^0?).

The last in the series of contributions on sources was given by *Osamu Fukumusa* comparing the performance of an ECR volume source (2.46 GHz, 1.5 kW launched from a magnetron through an annular slot antenna) with that of a filamented source of the same basic configuration (without caesium, driver and extraction region separated by a magnetic filter). In contrast to its advantage of relatively easy and inexpensive plasma generation the plasma parameters (of interest for efficient volume negative ion production) obtained with the ECR method appear to be somewhat disappointing. In ECR mode the filter does not seem to work as there is no difference in electron temperature ($T_e \sim 2$ eV) between the two regions. In case of a filamented discharge the electron temperature in the driver region is found to be substantially lower. Also, the electron density in the driver region is about a factor 10 lower for the ECR case than for a filamented discharge. How to proceed, to find out "why it does not work as expected" or to introduce caesium (as suggested by one of the "builders"), seems again a question of mentality and the research boundary conditions.

2.9 Negative Ion Acceleration

Turning towards the long and stony road of demonstrating 1 MV acceleration of a negative ion beam extrapolable to ITER neutral beam injection parameters (3 D^0 beams of 17 MW with 5 mrad divergence for 1000 - 10 000 s from primary D$^-$ beams of 1 MeV, 40 A, 20 mA/cm^2) *Kazuhiro Watanabe* presented the latest results from the 1 MV Test Stand at *JAERI*. The acceleration concept used is the same as envisaged in the reference design of the *ITER* injectors featuring an electrostatic **Multi Aperture Multi Grid** (*MAMuG*) system with the acceleration to 1 MV provided by means of 5 intermediate electrodes in steps of 200 kV.
High voltage conditioning of the system has been undertaken with great patience. Best values obtained so far are 920 kV without beam and 886 kV, 19 mA H$^-$, 1 s with beam. Extrapolation of perveance measurements (Child - Langmuir) in which up to 6 mA/cm^2 H$^-$ have been accelerated to 400 kV gives confidence that the required 20 mA/cm^2 D$^-$ may indeed be achieved at 1 MV.
Although these results are quite impressive without indicating any sign of principle obstacles it appears that the speed of progress is determined by difficulties encountered with the 1 MV large insulator made of fibre glass re-inforced epoxy. After having identified most of the (entirely technological) problems great effort is made to obtain a satisfactory technical solution for this "piece de résistance".

Although with a different concept the goal remains about the same for the *Cadarache* 1MV test facility: demonstrating a reliable 1 MV, 20 mA/cm^2 D$^-$ beam of good quality, as *Corinne Desgranges* reported. In the alternative *SINGAP* concept intermediate electrodes are avoided and beamlets after being extracted and pre-accelerated (50 keV) through a group of apertures are then merged and post-accelerated across one *SIN*gle *GAP* into a *SING*le *AP*erture. Best values obtained during the first beam campaign are 860 keV, 43 mA, 3.5 mA/cm^2 H$^-$, 1 s (without caesium seeding of the ion source) and 630 keV, 106 mA, 8.8 mA/cm^2 D$^-$, 1 s (with caesium) during the second campaign exploiting fully the nominal current capability (100 mA) of the HV power supply.

Relatively good beam optics (beam envelope divergence about 10 mrad) has been demonstrated for beams of quasi-circular and ribbon shape (the latter more relevant to *ITER* neutral beam injection but more difficult for *SINGAP* optics) already in the low energy range 250 - 500 keV. Although no particular optimization has been carried out it appears that the 3D modeling by means of the code from *Vector Fields* has been an indispensable guide on the way to these promising results.

More insight into what is happening inside the ion source might be obtained with the infrared diagnostics applied. In the example shown plasma non-uniformity in the source (without caesium) and beamlet displacement caused by the permanent magnets in the extraction electrode where clearly visible. It may be interesting to repeat these measurements with caesium.

The experiments of the second campaign have been limited to energies below 700 keV because of faults occuring in the 1 MV insulator bushing (the two top rings out of nine turned out to be conducting) also made of fibre glass re-inforced epoxy (but using a different fabrication process than above). Since voltage holding even slightly above the 1 MV level has been obtained relatively quickly it seems that these defects occur with usage during the course of time. This rises again the question whether, even with a reasonable design, the choice of material (inexpensive but eventually inadequate for very high voltages in vacuum because of its weaknesses, i.e. high secondary electron coefficient, degassing, carbonization) was not unfortunate and whether (in hindsight) one should not have "attacked" straight away the technological challenge of a ceramic 1 MV insulator - which is required for the *ITER* injectors anyway.

Masaya Hanada's contribution regarding electron stripping in the 3 electrode *MAMuG* accelerator of the 400 kV *NIAS* (*N*egative *I*on *A*cceleration *S*tand) facility was given by *Kazuhiro Watanabe*. Assuming that stripped particles and electrons are dumped in the gap where they are created the calculations arrive at a total particle loss rate of 44% (source pressure 0.3 Pa) with the highest heat load on the first acceleration grid and the lowest load on the third (ground) grid. The fact that the highest measured heat load is found on the second acceleration grid indicates that electrons (stripped and / or scattered forward by secondary emission) may be accelerated from one gap into the next. Extrapolating the numbers towards the *ITER* injectors predicts stripping power losses of the order of 1 MW which is probably not impossible to cool away but certainly not desirable. This shows again the importance of reducing the source operating pressure even further to 0.2 Pa or lower.

In his contribution on electron separation in the *NSNS* (*N*ational *S*pallation *N*eutron *S*ource) H$^-$ source *John Whealton* investigated the influence of the electron deflecting magnetic field in the extractor on the negative ion trajectories by means of 3D calculations. Both the magnetic field and the asymmetry in the plasma sheath due to gyrating electrons cause a steering of the ion beam. A good correction of this effect with a minimum of emittance growth (for further acceleration) is therefore required. In contrast to aperture offset steering (frequently applied in neutral injection systems) a satisfactory solution is proposed through a split electrode (with different potentials, top - bottom) which manages to do the job by nearly retaining the emittance without the deflecting field applied.

2.10 High Voltage Effects

This session was preceded by an extremely interesting contribution given by *Jens Peters* presenting the RF (1 - 2 MHz) H⁻ volume source developed for *DESY* (*D*eutsches *E*lektronen *SY*nchrotron). After single aperture (8 mm diameter) extraction in 130 µs pulses, 1 - 4 Hz repetition rate and 0.5% duty cycle the negative ions are to be pre-accelerated by an RFQ system. The discharge is generated by an internal two turn alumina coated antenna. The mode of operation (300 days a year, 24 hours a day) puts very high demands on reliability. The negative ion current densities (up to 160 mA/cm^2, 80 mA H⁻ current) obtained with a tantalum aperture insert ("collar") and without caesium are among the highest achieved anywhere. It appears that applying the tantalum collar increases the negative ion yield by more than a factor two. Further enhancement may be obtained by a focusing effect due to a negative bias potential on the collar.

Beam transport, steering and power handling in the *ITER* injectors was the subject of *Alexander Krylov*'s presentation. The long narrow duct (length about 10 m, usable cross section 0.5 m^W x 0.8 m^H) at about 10 m distance from the beam source puts very high demands on beam divergence (≥ 5 mrad) and alignment (accuracy ±2 mrad horizontally, ±4 mrad vertically). To achieve an acceptable geometric beam transmission (80%) together with an optimised deposition profile (note with different magnetic configurations the magnetic axis of *ITER* may move by 40 cm) requires a well designed beam steering system, individually for each of the 5 grid sections (vertically by inclination, horizontally by aperture displacement). Additionally, extensive magnetic shielding of the torus stray field (down to the order of 1 G) is necessary to avoid possible beam deflection and consequent excessive power loads on beamline components.
The calculations of the power distribution along the line show that at the entrance of the neutralizer a strucutred profile (peak power density ~100 MW/m^2) is still distinguishable resulting from the individual grid sections. Finally, this ends up in a merged beam profile with a peak power density of 50 MW/m^2 at the magnetic axis of the torus plasma. On the way, the walls of beamline components have to withstand between nominally 5 MW/m^2 and 10 MW/m^2 in case of some misalignment.

How important fundamental research can be for the design of large systems like the *ITER* injectors became once again clear in *Yukio Fujiwara*'s contribution on *R*adiation *I*nduced *C*onductivity (*RIC*) in insulating gases. In the reference design of the injector the beam source (at high voltage) is foreseen to be electrically insulated by pressurized gas (e.g. SF_6, CO_2 or dry air). That these gases might get more conducting than desirable under the γ- and neutron radiation of the *ITER* environment was first pointed out by *Eric Hodgson* of *CIEMAT* (*C*entro de *I*nvestigaciones *E*nergeticas,

*M*edioambientales y *T*ecnologicas - Madrid) in small scale experiments showing leakage currents of only some milliamps. However, extrapolated to 1 MV and the large gas volumes of the *ITER* injectors (source pressure vessel ~30 m^3) this could lead to rather unacceptable power losses which, in the worst case, would even be absorbed in the gas.

With this in mind, *RIC* measurements have now also been done in Japan using a ^{60}Co gamma ray source and high voltages up to 100 kV applied to electrodes immersed in the gas and separated by some centimeters. In these circumstances it is found that the *RIC* is proportional to the mass of the gas, the volume, the pressure and the dose rate. The *RIC* for dry air is about 5x lower than for SF_6 at the same pressure (but the breakdown limit also 3x lower). Taking the rather large spread in the calculated expected dose rate (0.2 - 2 Gy/s) still existing in the radiation modeling the measurements would lead to extrapolated leakage currents of 5 - 50 A for 21 bar dry air and 8.8 - 88 A for 8 bar SF_6. Although there is some hope that the extrapolation to 1 MV and to larger distances (order of 1 m) might yield currents which could be an order of magnitude less there seems more than enough reason to look at an alternative design using vacuum insulation (as we will see below this is underway for some time). Simultaneously, it would be reasonable to gather more *RIC* results at higher voltages and with larger seperating distances.

In his contribution on surface charging and secondary electron emission *Junzo Ishikawa* showed us why negative (and not positive ions) may be applied in areas like ion implantation and the fabrication of *ULSI* (*U*ltra *L*arge *S*cale *I*ntegrated) circuits. For negative ions the "balance" between impinging and leaving charges appears to be very favourable leading to a relatively low charging voltage which makes them (because of their penetration - good optics provided) an excellent tool for surface treatment. As an example, measurements in the beam energy range between 0 - 40 keV indicate that the surface charging on an (isolated) metal stays between 0 and +7 V and on an insulator between 0 and -5 V.

With an enthusiastic review of "20 Years Brookhaven" this Symposium finally got the birthday lecture it deserved. No-one was better suited to present this lecture than *Charles Schmidt*, the only person present who attended all the meetings from the very beginning. In fact, the history of negative ion beams "started" long before the first (international) meeting. It was as early as 1952 (for the tandem Van de Graaf accelerator at *Berkeley*) when negative ions became attractive as they let themselves be accelerated to twice the energy "for the same money" (the *"Swindletron"*).

Since then negative ion sources (surface plasma source, planotron, Penning) became more and more popular. Contributions regarding these subjects were presented on conferences dealing with ion sources "of all signs", like the one at *Berkeley* in 1969.

In the early 70's the *Novosibirsk* surface plasma source (*Dimov*) reached a certain performance (without and with caesium) which nearly instantly triggered a parallel development at *BNL* (*B*rookhaven *N*ational *L*aboratory, *Sluyters*). Growing interest from the field of neutral injection was one of the driving forces to launch an international symposium addressing virtually all of the items related to the production of negative ion based neutral beams (ion sources, acceleration, neutralisation etc.).

The first in the series of meetings "with the long name" was held in 1977, like the many which were to follow, at *BNL*. Popular negative ion production techniques were then (still) the surface plasma source, the rod duo-plasmatron, double charge exchange (in a sodium or caesium vapour cell), molecular cluster ion sources ("Super Swindletron"?) and, for the first time and extremely puzzling, the volume plasma source.

In 1980 the elaborate double charge exchange method was virtually gone, the clusters still there. Meanwhile negative ion production in "volume" was confirmed and (thanks to *Marthe Bacal*'s dissociative attachment model of highly vibrationally excited molecules) a more complete idea of the production mechanism was gradually established. The number of contributions related to Fusion was increasing.

The "bucket" source with magnetic multipole confinement (*Berkeley* and later also *Culham*) and the surface converter source (surface plasma source with a dedicated converting surface inside the discharge, independent from the cathode - the converter) appeared during the mid 80's. Although quite successful, the converter sources became less attractive during the late 80's because the negative ions they produced (rather efficiently) had a relatively high transverse energy (~10 eV c.f. ~1 eV in volume sources). As a consequence of the high demands on beam quality for Fusion applications and, let us not forget, for the *S*pace *D*efence *I*nitiative (*SDI*) the attention turned more and more towards volume sources especially after it had been discovered that the negative ion yield could be boosted by several factors through adding caesium to the discharge. Since then the "caesium seeded volume source" has been developed vigorously and has become the prefered solution in all of today's negative ion based neutral injectors, the *ITER* ones included.

In conclusion, it appears that great progress has been made in the development of negative ion sources and beams despite the fact that a profound understanding of all the fundamental processes involved lacked (and still lacks) somewhat behind. With its openness and conciseness the "Brookhaven Symposium" also has been a great success providing the most complete documentation of the field.

Henk Hopman continued the session with a contribution regarding the *S*econdary *E*lectron *E*mission (*SEE*) from insulators under electron bombardment, a subject which is not only of fundamental interest for the performance of high voltage insulators, e.g. like the large 1 MV bushing of the *ITER* neutral injectors. Measurements of the *SEE* coefficient, δ, as a function of primary energy (0 - 2 keV) on mica show a maximum δ of about 5 accompanied by a surface charge which is positive (in contrast to *Junzo Ishikawa*'s result on insulators above).

That δ may be reduced by about a factor 2 (which possibly can improve the insulator performance) was shown in measurements on alumina which had been surface treated by high dose ion implantation. The finding that the insulator surface gets slightly conducting by the implantation treatment could be another positive point as it would help to improve the uniformity of the voltage distribution across the insulator (which again could enhance the performance or give a welcome reduction in size).

Very interesting indeed was also the work on diamond (electron affinity -0.5 eV, i.e. above vacuum level!) which in principle should be a "fantastic" (negative ion) surface converter material. Until now the measured conversion efficiencies remain however rather modest (~ 5%). *SEE* measurements have shown that the value of δ depends very sensitively on surface preparation. A small dose of irradiating electrons may increase δ by an order of magnitude which should be more than enough reason to try the material in a plasma environment.

As the expected participation from *LBL* (**L**awrence **B**erkeley **L**aboratory) finally turned out not to materialize a summary of the *Berkeley* negative ion source activities was given by *Jim Alessi*. Together with *BNL* and *ORNL* (**O**ak **R**idge **N**ational **L**aboratory) *Berkeley* collaborates in the development of the negative ion beam for the *NSNS* project (see *John Whealton*'s contribution above) by supplying the ion source. Envisaged is a pulsed RF source (2 MHz, 6% duty cycle) with internal antenna and a movable (magnetic) rod filter. With an RF power of 40 kW 35 mA H⁻ is now available without caesium in the discharge (which corresponds to a current density similarly high as the one reported by *Jens Peters* above). Experiments with an insert in the plasma electrode aperture ("collar") have indicated an optimum collar temperature of 200 ^{0}C for the best negative ion yield.

A second activity at *Berkeley* is the development of a cylindrical multicusp converter H⁻ source for *LANL* (**L**os **A**lamos **N**ational **L**aboratory). Currents up to 43 mA have been realized with further optimization in progress.

In his overview of the *ITER NBI* (**N**eutral **B**eam **I**njection) systems *Ron Hemsworth* brought us up to date with the (design) status of the three main injectors (for heating and current drive, for some of the parameters see *Kazuhiro Watanabe* 's contribution above) and the diagnostic beamline.

The recent events concerning the heating beamlines may be summarized as follows. The large gate valve seperating each of the beamlines from the torus and thus allowing injector maintenance or operation independent of the *ITER* machine has been omitted for cost and maintenance reasons. The Fast Shutter to reduce (tritium) gas flow from the torus into the injector has been taken out of the beamline vessel and forms now an independent module which can be removed (remotely) for possible maintenance. The beamline calorimeter design has been fixed.

Triggered by the alarming RIC measurements in insulating gases (see *Yukio Fujiwara*'s contribution above) the design of a vacuum insulated beam source has made rapid progress not only avoiding the *RIC* problem but also yielding several additional benefits. In this design the large alumina insulator column (3 m outer diameter, mounted horizontally) enclosing the accelerator structure (only allowing for axial pumping) is replaced by post insulators which (now entirely in vacuum) permit a certain amount of transverse pumping (which again is suspected to reduce stripping). The gas / gas 1 MV bushing of the reference design is changed for a ceramic gas / vacuum (transmission line / source vessel) bushing of "only" 2 m outer diameter mounted vertically ("easier" to manufacture than the 3 m one and featuring only compressional forces on the alumina). As we heard already in *Alexander Krylov*'s talk, the beam steering has been adapted to cater for the changes in the position of the magnetic axis required for some of the *ITER* plasma scenarios.

The project of the diagnostic beamline (5 MW H^0 produced from a primary 125 kV, 120 A H⁻ beam modulated with 5 Hz) allowing charge exchange spectroscopy measurements of the thermal helium "ash" in *ITER* also has come into shape. Although lower in energy this beam may be no less difficult to realize than its "bigger brothers" because the boundary conditions (125 kV for maximum cross section, perpendicular injection for good penetration) demand a beam footprint "as small as possible" (~200 x 200 mm^2) with the consequence of quite high peak power densities. Also, because of the "lower" energy and the substantially higher number of beam particles the issues of duct re-ionization and beam blocking become more important.

The *JAERI* negative ion programme not only covers almost all the items to be resolved for the *ITER* injectors but also looks after the "good health" of the *JT60-Upgrade* 500 kV *NBI* system. In his talk *Kazuhiro Watanabe* concentrated mainly on two subjects, i.e. continuous operation with a caesiated source and experience with the *JT60-Upgrade* injector. In a demonstration experiment the *JAERI* Long Pulse Negative Ion Source has been run continuously for 140 h after initial injection of 600 mg caesium into the discharge. Requiring a caesium consumption of <3 mg/h the negative ion yield remained stable without additional seeding. With a value of 6 mm (starting from 2 mm) the thickness of the frame cooled plasma electrode envisaged for long pulse operation has been optimized, giving the same H$^-$ yield but less electrons (electrodes thicker than 6 mm spoil the beam optics as we learned from *Rusty Trainham*'s contribution above).

While doing important plasma experiments the *JT60-Upgrade* negative ion based neutral injector is steadily running up to its specifications (500 kV, 20 A D$^-$). Obtained so far are 400 kV, 13.5 A D$^-$ and 350 kV, 18.4 A H$^-$, the latter representing an injected H^0 power of 3.2 MW (through a duct opening of 50 cm^W x 60 cm^H). Although the sources are capable of operating at 0.1 Pa an optimum pressure of 0.25 - 0.3 Pa is found. The negative ion yield is a factor 1.4 higher in hydrogen than in deuterium (isotope effect). In deuterium the ratio of negative ion to electron current is >2. To improve the plasma uniformity the filaments are regulated individually. The caesium consumption is about 30 mg/h which does not result in any adverse effects on voltage holding. For the next experimental campaign (to start in September 1997) improvements made earlier to only one source will now be accomodated to the second one which certainly will yield some more exciting results.

The European negative ion activities are distributed among *DRFC Cadarache* (*D*épartement de *R*echerche sur la *F*usion *C*ontrollée - 1 MV *SINGAP* acceleration, D$^-$ source collaboration with *JAERI*), *IPP Garching* (*I*nstitut für *P*lasma*p*hysik in collaboration with *Cadarache* - RF source development), *École Polytechnique Palaiseau* (fundamentals of single aperture beam) and *CIEMAT* (insulating materials), as *Ron Hemsworth* reported.

Results and status of the various activities may be summarized as follows. With comparatively little effort the 1 MV *SINGAP* experiment has obtained very promising results (860 keV, 9 mA/cm^2 D$^-$) producing a high energy D$^-$ beam of good quality with the relevant parameters by all means extrapolable to *ITER NBI* specifications. Advantages are a substantially simpler acceleration system and possibly a lower amount of stripping, not only in fraction of particles but also in power loss.

Very satisfactory results have also been obtained with the "*Kamaboko*" source on the *MANTIS* test bed (collaboration with *JAERI*). *ITER NBI* design current densities of 20 mA/cm^2 D$^-$ have been demonstrated with an improved magnetic filter reducing the fraction of co-extracted electrons below half the negative ion current. Long pulse experiments have now arrived at pulses of several minutes testing different designs of water cooled plasma electrodes with thermal barriers to allow continuous operation at the optimum surface temperature of 300 °C.

The emittance measurements on the fundamental single aperture negative ion beam experiment have started to deliver the first results (see *Claire Michaut*'s contribution). So has the RF negative ion source project at *Garching* (contribution of *Peter Frank*). The survey of insulating materials for the large 1 MV insulator of the ITER injectors has identified porcelain C221 as a suitable alternative to the envisaged alumina. This porcelain has not only the advantage that it is already commonly used in the insulator industry (thus possibly cheaper to produce) but it also has the same thermal expansion coefficient as titanium envisaged for the metal parts of the large insulator (thus no thermal expansion stresses).

Although still in the proposal stage the investigations concerning the joint European project *JENIPHER* (**J**oint **E**uropean **N**egative **I**on **P**roject for **H**igh Ene**R**gies) have already resulted in a common specification for a beam source module of 350 keV, 20 A D$^-$ and 360 mm^W x 800 mm^H extraction area which will equally suit machines like *ASDEX Upgrade, Wendelstein VII X, Tore Supra* and possibly *JET*. Depending on the beam quality achievable at this energy *SINGAP* or *MAMuG* acceleration will be used. The envisaged pulse length is >100 s. As high energy *NBI* still appears to be the most credible heating and current drive method in large dense plasmas we can only hope that there will be enough will to bring this interesting project to the success it deserves.

The panel discussion on *Neutral Beam Injectors and Other Negative Ion Applications* was moderated by *Jim Alessi*. The panel members (again in alphabetical order) were: *Samar Guharay, Ron Hemsworth, Junzo Ishikawa, Alexander Krylov* and *Kazuhiro Watanabe*. To stimulate discussion experts in the panel were asked to give a short asessment of their familiar application area, i.e. spallation source (*Jim Alessi*), micro beams and surface treatment (*Samar Guharay* and *Junzo Ishikawa*), neutral injection (*Ron Hemsworth*). This was further enriched by a short contribution by *Nikolai Semashko* presenting the Russian plasma neutralizer programme for *ITER NBI* application.

The points raised during the successive lively discussion are summarized as follows. New accelerator users of H$^-$ are the spallation neutron sources (*NSNS* in America and *ESS* in Europe). The negative ion source development (high discharge power, high current density, pulsed operation) for this application has made impressive progress and the parameters obtained are not far off the required specifications.

H$^-$ beams seem predestined for a whole variety of surface treatments and implantation because of their negligible surface charging. The high brightness micro beams required for electronic chip production pose interesting problems related to space charge neutralisation and very high quality optics. Many areas in this field are expanding rapidly.

The design of the (rather larger) *ITER* injectors gives confidence that such systems will work satisfactorily (like their positive ion based ancestors) despite their complexity and the technological challenges involved. In contrast to the downstream beam components which pose relatively little problems the development of the beam source (low pressure - long pulse ceasiated operation, vacuum high voltage insulation, remote maintenance) and the 1 MV insulator (voltage holding in a radiation environment, size far larger than ever built) need some more attention. "Advanced" components like the plasma neutralizer are very "exciting" to develop but the merit has to be weighed carefully against the effort.

Experience in many laboratories has shown no detrimental effect of caesium on voltage holding or negative ion beam extraction. Even at *DESY* where the use of caesium has been ruled out right from the beginning there is no real "hard evidence" against it. Meanwhile the "volume" production there has reached such a high level that the (anticipated) "problem" has become irrelevant.

The question why "some do and some don't" need caesium remains difficult to answer in detail. However, apparently all the high current densities produced without caesium seeding, may it be "Volume" or "Surface", result from single small aperture pulsed sources of small volume with high specific discharge power and relatively high gas pressure offering negative ion currents up into the 100 mA range with a relatively large fraction of co-extracted electrons. In contrast, neutral injection sources are supposed to produce a continuous high energy beam of many 10's of amps without excessive losses in the accelerator and with a certain economy in a large source. This again requires just the opposite, i.e. large area multi-aperture extraction, resonable specific source power, low gas pressure and a minimum of co-extracted electrons. What else is there to do - than put caesium?

... *"I want to see 'that beam' going down all the way right through the middle of 'that duct' ..."*
Ken Ehlers, "Brookhaven '86"

3. CONCLUDING REMARKS

This wish expressed rather emotionally "after hours" contains quite a high dose of self criticism (directed towards all of us) meaning that, at the time, despite the substantial experimental and modelling effort persisting throughout many years a negative ion based neutral injector seemed still far out of sight (note that at this time and presented at that conference, the *JAERI* people only just had started experiments reversing the extraction potential on a multipole source demonstrating the first multi-aperture, multi-amp extraction of at least some negative ions). Today, we may ask ourselves two questions: "has there been progress" and, if so, "what is the recipe for exceptional progress"?

Looking back over 11 years of development we can answer the first question with a clear "yes". And Ken's wish has no doubt come true as we have today (at *JAERI*) a working 500 keV negative ion based *NBI* system which has already injected several MW of neutrals and continues to do important experiments. Not only that, with the *NIFS* 180 keV injectors we have a second large system in the commissioning phase. But there is more! The detailed design of a system to deliver no less than 50 MW of 1 MeV neutrals to ITER is as good as completed and the supporting R&D results give great confidence that such a system will work and achieve the required specifications (note that most of the important parameters are close to being achieved or have already been demonstrated separately - which does not mean that this will go without the necessary struggle, as we all know).

With answering the second question we somehow touch the domain of personal opinion. But I think many of you will agree with the following. Expressing it shortly, there seem to be three things necessary for exceptional progress and a good chance of success: a *customer*, a good *strategy* and the *resources*. As typical examples how successful *customers* (note that this is only a selection, as the complete list of happy *NBI* users is so long that it would cause space problems to mention everybody) have stimulated *NBI* progress (or still do) we may mention *JET* (160 keV, 20 MW, positive based), *JT60-Upgrade* (500 keV, 10MW, negative based) and *ITER* (1 MeV, 50 MW, negative based, not yet a running machine but already in the design phase having triggered tremendous R&D progress). Where would *NBI* be today without these projects?

For examplary *strategy* there are two different approaches worth mentioning. The first, nearly being history and unfortunately without a follow-up, again *JET* who managed to push the positive ion based *NBI* power to the multi-megawatt level (accompanied by advances in many areas of technology) within just a couple of years by employing two competing design teams. And the second, the unequalled Japanese negative ion programme, a typical example for a solid long term and persisting strategy (many of us may remember the nicely structured programme already presented at "Brookhaven '86) which definitely does pay off now making available all its advantages for *ITER* (without *ITER* it would have been *FER*, the *F*usion *E*xperimental *R*eactor).

The *resources* are mainly a consequence of the two points already mentioned. Most of the time they are already available somewhere and they do not need to be especially raised. It is rather a matter of priorities and will to channel them into a desirable project. Therefore, in the cases of *JET* and *JT60-Upgrade resources* did not pose a particular problem. In contrast, *ITER* may have to struggle a bit harder as a consequence of its size and its rather complex organisational structure.

Although important and probably the largest, *NBI* is of course not the only application of negative ions. It is a pleasure to see that they receive growing interest and ... *new customers*, be it in the accelerator field and as a building block of a spallation neutron source or be it in the field of diagnostics (*PIXE, SIMS*) and as micro beams for a whole range of surface applications. Especially the micro beams for electronic chip production and possible medical applications may benefit from the presence of "serious" (industrial) *customers* which could generate progress even more spectacular than in the field of neutral beams. Finally, the judgement about the complete absence of a contribution related to *SDI* at this Symposium is left to anybody's conscience (thank God for that!).

The last alinea of this Summary is reserved for words of thanks now appearing in a more conventional place than in the oral, in situ version where they were placed pragmatically at the beginning, to counteract expected time problems. Thank you, *Véronique, Anne* and *Natasja* for your charming presence and the work which made this Symposium run so efficiently and pleasantly. And thanks again (of course) to *Claude Jacquot* who managed so well to expatriate the "Brookhaven Spirit" to Europe. A spirit which is well worth and in our interest to be kept going. Therefore, see you all again at "Brookhaven '00".

List of Participants

Alessi, James G.
Brookhaven National Laboratory
PO Box 5000
Upton, NY 11973-5000   USA
phone: 516-344-7563
FAX: 516-344-5011
email: alessi@bnl.gov

Bacal, Marthe
Laboratoire de Physique des Milieux Ionises
Ecole Polytechnique
F-91128 Palaiseau Cedex   FRANCE
phone: 33-1-69-33-32-52
FAX: 33-1-69-33-30-23
email: bacal@lpmi.polytechnique.fr

Baksht, Fedor G.
A. F. Ioffe Physical-Technical Institute
26 Polytechnicheskaja
R-194021, St. Petersburg   RUSSIA
phone: 7-812-247-86-40
email: aleb@plasmadiv.ioffe.rssi.ru

Fukumasa, Osamu
Department of Electrical & Electronic Engineering
Faculty of Engineering
Yamaguchi University
Tokiwadai 2557, Ube 755   JAPAN
phone: 81-836-35-9463
FAX: 81-836-35-9449
email: fukumasa@plasma.eee.yamaguchi-u.ac.jp

Guharay, Samar K.
Institute for Plasma Research
University of Maryland
Energy Research Building 223
College Park, MD 20742   USA
phone: 301-405-5013
FAX: 301-314-9467
email: guharay@plasma.umd.edu

Hayashi, Daiyu
Department of Electronics, Kadota Lab.
Nagoya University
Furo-cho, Chikusa-ku
464-01 Nagoya   JAPAN
phone: 81-52-789-3940
FAX: 81-52-789-3932
email: d-hayashi@echo.nuee.nagoya-u.ac.jp

Hershcovitch, Ady
Brookhaven National Laboratory
PO Box 5000
Upton, NY 11973-5000   USA
phone: 516-344-4531
FAX: 516-344-5954
email: hershcovitch@bnldag.bnl.gov

Hopman, Hendrik J.
FOM Institute for Atomic and Molecular Physics
PO Box 41883
NL-1009 DB Amsterdam
THE NETHERLANDS
phone: 31-20-608-1234
FAX: 31-20-668-4106
email: hopman@amolf.nl

Ishikawa, Junzo
Department of Electronics
Kyoto University
Yoshida-honmachie, Sakyo-ku
Kyoto 606-01   JAPAN
phone: 81-75-753-5325
FAX: 81-75-751-0297
email: ishikawa@kuee.kyoto-u.ac.jp

Katsch, Hans-Michael
Inst. f. Laser & Plasmaphysik
Universitat Essen
Universitatsstrasse 5
D-45117 Essen   GERMANY
phone: 49-201-183-3130
FAX: 49-201-183-2120
email: katsch@uni-essen.de

Krylov, Alexander
ITER-JCT
801-1 Mukouyama, Naka-machi
Naka-gun, Ibaraki-ken, 311-01
JAPAN
phone: 81-29-270-7724
FAX: 81-29-270-7505
email: krylova@itergps.naka.jaeri.go.jp

Lakatos, Andreas
Institut für Angewandte Physik
Universität Goethe
Robert Meyer Strasse 2-4
D-60054 Frankfurt   GERMANY
phone: 49-69-1-7982-3470
FAX: 49-69-1-7982-8510
email: lakatos@em.uni-frankfurt.de

Maaser, Andreas
Institut für Angewandte Physik
Universität Goethe
Robert Meyer Strasse 2-4
D-60054 Frankfurt   GERMANY
phone: 49-69-1-7982-3470
FAX: 49-69-1-7982-8510
email: maaser@iap.uni-frankfurt.de

McNeely, Paul
Plasma Research Laboratory
School of Physical Sciences
Dublin City University
Glasnevin, Dublin 9   IRELAND
phone: 353-1-704-5948
FAX: 353-1-704-5951
email: pmcn@alice.physics.dcu.ie

Michaut, C.
Laboratoire de Physique des Milieux
Ionises
Ecole Polytechnique
F-91128 Palaiseau Cedex   FRANCE
phone: 33-1-69-33-32-34
FAX: 33-1-69-33-30-23
email: claire@lpmi.polytechnique.fr

Nishiura, Masaki
National Institute for Fusion Science-Gifu
322-6 Oroshi-cho, Toki City
Gifu 509-52
JAPAN
phone: 81-572-58-21-96
FAX: 81-572-58-26-19
email: nishiura@LHD.nifs.ac.jp

Peters, Jens
DESY/MIN
Deutsches Elektronen-Synchrotron
Notkestrasse 85
D-22067 Hamburg   GERMANY
phone: 49-4089-98-2495
FAX: 49-4089-98-4364
email: petersj@pktr.desy.de

Riz, David
Laboratoire de Physique des Milieux
Ionises
Ecole Polytechnique
F-91128 Palaiseau Cedex   FRANCE
phone: 33-1-69-33-32-39
FAX: 33-1-69-33-30-23
email: riz@lpmi.polytechnique.fr

Sherman, Joseph D.
Los Alamos National Laboratory
LANSLE-2, PO Box 1663
Los Alamos, NM 87545
phone: 505-667-3511
FAX: 505-665-2509
email: jsherman@lanl.gov

Simonin, Alain
DRFC
Commissariat à l'Energie Atomique
CEA/Cadarache
F-13108 St. Paul lez Durance
FRANCE
phone: 33-4-42-25-46-60
FAX: 33-4-42-25-49-90
email: simonin@gemini.cad.cea.fr

Trainham, Rusty
DRFC
Commissariat à l'Energie Atomique
CEA/Cadarache
F-13108 St. Paul lez Durance
FRANCE
phone: 33-4-42-25-62-41
FAX: 33-4-42-25-49-90
email: trainham@drfc.cad.cea.fr

Tsumori, Katsuyoshi
National Institute for Fusion Science-Gifu
322-6 Oroshi-cho, Toki City
Gifu 509-52
JAPAN
phone: 81-57-258-2206
FAX: 81-57-258-2622
email: tsumori@phlkiku.nifs.ac.jp

Author Index

A

Akiyama, R., 93
Asano, E., 93
Asano, S., 93

B

Bacal, M., 3, 12
Baksht, F. G., 12
Beller, P., 133
Boilson, D., 113
Bucalossi, J., 81, 187, 237

C

Curran, N., 113

D

Desgranges, C., 187, 237
Di Pietro, E., 199, 227
Dlugach, E., 199
Döbele, H. F., 77

E

El Balghiti-Sube, F., 3
Elizarov, L. I., 3

F

Feist, J. H., 119
Frank, P., 119, 237
Fujiwara, Y., 105, 179, 205
Fukumasa, O., 54, 165
Fumelli, M., 187, 237

G

Gleeson, M. A., 37
Gotoh, Y., 217
Guharay, S. K., 158

H

Hanada, M., 179, 199, 205
Hayashi, D., 123
Heinemann, B., 119, 237
Hemsworth, R. S., 187, 199, 227, 237
Hodgson, E. R., 237
Hopkins, M. B., 113
Hopman, H. J., 221

I

Inoue, T., 179, 199, 205, 227
Ishikawa, J., 217
Ivanov, V. G., 12

J

Jacquot, C., 105, 119, 237

K

Kadota, K., 123
Kaneko, O., 93, 123
Katsch, H. M., 77
Kawamoto, T., 93
Klein, H., 133
Kleyn, A. W., 37
Koppers, W. R., 37
Kraus, W., 119
Krause, W., 237
Krylov, A., 199, 227

L

Leung, K. N., 244

M

Maaser, A., 133
Massmann, P., 187, 237
McNeely, P., 113
Michaut, C., 81, 237
Miyamoto, K., 105, 179, 199, 205
Miyamoto, N., 179, 205
Mosbach, T., 77

O

Ohara, Y., 179, 205
Oka, Y., 93, 123
Okumura, Y., 105, 179, 205, 227
Okuyama, T., 93
Olsen, D. K., 41
Osakabe, M., 93
Ott, W., 237

P

Paméla, J., 62, 187
Panasenkov, A., 199
Penningsfeld, H.-P., 237
Peters, J., 172
Probst, F., 119

R

Raridon, R. J., 41, 244
Riz, D., 62, 81, 105, 237
Ryabtsev, A. V., 139

S

Sasaki, K., 123
Schmidt, C. W., 254
Shchedrin, A. I., 139
Simonin, A., 187, 237
Soloshenko, I. A., 139
Speth, E., 119, 237
Stäbler, A., 237
Suzuki, Y., 93

T

Takanashi, T., 93
Takeiri, Y., 93
Tanii, M., 227
Tontegode, A. J., 3
Trainham, R., 105, 119, 237
Tsuda, N., 165
Tsuji, H., 217
Tsumori, K., 37, 93, 123

V

Vender, D., 113
Verhoeven, J., 221
Volk, K., 133
Volmer, O., 237

W

Watanabe, K., 179, 205
Watson, M., 227
Weber, M., 133
Whealton, J. H., 41, 244

Y

Yoshino, K., 54